Compassion, by the Pound

COMPASSION, BY THE POUND

THE ECONOMICS OF FARM ANIMAL WELFARE

F. BAILEY NORWOOD
AND JAYSON L. LUSK

OXFORD
UNIVERSITY PRESS

OXFORD
UNIVERSITY PRESS

Great Clarendon Street, Oxford OX2 6DP

Oxford University Press is a department of the University of Oxford.
It furthers the University's objective of excellence in research, scholarship,
and education by publishing worldwide in

Oxford New York

Auckland Cape Town Dar es Salaam Hong Kong Karachi
Kuala Lumpur Madrid Melbourne Mexico City Nairobi
New Delhi Shanghai Taipei Toronto

With offices in

Argentina Austria Brazil Chile Czech Republic France Greece
Guatemala Hungary Italy Japan Poland Portugal Singapore
South Korea Switzerland Thailand Turkey Ukraine Vietnam

Oxford is a registered trade mark of Oxford University Press
in the UK and in certain other countries

Published in the United States
by Oxford University Press Inc., New York

British Library Cataloguing in Publication Data
Data available

Library of Congress Cataloging in Publication Data
Data available

Typeset by SPI Publisher Services, Pondicherry, India
Printed in Great Britain
on acid-free paper by
Clays Ltd, St Ives plc

ISBN 978-0-19-955116-3

1 3 5 7 9 10 8 6 4 2

Bailey dedicates this book to his grandmother, Virginia Norwood, who believed he was somebody long before he was.

Jayson dedicates this book to his father, Raymond Lusk, who taught him the value of hard work (much of it with farm animals) and intellectual curiosity.

ACKNOWLEDGMENTS

While conducting research for this book, we traveled to many farms and had conversations with numerous farmers, animal advocacy groups, livestock industry groups, and ordinary consumers. We are grateful to everyone who took the time to talk with us and help us to better understand the farm animal welfare debate. This includes Don Lipton and Mace Thornton of the American Farm Bureau Federation and Paul Shapiro of the Humane Society of the United States. Other organizations have asked us to give presentations on our work, which has allowed us to interact with a variety of people and helped us to better understand the positions staked out by various parties in the farm animal welfare debate. These organizations include: the American Farm Bureau Federation, the Alabama Farm Bureau, the Oklahoma Farm Bureau, the National Institute for Animal Agriculture, the University of Nebraska, Animal Legal Defense Fund, Oklahoma State University, Alberta Agriculture and Rural Development, Oklahoma, Missouri, Arkansas (KOMA) Annual Cattle Conference, University of Arkansas, The Ohio State University, Centre for Research on Agro-Food and Development Economics in Spain, the French National Institute for Agricultural Research (INRA), North Carolina State University, the Beef Cattle Institute, and the International Symposium on Beef Cattle Welfare. Many of the scientific results in the book were derived from economic experiments that were funded by the United States Department of Agriculture (NRI Competitive Grant OKLo2662), and we thank them for their funding and support.

The farm animal welfare issue is a contentious topic, and we could only conduct objective research if we knew our administration would support us, and would prevent interest groups who might want to suppress our work from doing so. We are proud to work for an administration who is devoted to objective research and the dissemination of knowledge, and an administration that does not allow politics to interfere with the search for knowledge. Specifically, we thank our Department Head, Mike Woods; our Associate Dean of Research, Clarence Watson; and our Vice President and Dean, Bob Whitson. Rob Prickett and Lacey Seibert conducted their Masters' thesis on the farm animal welfare issue, and some of their findings played a pivotal role in this book. Two anonymous reviewers from Oxford University Press provided helpful suggestions, and we appreciate their thoughts, as well as those from Georgia Pinteau and Aimee Wright at Oxford University Press.

A final expression of gratitude is reserved for our families. This book required long hours of work, and we are fortunate to have wives who tolerate our obsession with research and writing. We also thank our children, who remind us that there is such a thing as working too hard.

CONTENTS

LIST OF FIGURES

LIST OF TABLES

CHAPTER I

Economics and the Farm Animal Welfare Debate

"The economic approach isn't meant to describe the world as any one of us might want it to be, or fear that it is, or pray that it becomes—but rather to explain what actually is. Most of us want to fix or change the world in some fashion. But to change the world, you first have to understand it."

Steven Levitt and Stephen Dubner in *SuperFreakonomics* (2009)

We live in a remarkable age. Never have so few people fed so many. For the past 10,000 years, most humans lived and worked on some sort of farm, but today that number is less than 1 percent.[1] Farmers today are incredibly productive. In fact, the poor among us are threatened by both hunger and obesity—what a combination! Perhaps because so few people are now involved in production agriculture, the public is showing a renewed interest in food. Consumers are visiting local farmers' markets, paying hefty premiums for organic foods, turning on the *Food Network*, and buying bestselling books about food and agriculture. When interest in food is aroused, people are often surprised by what they find. For many, the only farms they have seen are those depicted in children's books. People are often shocked to find that the story-book farms are nowhere to be found.

When people's romanticized notions of an agrarian lifestyle meet with the realities of the modern industrial farm, the result is often a plea for a return to antiquated production methods. But if consumers are disenchanted with the way farm animals are now raised, farmers are mystified that those so disconnected from production agriculture presume to know so much about how to run a farm. Then the exchange of insults begins. Some consumers claim farmers have lost their humanity by confining hogs and chickens to small cages, and some farmers accuse consumers of being food-elitists or even ignorant. Insults lead to action. Animal advocacy groups have pursued ballot initiatives and the courtroom to outlaw certain production practices and have pressured food retailers into purchasing food from more "humane" farms. In turn, farmers and ranchers have used political clout to block legislation and referenda.

The result is a brewing controversy between animal activist groups, farmers, and consumers that is currently being played out in ballot boxes, courtrooms, and in the grocery store. These developments have led some prominent writers to argue that the relationship between humans and animals is being fundamentally rethought.[2] We are at the precipice of perhaps the greatest challenge facing the livestock industry in recent history—the farm animal welfare debate.

You have no doubt noticed that buying eggs is not as easy as it once was. Grocery stores still carry regular white eggs, but also sell brown eggs, cage-free eggs, free-range eggs, organic eggs, and even Omega 3 eggs. Which eggs should you buy? How much should you be willing to pay? Californians recently had a chance to vote on a state-wide referendum to ban the production of regular cage eggs. If you had the chance to vote on a similar referendum, where would you seek honest, objective information? At present, there is no clear answer.

Sources such as the American Meat Association and People for the Ethical Treatment of Animals (PETA) have vested interests in your vote and your shopping choices. Although they rarely lie, all special interest groups are adept at sensationalizing and leaving out inconvenient truths that fail to fit their agenda. Instead, one could turn to the numerous writings of philosophers or ethicists, almost all of whom champion improved animal conditions for farm animals. Like the special interest groups, these writers communicate valuable information, but they tend to neglect arguments that would weaken their case. However, writings by ethicists and philosophers, though perhaps biased at times, is helpful in clarify one's views on animal welfare.

Ultimately, however, reading the writings of competing ethical philosophers is unsatisfying because, *"the philosophical discourse on animal rights is inherently inconclusive."*[3] Ethical and philosophical arguments are of little practical use in evaluating the consequences of the estimated 50 to 60 pieces of legislation regarding animal welfare that are introduced in the US Congress each year or the numerous matters considered by the European Union and their Scientific Committee on Animal Health and Welfare (SCAHAW).[4] Although much about farm animal welfare has also been written by veterinarians, biologists, and animal scientists, it too is inadequate in fully illuminating the consequences of the regulatory and market-oriented initiatives related to farm animal welfare.

What is missing in the animal welfare debate is an objective approach that can integrate the writings of biologists and philosophers while providing a sound and logical basis for determining the consequences of farm animal welfare policies. What is missing in the debate? *Economics.*

We are at an impasse when dealing with competing philosophies or moralities because we are dealing with issues of right and wrong. The economic approach is instead to treat peoples' views on animal welfare as *preferences*. This approach allows us to measure trade-offs between two competing views from the perspective of individuals and society as a whole. Implicit in such an approach is the need to discover exactly what it is people want; indeed, we spend a great portion of this book reporting on research aimed at finding out exactly that. We do not argue whether certain agricultural practices are right or wrong, nor do we try to persuade you to adopt a particular system of moral beliefs. Rather, the economic approach involves asking you, as citizens, how *you* think animals should be raised. This information is then used to help determine the consequences of the actions pursued by you and others in society. For example, in Chapter 8 we use the tools of economics to describe how changes in one's diet impact on the lives of farm animals. These tools provide some results that are obvious. Other results, however, are counter-intuitive, and it is these results that demonstrate the value of applying reasoning and logic to the farm animal welfare debate.

Because of the immense literature on animal welfare and rights, it is important to spend a few moments clarifying some of the assumptions of the economic approach we advocate in this book. The first assumption was mentioned in the preceding paragraph: although moral reasoning has its place, philosophical arguments are often unable to settle any two conflicting moral views. For example, if we must choose between providing chickens with more room to move and protecting them from injuring one another, which priority should we choose? *The most elegant and thoughtful philosophy cannot answer this question*, and so in a pluralistic and democratic society, it is prudent to turn to the people themselves to resolve the dispute. Throughout this book we frequently turn to regular citizens and ask about their preferences for animal care.

Second, *the vast majority of people have no intention of becoming vegetarians*. Most surveys suggest that about 98 percent of Americans and 94 percent of Europeans eat meat.[5] Although some argue vehemently that eating animals is wrong, the reality is that meat eating and, therefore, animal rearing for food, is likely to be with us for some time. However, just because people eat meat, this does not imply that they favor any kind of torture of animals, nor does it imply that they will not pay higher prices to allow for better animal care. Thus, the third assumption is this: *almost everyone cares about the treatment of farm animals, to some degree*. People are, generally speaking, 'compassionate carnivores'. We care about animals *including* the animals that we use to provide us with meat, a fact that means people must make trade-offs between competing sets of concerns and preferences. Consider the recent behavior of a well-known American football quarterback, Michael Vick, who was accused of killing dogs by hanging and drowning. Obviously, Vick was not compassionate towards animals, but the fact that the story made headlines suggests most people do care. Further, simply because PETA videos some farmers being cruel to their animals, this does not imply that many (or even most) farmers do not care about the well-being of the animals they own.

Fourth, in this book, *we argue that the study of animal welfare is, for all practical purposes, the study of farm animal welfare*. It has been estimated that of all animals that have at least some contact with humans, 98 percent are farm animals.[6] There are some important and interesting issues related to the use of animals for research purposes,[7] and puppy mills, for example, are detested by most pet owners, but if one is interested in improving the lives of the vast majority of animals, it is clear that the farm is the place to start.

Fifth, economics focuses on *trade-offs*. We all tend to prefer more food to less food, more clothes to fewer, and bigger houses to smaller houses, but we have limited resources. We cannot have it all. The question is how much and what kind of food we eat and how many and what kind of clothes we wear, given our income constraints. You may decide to have a bigger house, but this means you must have fewer (or less expensive) shoes, for example—you cannot have both unless you get a pay raise, go into debt, or stop consuming some other goods. There might even be a part of us that cares deeply enough about animals that we would prefer to stop eating meat, but then, eating a juicy steak is immensely pleasurable. We can choose to improve the lives of farm animals, but we must give something up to accomplish this mission. As economists, we are not content to rely on the good intentions of animal activists or livestock industries, but rather we look to analyze the *outcomes* their actions produce, and what we, as citizens, must give up to achieve these outcomes. Moreover, we do not simply assert that generally people make trade-offs; we want to measure these trade-offs. And it is not a question of *whether* people have to give something up to improve animal welfare, but rather *how much* they are willing to give up.

The economic approach also asserts that *individuals tend to be most happy when they are free to make their own choices in the marketplace*. The market price of food reflects the cost of all the inputs used to produce the food as well as the value the consumers place on the food they buy relative to other goods they could have bought with the same money. Market prices reconcile the competing costs and values, and so they help people decide how to divide their limited time and resources, to ensure that these are allocated in a way that reflects the buyers' values. Markets set prices, but however quickly we may turn to the government to resolve social problems, the government is not particularly adept at allocating resources in a way that increases the size of the economic pie. Government bureaucrats, no matter how kind or benevolent, simply do not have the information needed to allocate goods the way we do, as individuals, by responding to the incentives created by the market prices we face. Moreover, markets allow experimentation and failure. This is important: if a private business introduces a new product that is not desired by the public, it soon disappears. (Think of the McLean burger—if you've never heard of it, Google it!). By contrast, bad policies are rarely eliminated. Finally, markets decentralize power. If you do not want one special interest group, such as PETA or the pork industry, to control how farm animals are raised, you might like market outcomes to do so, these being controlled by large numbers of individual consumers and firms.

This is not to suggest that additional government regulation might not improve the lives of humans and farm animals. There are some peculiarities of the animal welfare problem that suggest market outcome controls may be less than desirable in some situations compared to some regulated alternative. But saying government action *could* benefit society does not mean that it *will*.

Although this is a book about farm animal welfare issues, we posit that studying what *people* want is ultimately what matters. Even in the unlikely event that we could all agree that the goal is to maximize the well-being of farm animals, we humans are the ones that must decide how to make it happen. Research on animal biology and physiology can help us better understand what makes animals happy, but humans must take the steps to bring about their happiness. If animal rights activists were given complete and absolute power over farm animals, for example, they would still face some tough choices. They could choose to take all existing farm animals and place them in a wildlife preserve, away from the "greedy" omnivores who wish to place the animals in cages and eat them. Still, a host of questions would arise requiring difficult trade-offs. Should people remain responsible for feeding the animals? Should we try to prevent wolves from eating piglets? Are the animals to be allowed to breed, and if so, what should we do when their population exceeds the capacity of the preserve to feed them? Should we let natural diseases run their course in the animal population or should we intervene? If two bulls are fighting, should they be stopped before one is hurt? What do we do with a wild bull that gores a human? The point of these questions is not to mock animal rights advocates, but to point to the fact that regardless of whether humans eat them or not, farm animals are dependent on humans and their choices. The complete abolition of livestock as property does not make the choices easier, unless one's definition of abolition includes the complete extinction of farm animals. However much animal rights activists associate themselves with the abolitionist movement of the nineteenth century, livestock can never really be liberated. Livestock must be owned and cared for or they will become practically extinct.

Of course, not everyone agrees that the goal in life is to maximize the well-being of animals. Still, animals matter because people care about them. Because the happiness of animals is tied to the happiness of humans, improving the welfare of humans means finding out what we want for animals.

The Journey

If you are looking for easy answers to questions about farm animal welfare, this book is not for you. What we can offer is a journey into the most controversial issue facing modern livestock agriculture. We offer a different perspective on the farm animal welfare debate than that which is typically offered by ethicists or veterinarians—because we look at the issue through the lens of economics. This lens is focused largely on animal welfare issues in the US. We focus our attention here, not because the issue is uncontroversial or irrelevant in other parts of the world, but because this is where our expertise lies. Moreover, our unique consumer research was conducted with US consumers. From time to time we will delve into animal welfare issues in Europe and the rest of the world, but the reader is forewarned that our discussion and results are largely US-centric.

We have little doubt that you will sometimes disagree with our thoughts on particular issues. Writing about animal welfare is a bit like writing about politics or religion—it is virtually impossible to keep from offending the reader. We have done our best to tell the whole story, giving all views their proper due, but we have not refrained from taking positions when the data and logic warrant it. We have no formal affiliation with the animal production industries or with animal advocacy groups. Strangely, for authors of a book on farm animal welfare, we did not choose to write on this topic out of an intense concern about farmers or farm animals. We are passionate about science, economics, and the truth, and we were drawn to this topic because we felt the controversy needed a heavy dose of honest, objective, and *dis*passionate information.

While reading the book, keep in mind that our goal is not to advocate a particular stance on how farm animals should be raised. If, when the final page of the book is turned, you know a little more about farm animal welfare, and if you have been encouraged to think more deeply about the consequences of your actions and those of the government, then we will consider our efforts to have been a success.

CHAPTER 2

A Complex Relationship

A Natural and Cultural History of Humans and Their Livestock

"But there came a day when there was no rice left and no wheat left and there were only a few beans and a meager store of corn, and the ox lowed with its hunger and the old man said,
'We will eat the ox, next.'

Then Wang Lung cried out, for it was to him as though one said, 'We will eat a man next.' The ox was his companion in the fields and he had walked behind and praised it and cursed it as his mood was, and from his youth he had known the beast, when they had bought it as a small calf. And he said,
'How can we eat the ox? How shall we plough again?'"

Pearl S. Buck in *The Good Earth* (1931)

The First Herdsmen

Humans are not the first species to engage in livestock farming. In Malaysia there is a species of ant known as *Dolichoderus cuspidatus*. These interesting ants do not hunt for food. Besides scavenging for the occasional dead insect, the ants' only source of food is the mealybug (*Malaicoccus formicarii*), which they "farm." Mealybugs feed on sap and leaves from trees. Like a shepherd leading a flock to green pastures, each day the ants carry the mealybugs to a feeding location up to 20 meters away from the nest. As mealybugs eat, they emit honeydew droplets, which are harvested by the ants for food. A typical nest contains a queen, 10,000 worker ants, 4000 larvae and pupae, and more than 5000 mealybugs. The ants are highly protective of the mealybugs and will fight to the death to keep away predators.

Welcome to *Livestock Farming 101*. The mealybug and the ant have formed a symbiotic relationship where each species critically depends on the other. This relationship has evolved due to the interactions of the two species. Over time, the ant's interaction with and care of the mealybug began to alter the mealybug's genetic makeup—in an insect version of animal domestication. Today, the mealybugs' genetic makeup has been altered by the ants to such an extent that they cannot survive alone.[1] Moreover, the species' lives are so intertwined that the ants cannot live without the mealybugs either.

This story may sound familiar. It resembles the story of humans and their livestock. Thousands of years ago humans began controlling and caring for wild animals and through their actions altered the animals' genetic makeup. Domestication occurred, and produced the cattle, pigs, and chickens of today, on which we rely extensively for our food. Like the ant–mealybug story, the human–livestock story is symbiotic. Livestock require our daily care and attention, and they would fare poorly without our fences and feed buckets.

However, there are a few salient aspects of the human–livestock story that distinguish it from the ant–mealybug story. Human and livestock relationships did not emerge through biological evolution but as a result of the intentional actions of humans. Humans *chose* to raise farm animals, and emphasis is placed on the word *chose* because that choice is still ours to make. The ants cannot survive without mealybugs but we can survive without our livestock. It is this *choice* of whether and how to raise farm animals that is the focus of this book.

The Many Faces of Humans

History reveals the human to be a flexible creature. Her ability to be compassionate or cruel, emotional or logical, is well-documented. This includes her relationship with farm animals—and to seriously consider the topic of farm animal welfare is to confront a complex topic. There are easy ways out. One can simply purchase food without any consideration towards farm animals, or one can shun all animal products in the belief that raising animals for human use is always cruel. Both reactions are naive. To engage in the farm animal welfare debate and thence emerge with a clear, unambiguous conclusion is almost impossible. But making decisions based on a well-informed grasp of the complexities involved will surely produce a better world for humans and animals alike.

The difficulty in deciding how to treat animals is revealed in the myriad ways humans have treated farm animals throughout history. Consider Figure 2.1 describing the many faces of man. The top quote from a Roman historian describes the horribly cruel manner in which some sows were treated in order to provide tasty food. The second quote describes the actions of a French philosopher's entourage, who believed animals to have no soul, conscience, or ability to

> *"They thrust red hot irons into their living bodies so that...they may render the flesh soft and tender. Some butchers jump upon or kick the udders of pregnant sows, that by mingling the blood and milk and matter of the embryos..."*
>
> – Plutarch, Roman Historian (46–120 AD)
>
> *"They administered beatings to dogs with perfect indifference and made fun of those who pitied creatures as if they felt pain. They said the animals were clocks; that the cries they emitted when struck were only the noise of a little spring that had been touched, but that the whole body was without feeling."*
>
> – description of Rene Descartes' Entourage (1596–1650)
>
> *"A ruthless murderer's lack of remorse in taking human life began in a past life of simple disregard for the lives of others as seemingly inconsequential as animals or insects...Upon achieving an impartial attitude toward all sentient beings, you should meditate on love, the wish that they find the happiness they seek."*
>
> – The Dalai Lama (present day)
>
> *"Cow killers and cow eaters are condemned to rot in hell for as many thousands of years as there are for each hair on the body of every cow they eat from."*
>
> – Ancient Hindu Scripture

Figure 2.1 The many faces of humans

Sources: Spencer, 2000; Dalai Lama, 2001; Hills, 2005.

feel pain or emotion. In sharp contrast, the current Dalai Lama encourages us to treat animals with kindness and with the same respect as we would treat humans. The final quote elevates a common farm animal above the status of humans, deifying the cow and condemning to hell anyone who eats beef.

As a species, we are apparently quite flexible in our views about farm animals. To understand how modern societies treat farm animals today, we must first look back in time and witness the long history between humans and their livestock. We will look at how humans and their animals have suffered through difficult times, and how humans have now reached a period of luxury and comfort never before experienced. The question we would like to pose is this— should we now extend some of this improved comfort to livestock as well?

Down from the Trees

Eating meat is as natural to humans as walking on two feet. Humans were always omnivores, and omnivorous primates existed before hominids (great apes) arrived.[2] The diets of most hominids consisted of fruits, leaves, and grains, but they were flexible and developed a taste for meat first through scavenging and later through clever methods of hunting. Today, even chimpanzees hunt smaller monkeys and other prey, sometimes consuming 200 grams of meat in one day. However, just because chimpanzees or early humans ate meat does not necessarily mean we humans should continue to do so now. After all, chimpanzees (and modern humans) are known to carry out genocides![3] We explore the past not to justify but rather to better understand our current culture and actions.

Most scholars think that humans, or *Homo sapiens*, appeared around 200,000 BC. *Homo sapiens* journeyed out of Africa to populate the world. The first *Homo sapiens* exhibited little proclivity for agriculture: for many thousands of years humans fed themselves by hunting, gathering, and scavenging. Food was eaten raw at first, but humans eventually learned to cook food—perhaps by trial and error or from observing other species, as evidence suggests that cooking predates humans. The act of cooking increases the nutrient availability of meat. Some biologists believe this played a role in increasing the brain size of hominids, and is one reason why hominids evolved into the more intelligent species called the "human."[4] Meat was initially roasted over an open flame. Boiling came later and in many ways was more efficient, as it preserved calories (i.e., fat) as well as tenderizing the meat. The clear benefits of boiling led humans to utilize the practice even before the pot was invented. Some tribes may have used turtle shells; many tribes likely boiled meat in an animal stomach hung over the fire. Eventually leather replaced animal stomachs around 13,000 BC, and then clay pots emerged.[5]

As cooking improved so did meat storage. Our ancestors probably noticed how the meat of dead animals was easier to preserve during cold winters, and learned to bury meat under ice. Humans have lived most of their existence in the middle and late Paleolithic Age, which ended in 10,000 BC. Yet before this period, in the early Stone Age, humans learned to dry and smoke meat. Despite all the food advances made during the Stone and Paleolithic Ages, starvation was a persistent threat.

A Seed is Planted

Enter the Neolithic Revolution, when humans slowly assumed the role of farmer. Farming emerged from hunter-gatherer societies, where women assumed the primary role of the seed gatherer. As seeds were brought back to the cave or hut, people would notice the plants beginning to grow close to where the seeds were kept. Moreover, these humans would have observed seed-bearing plants growing in refuse piles, emerging from the seeds that had not been digested. The non-random appearance of new plants would have suggested to people that something important was happening. More than a few humans would have had an "aha" moment, realizing that a seed placed in the ground would yield a plant in time. Farming evolved from this realization.

The delicious cowpeas, snow peas, and sweet peas on our dinner plate today bear almost no resemblance to their ancestors from pre-domestication. The process of domestication causes dramatic changes to the genetic makeup of the plant. In fact, the very definition of the term *domestication* is "the genetic alteration of a plant or animal caused by human decisions over which offspring are permitted to reproduce." For example, humans were more likely to harvest larger seeds because they would yield more food when they were subsequently planted. As a result, fields began to produce crops that yielded larger seeds than the crops of earlier years. For example, the ancestor of modern corn was no longer than your thumb; today corn cobs can reach from your elbow to your wrist.

Likewise, humans embarked on a similar voyage with wild animals. No one knows exactly how animals were domesticated, but the processes likely happened in different ways and at different times with different animals. Nevertheless, it is useful to conduct a few "thought experiments" to imagine how such domestication might have taken place.

Imagine the following fictional, but plausible scenario. A clan of hungry humans were hunting for wild goats. They came upon a valley with only one point of entry and one point of exit. In it they found a group of goats that they were able to corner, making an easy kill. Once the hunters had caught and killed as many goats as they needed, they had the time to notice that this valley had a water source and plenty of forage. A more thoughtful member of the hunt suddenly had the idea, "Why not block the exit and keep the remaining goats inside, thus keeping them continually available of future hunts?" This enclosure would then have resembled a modern farm, where goats are kept in pastures and are contained within fences.

At this stage in the animal's history, the goat would not yet be classified as domesticated—because these goats were simply a free product of nature. But, over time it is easy to see how, even if unintentionally, humans would determine which goats reproduced. Aggressive males, for example, would be killed off first so that subsequent hunting would be easier. The less-aggressive males would then be left to mate, producing offspring that were also less aggressive. These animals would then be easier to tame, being less aggressive. The humans might soon have begun to live permanently with the herd to protect it from predators, and the animals' offspring would have become familiar with the humans from birth, making it easier for the humans to control the animals and determine which animals would be allowed to breed. Consequently, over time, animal domestication would have occurred and wild animals would have been slowly transformed into livestock animals.

A number of other stories could be thought up to explain how domestication might have occurred. Consider another example of how the human–livestock relationship might have begun. Some human populations were sizable and used latrines to hold their excrement. Many animals, most notably the wild hog, would see the latrine as a food source. It is a distasteful thought, but even the modern-day pig retains a proclivity for eating feces. In fact, in the not-too-distant past, some farmers relied on cow patties for pig feed. A 1928 livestock production manual bluntly asserts, "corn salvaged from cattle droppings is clear gain. Experimental results show that for every bushel of corn fed to cattle, enough feed is recovered by swine following them to produce one to two pounds of pork."[6] It is possible that the pig was domesticated through being drawn to these early latrines or the human's trash dumps, the pig's diet closely resembling that of humans. Perhaps as some of these hogs grew more comfortable living near humans, they slowly became domesticated.

In different regions and at different times, different animals were domesticated. Sheep and goats were domesticated in Southwest Asia around 9000 BC. Shepherds are known to have tended sheep herds in modern-day Iraq and Romania during this time. About 2000 years later (although it may have happened earlier, the science behind such dating being inexact) the pig was domesticated in China. Pigs are prolific breeders, can eat the same foods as humans, and in ancient times were small animals. As a result, by 6000 BC pigs are thought to have existed in every Chinese household, eating household scraps and, unless reserved for breeding, being slaughtered before the age of one year. The pig's contribution to the ancient Chinese diet was so essential that the Chinese word for "meat" and "pork" remain synonymous. Quite separately, the pig was also being domesticated in South America, but not until about 3500 BC.[7]

At least as early as 7000 BC, an animal called an auroch (which looked much like a modern-day ox) was domesticated in the area that is now known as India and Pakistan, and evolved into the modern-day cow. The wild auroch survived for thousands of years, and the last recorded one was killed in the Netherlands, around 1600 AD.[8] Turkeys were domesticated in Central America in about 3500 BC. Less is known about chicken domestication, but it was probably earlier than 3000 BC, when Asians began to keep and breed wild fowl. In ancient Egypt, around 6000 BC, the skill of incubating eggs was well-known and widely practiced.[9] The dog was the first domesticated animal, coming under human influence before 10,000 BC, and while they were valued for their assistance on the hunt their meat also helped provide nourishment. Even a species of fish (carp) was farmed before the BC era ended. Horses, camels, and other animals were also domesticated well before AD 1.

As one historian notes, "humans are not born farmers,"[10] and at the beginning animals were kept on an *ad hoc* basis—if an animal or group of animals happened to be available and happened to suit an individual or group of human's needs at a particular time. Almost every ruminant animal has been held captive by humans at some point, but only a few of those ruminants have gone on to become domesticated livestock.[11] Not every animal can be domesticated. In his fascinating book, *Guns, Germs, and Steel*, Jared Diamond outlines the traits necessary for successful animal domestication on a large scale. Animals capable of domestication must: be comfortable living in herds, be herbivores (omnivores can be acceptable), grow quickly, breed in captivity, be relatively docile and gregarious, have a hierarchical social structure (allowing humans to dominate them), and interact amicably with other herds of the same species. Domesticated animals are generally kept in fairly confined groups, living in close proximity to humans. The horse was domesticated while the zebra was not, probably because the zebra is horribly ill-tempered.[12] To thrive in groups, animals must be able to breed in groups; cheetahs, for example, will not breed in groups. Carnivores would be too expensive and time-consuming to feed (as other animals would have to be hunted or raised to provide them with meat), which led to herbivores who would eat grass being favored.

Not Just Food

Livestock provided humans with a food source, but they were much more than this. The human–livestock interaction has shaped the face of the modern world. A wild animal must possess certain traits to be suitable for domestication, as discussed above. Only a few did, and they were only found in some parts of the habitable world. The existence of domesticate-able animals in certain parts of the world ultimately decided which cultures would dominate in the world. Asia was blessed with most of the animals that would eventually develop into the most important livestock animals: sheep, goats, pigs, horses, cows, chickens, and donkeys. And because Europe was connected to Asia by land, these same animals eventually became available to the Europeans as well. The Americas, separated from Asia and Europe by water, only had access to llamas, guinea pigs, dogs, and turkeys. They did have bison, but however much these might outwardly resemble cattle, they differed in their emotional makeup, experiencing stress-related disorders in confinement, as early English settlers who tried to domesticate the bison soon found out. Consequently, ancient humans in the Americas did not farm animals on a scale even close to that seen in Europe and Asia.[13]

The location of "domestication-friendly" animals had further important implications in the course of human history. One factor was that most of man's major diseases originated in their livestock. When Europeans sought to conquer the Americas, it was not their bullets that gave them the winning advantage over the indigenous cultures; it was their immunity to smallpox and other diseases. Of course, the sophisticated weaponry of the Europeans helped in battle, but it could not overcome the European's lack of numbers. When Cortes landed in present-day Mexico in 1519, he was unable to conquer the Aztecs. However, when a slave arrived in Mexico from Cuba infected with smallpox, almost half of the Aztecs were killed, and it was their decimated numbers that made Cortes' second assault successful.

Smallpox was a disease that evolved from a pathogen that infected cattle. The Europeans' daily encounters with cattle had led to them being immune to smallpox, whereas the Aztecs, who had no cattle, had no consequent immunity. When discussing European ventures into the New World, Diamond asserts, "Far more Native Americans died in bed from Eurasian germs than on the battlefield from European guns and swords."[14] The ownership of livestock in Europe not only provided Europeans with labor and food, but also a weapon that allowed them to conquer and settle new lands. The face of the world was changed by the ownership of animals. Humans altered the genetic makeup of animals through domestication, but animals altered our genes in turn. For example, were it not for the domestication of livestock and a desire for the nutritious milk that mammals provide, almost all adult humans would still be unable to digest milk. As we reconsider our relationship with animals, let us ponder their role in our history. Let us go back to the beginning of human civilization and discover why we chose to farm livestock rather than remaining hunter-gathers. How did our *complex relationship* with farm animals emerge?

Why We Farm

Agriculture—the domestication of plants and animals—arose independently in nine different regions, and while some societies remained hunters and gatherers, others adopted a new agricultural life. Agricultural societies experimented with plants and animals, discovering new ways of managing their biological assets to improve their lot. People learned that the milk produced by livestock was tasty and nutritious. Animal do not die from being milked, and they can be milked daily, providing a daily source of nutrition to humans. Without refrigeration, people stored milk in animal stomachs, which contained natural enzymes and bacteria, leading to the first forms of cheese, yogurt, and butter. While some of these dairy products proved to be dangerous to consume, others were not. In some regions of the world, a form of cheese was eaten even before humans had begun to make pottery.[15]

It is tempting to think that agriculture began a revolution, and that within a few decades humans never looked back on their seed-gathering days. The reality is more complex. Agricultural societies competed with hunter-gatherer societies, and many of the latter existed until modern times. Where wild edible plants abound, it is easier to gather these than to go to the effort of planting and growing your own crops. Although hunting may seem like hard work, maintaining livestock is also strenuous, having to spend time building fences or, for example, herding sheep. Hunter-gatherer societies maintained a vibrant existence until recently, and some exist still. The Bushmen of the Kalahari are an example. When a present-day researcher asked them why they do not become farmers, they replied, "Why should we plant, when there are so many mongomongo nuts in the world?"[16]

The inhabitants of North Sentinel Island, in the Bay of Bengal, still thrive as hunter-gatherers and avoid contact with other people. They have kept to themselves for some time; according to their genetic profile, they have remained isolated for 60,000 years! For a long time agriculture had disadvantages that caused societies to remain hunter-gatherers. Agricultural societies' reliance on a smaller subset of food led to vitamin deficiencies and in some cases to a general decline in health. In the first stages of agriculture, farmers were less healthy than their hunter-gatherer counterparts. Even when they had access to the same number of calories, their diets consisted of less protein and fat and lacked certain vitamins and minerals. The first farmers were shorter, had more fragile bones, fewer teeth, and suffered more when the diseases that plagued their livestock began attacking them.[17]

From around 8000 BC until today the use of agriculture has steadily developed worldwide, perhaps more out of necessity than anything else. For various reasons, people have moved to live closer together over time, creating many more instances of densely populated areas.[18] Successful hunter-gatherer clans require large amounts of land per person. Once people are forced to live in closer proximity to one another, they must extract a greater amount of nutrition from each acre of land in order for everyone to survive—and it is here that agriculture provides a clear advantage over the hunter-gatherer lifestyle.

Earlier we told a fictitious story of a clan of erstwhile hunters who discovered that it was easier to raise goats than to hunt them. As animals became domesticated, and therefore easier to manage, the advantages to livestock farming increased. The obvious advantage of keeping livestock over hunting is that it has the possibility of providing greater amounts of meat for equal amounts of effort. Another benefit of livestock farming over hunting is meat storage. Preserving meat was difficult in 8000 BC, and preserving it poorly was, and still is, of course, deadly. For example, many of these early hunters obtained meat from hunts of migratory large animals, such as buffalo or mammoth. These hunts were infrequent, but yielded much meat, more meat than the humans could possibly eat before it spoiled. If this meat could have been stored, then hunting may have proven more efficient than raising animals. Although early humans had some methods for preserving meat (e.g., drying) it was imperfect, and it was risky. There is no better way of preserving meat than by keeping the animal alive, close, and available for "harvest" at any time—and this was one of the great advantages of farming over hunting.

For many ancient societies (and some poorer societies today) domesticated cattle were valued more for their labor than for their milk or meat. Many ancient farming communities were located near water, and early humans utilized natural and man-made mechanisms to irrigate and fertilize crops. For example, ancient Egyptians experienced Nile river floods between July and November. When the water receded, the Egyptians would sow seeds in the land, which was now moist from the flood water and rich with nutrients that had been carried downstream. These lands were open and generally treeless, making it easy for a plow to tear through the soil. It was in these areas that humans used cattle to pull a plow. There was a preference for larger, stronger cattle over cows that provide large quantities of milk and meat. As time progressed other animals were strapped to the plow, including the ass and the horse (the horse could not pull a plow efficiently until later, when a specific type of harness and horseshoe were developed). Animals were utilized for labor in other areas, such as making flour and providing power for irrigation machines. The plow pulled by today's modern tractor is quite similar to the plow pulled by the cow thousands of years ago: it all began with man and his cow.[19]

Livestock helped fertilize fields. While some farmers planted seeds near the river banks, others cleared forests to provide cropland. With their primitive tools, humans were unable to remove large tree trunks or roots. Consequently, such forest clearings being ill-suited for the animal-drawn plow, the land was turned by hand instead. Without river floods to fertilize the land, cleared forests would have had to be abandoned after a few years due to depleted nutrients— animal manure used as fertilizer changed this.

Consider the European and Mediterranean settlements in the BC era as humans were becoming better farmers. As populations increased, forests suitable for clearing and farming would have become scarce. Every parcel of land was assigned to one of three categories: *saltus*, *ager*, or *silva*. *Saltus* were hilly, deforested areas that could only be farmed until the nutrients were depleted. Over time, the *saltus* could no longer be planted with crops, and would be used to accommodate livestock who could live on the thin grass that grew. The *silva* were areas too difficult to deforest; such as hilly land or swamps. While humans would find *silva* a useful source of wood, the land was otherwise largely untouched and left in native form.

The *ager* was where most food was grown. Below the *saltus* and the *silva* were valleys that received rain and nutrients from the hills above. The land was easily cleared of forest and free of trees. Though the *ager* would have initially been extremely fertile, many years of repeated planting would deplete nutrients (e.g., nitrogen, phosphorus, and potassium). The *ager* would be in need of care, and livestock provided the solution.

Humans learned to fertilize the *ager* by grazing livestock on the *saltus* during the day and housing them on the *ager* at night—where the excrement from foraging would be transported to the *ager*, the cropland. The process effectively transported nutrients from the *saltus* down to the ager, providing fertilization. Humans became resourceful in how they managed livestock for fertilizer. The plants on the *saltus* grew better during certain seasons of the year, and its ability to support animals varied accordingly. People adapted to this situation by managing cattle so that they were taken to the *saltus* when they were in most need of feed and the *saltus* had the most plants to feed them with. Moreover, during the seasons when the *saltus* was less productive, surplus cattle would be sent to other areas to forage. Today, we fertilize farmland by using nitrogen from the atmosphere and mineral deposits from deep within the earth, and we no longer need to rely on the manure provided by livestock. In the modern age, agriculture is used to feed over six billion people, in large part because of the inexpensive fertilizers humans have developed over time. However, it all began with the first farmers and their cows.[20]

Agriculture allowed humans to live in more densely populated areas, forming large and complex societies. Our great wealth today is attributable to our ability to engage in mutually beneficial trade with each other. To be self-sufficient is to be poor. Think about it: what if you were only able to consume the goods and services that you yourself produced? Virtually everything we consume is produced by other people.

In the years 8000–2000 BC some people lived in areas suitable for grain production and others lived in areas more suitable for keeping livestock. Equally, some humans were naturally more talented in grain production and others in raising livestock. Managing the movement of cattle between the *saltus* and the *ager* would likely have been handled via a trade of services. Let us again make up an imaginary story of how this might have been handled. Let us imagine one man, who we shall call Abel, who specializes in raising livestock, and another, who we shall call Cain, who mainly produces crops, but who keeps some livestock as well. Cain needs more livestock than he owns to graze the *saltus* during the productive seasons, in order to transport more nutrients down to the *ager* where his crops grow. There is a clear opportunity for mutual benefit here. Abel could lend Cain a portion of his livestock during the productive season; this would mean that Abel would receive free grazing for his livestock, and Cain's fields would receive maximal fertilization. Both would be better off as a result of the trade. Both would have more meat and grain, respectively, than they would have had otherwise.

One can easily imagine other opportunities for trade that emerged as agriculture and human societies developed. For example, some people would have been better at making *ards* (the first plows) than others. A person who specialized in making *ards* would make better *ards* in less time and with fewer resources than someone who made their own *ard* themselves, once every two years. It is therefore to everyone's advantage for the *ard*-making expert to concentrate on making *ards*, providing them for a group of farmers, who could then in turn compensate the *ard*-maker with a portion of their harvest. The human is a bartering animal, a social animal. This trait has served us well, and agriculture gave humans their first opportunity to trade on a large scale.

Agriculture allowed, and perhaps caused, humans to live in densely populated areas and in close contact with one another, making it both more necessary and profitable to engage in trade. This is an often overlooked benefit of agriculture. However, trade was important long before the advent of agriculture. Indeed, the human penchant for trade is thought to be one of the reasons why humans thrived while *Neanderthals*, who apparently did not trade, perished. In fact, some scientists believe that the human propensity for trade was as important to the species' survival as was the development of agriculture.[21]

Trade of goods provides humans with many benefits, but it is the exchange of information, knowledge, and technology that truly allow humans to progress. Better ideas on how to grow food, build houses, and make clothes have emerged from the entrepreneurial spirit of humans. People steadily improve upon the ideas of others through the exchange of ideas and goods. Large, stable populations facilitate the creation and dissemination of knowledge. The opportunity provided by agriculture to feed dense populations is of monumental importance: it provided the wherewithal for societies to remain together, producing innovation and culture.[22]

Agriculture became increasingly widespread in the late bc era, allowing humans to live close to one another and form tight, complex, social bonds. In many areas, agricultural societies formed large armies that easily overtook other less well-organized hunter-gatherer groups. The use of agriculture slowly spread across the globe. Agricultural methods were more easily expanded eastward and westward, latitudinally, from the Mediterranean to Central Europe, and on to the Middle East. Plants and animals that were raised in one area could not easily be raised in another area that was significantly hotter or colder, and this prevented expansion northward and southward. Seeds from grains raised in modern-day Iraq could not be transported to and successfully cultivated in Siberia, for example. Hogs domesticated in Asia would suffer greatly if they were transported to the desert regions of Africa—because pigs cannot sweat, and so they need water or mud to wallow in when temperatures are hot.

However, if seeds and animals were migrated slowly northward or southward, they could adapt over time, just as humans' skin color has slowly evolved in response to the different sun exposure levels at different parts of the planet. Animals that survived the hotter or colder conditions and lived to breed would produce offspring that would also be likely to survive, and so on. If moved northward slowly, for example, the animals' genetic composition would eventually change, leading to thicker hair, more fat, and so on, through natural selection.

Cattle, pigs, and chickens spread rather quickly across Eurasia because much of the continent lies on roughly the same latitude, and thus has similar climate. Llamas were domesticated in South America but never made it to North America, and the corn planted by Native Americans was never adopted by Peruvians. It was not the physical distance between North and South America that was so important, but the differences in a climate.

As livestock spread across regions their genetic profiles changed. All cattle have a common ancestor in the auroch, but they branched into two different subspecies. *Bos taurus* cattle have a thick coat, do not have a hump, are fatty, and cannot sweat. These cattle were raised in Europe and modern cattle breeds like the Angus, Hereford, and Holstein are *bos taurus*. By contrast, the *bos indicus* subspecies originated in India. In the hot Indian climate the cattle either developed or retained the ability to sweat. Moreover, *bos indicus* cattle have a thin coat of hair, are lean, and have a pronounced hump. These differences are important today. Because the *bos taurus* cattle are fattier they taste better; it is the small deposits of fat within the muscle, called marbling, that gives steak its great taste. This is why upscale restaurants often boast about selling Angus beef. *Bos indicus* cattle are leaner and less tasty, but they still exist today because they can better survive in warmer, drier climates and make perfectly good ground beef.

The spread of agriculture brought about a change in how food was obtained, and allowed more people to survive on the same area of land. As a result, the world population exploded, increasing from three million people in 10,000 BC to 100 million in 3000 BC (see Figure 2.2). Starting in 1000 BC, the population began growing in a clearly exponential manner. The particularly large growth rate from 1500 AD to today is a testament to the wonders of technology and political arrangement, which are more conducive to human life and growth. Though this tremendous surge in human populations was the result of a myriad factors, it is undeniable that the labor and food livestock provided was of paramount importance. The chances are you would not be alive today had

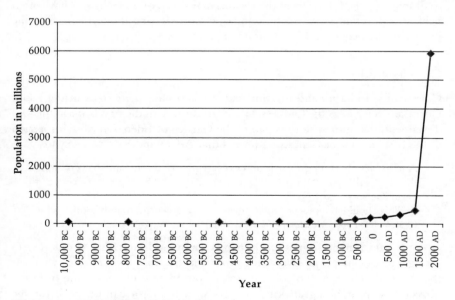

Figure 2.2 Historical world population

Source: U.S. Census Bureau (2010).

man not domesticated livestock. In short, farming proliferated throughout the world. Just as the farmer-human displaced the hunter-human, the domesticated animal began grazing the land of their wild ancestors. Side by side, people and their livestock built empires. Cattle began pulling plows in 3000 BC. As the Nile receded from the yearly floods, the Egyptians threw their seeds on the ground and drove the swine over it to help push the seeds down.[23] Men battled on their horses and slept in the same house as their pigs. Animals evolved through their domestication and human culture evolved with it. Together, man and animal trudged through a difficult and unforgiving world, developing new social norms and cultures that would make their communities more successful.

People and their Gods

The varied histories people have lived with different animals have resulted in a surprising diversity in cultural attitudes toward animals (as we first noted in Figure 2.1). A resident of Delhi, India, is much less likely to eat beef for dinner than a resident of Dodge, Kansas. It is no coincidence that the Indian resident is likely to be Hindu and the Kansas resident is likely to be Christian. To understand why the cow is worshiped in India and slaughtered in Kansas requires an understanding of cultural and religious views on animals. As we shall soon see, the followers of the various major and minor religions have often held very different views about animals, often reflecting the differences in the history of the people and their varied animals.

Outside of Eden

Christianity, Judiasm, and to some extent Islam have their roots in the books Christians refer to as the Old Testament. According to the first book of the Old Testament, the human story began in the Garden of Eden. On close inspection of the first chapter of Genesis it appears that Adam and Eve were vegans.

> And to every beast of the earth, and to every fowl of the air, and to every thing
> that creepeth upon the earth, wherein there is life, I have given every green
> herb. (Genesis 1:30)
>   This should be interpreted to mean: every green herb and nothing else.
> (Jean Soler, *Food: A Culinary History*, 1996)[24]

After banishment from the Garden, Genesis indicates that people began eating meat—but people were not given the freedom to consume whatever they wished. The most notable examples from these traditions come from the Mosaic law, prohibiting all kinds of foods, including a ban on pork for the Israelites, a prohibition that remains today in the Jewish and Muslim religious. The reason given in the Bible for this law is that pigs do not chew their cud.

Leviticus 11:3 says, "Whatsoever parteth the hoof, and is clovenfooted, and cheweth the cud, among the beasts, that shall ye eat." Thus, Jews could eat a cow but not a camel, whose hoofs are not parted, or a pig, who does not chew cud. It is not exactly clear why this limitation should have been placed on the Jewish diet, but a number of hypotheses exist.

Some historians argue that Jews prefer to eat only animals that eat plants (herbivores), because they are closer to the vegetarian ideal God had originally intended. All animals that are ruminants are herbivores. Ruminants are easy to spot because they chew cud—cud being grass that has been regurgitated to be chewed for a second time—making them a double-vegetarian of sorts and extra clean from the rabbi's point of view. The hog was perhaps deemed unclean, in part, because it was an omnivore, and not the vegetarian ideal.[25]

Another hypothesis is that such traditions emerged from the insight that certain foods are safer to eat than others. Pigs and humans share many common parasites and diseases that can easily be transmitted—the "Swine Flu" or H_1N_1 virus is a current example. Another example is Trichinella (a parasitic round-worm), which was once common in pork and is deadly to humans. Some historians dismiss this explanation because it does not explain why other cultures (e.g., Chinese) did not also refrain from eating pork. Finally, we cannot rule out the possibility that there is no *logical* reason for the prohibitions other than tradition—the Mosaic laws could have resulted from stories of divine revelation or simply from arbitrary rules developed long ago. In either case, the law would be subsequently followed simply for being an integral part of the religion and culture.

Christianity and the New Testament outlined a new set of rules relating to farm animals. From the onset of Christianity the old Jewish norms on the uncleanness of pigs and other foods were set aside as two of the most prominent and influential apostles, Paul and Peter, both shunned the old eating rules. Later, the Catholic Church, which became quite powerful in the Middle Ages, organized a set of guiding principles regarding meat consumption. Meat fasts were required by the Roman Catholic Church. Although fasting is common in many religions, the Roman Catholic Church specifically targeted meat. When the church bans were fully enforced Christians had to abstain from certain animal products on Wednesdays, Fridays, Saturdays, and all the days of Lent. This constituted half the days of the year! Sometimes meat was interpreted to include all animal products, and sometimes exceptions were made, such as fish.[26]

Reasons for the religious prohibitions on meat might be found in the conflu-ence of social status and personal character in the Middle Ages. At this time, meat consumption was associated with strength and power. The wealthier noble classes ate far more meat than the peasants, and they often roasted meats which allowed much of the meat fat to escape. Peasants, when they had access to meat, boiled the meat to make sure every morsel was consumed. Meat was such a strong identifier of status that if a person of power was punished they were often forced to abstain from meat, sometimes for life.

There were three general classes of people in Middle Ages in Europe: the nobles, the peasants, and those affiliated with the Church. While the church leaders were often as obsessed with power as were the nobles, their strategy for gaining power was very different. Entering the church was supposed to imply that one had renounced the world and any quest for power. The monastery demanded humility of its subjects. Thus, to signal a rejection for the thirst for power many monks abstained from eating meat (except when sick).[27] "...renouncing meat—a symbol of violence and death, physicality, and sexuality—was a cardinal point of monastic spirituality from the dawn of Christianity" (Montanari, 1996: 183).

This view of meat cultivated in the early Christian churches gained acceptance over the centuries. Many modern Catholics still abstain from meat on certain days of the year as a result of the church's ongoing respect for vegetarianism, both in remembrance of the Garden of Eden and due to the role of meat as an expression of social identity and power in centuries past.

Within a century of Jesus' death there emerged competing "Christian" sects, each with their own doctrines and interpretations of the gospels. Many of these sects became later knows as heretics: these included the Paulicians, Massalians, Bogomils, and Cathars. They were *Christians* in the sense that they based their beliefs on the life of Jesus and the gospels; as historian Michael Frassetto writes, "Devotion to the scriptures and the life of Christ and the Apostles was promoted by all the leading heretics."[28] A common theme among many heretic sects was adherence to asceticism. The earthly and material things were thought to be vulgar; true treasure was in heaven. Some sects had a dualistic conception of the world, where it was believed that the physical—the earth—was the domain of the devil. Shunning the physical world and thus the devil's lies would bestow life in heaven. Consequently, many ascetic sects abstained from wine and meat, and some sects even forbade the consumption of milk, cheese, and eggs.

Some of these sects died out, some persisted, and other new ones developed. The existence of sects throughout the Dark Ages and Middle Ages is partially attributable to heterogeneity in people's preferences for religious intensity, as, for example, manifested in the demand for asceticism. Although some groups worked within the Catholic Church for reform, for some the demand for asceticism was too great and new sects would emerge. For example, when people with a strong preference for asceticism believed the Catholic Church had become too worldly, new sects would develop.[29] The presence of sects became particularly problematic for the Catholic Church in the Middle Ages as the Church sought to continue its monopoly on religious beliefs by squashing heresies.

Desiring to maintain their monopoly position as Europe's spiritual authority, the Catholic Church often sought to squash the heretics through a series of Inquisitions. Those under investigation were asked to assert their devotion to the Catholic Church, and one observable factor seen to separate heretics from Catholics was the eating of meat. One accused heretic during the Medieval Inquisition pleaded the accusation was wrong by stating, "I am not a heretic, I have a wife whom I love, I have children, I eat meat, I lie and I take oaths, I am a good Christian."[30]

The vegetarianism (and sometimes veganism) of the heretics did not emanate from respect for animals necessarily, but rather from a disrespect to worldly pleasures. As we have seen, some Catholic monks would abstain from meat for this reason. There is evidence that some heretical behavior was driven by compassion for animals. The Bogomil sect, for example, did not condone violence toward animals. Nevertheless, empathy and respect for animals was not the major motivator for heretics. There is no compelling evidence to suggest the heretics were more concerned for the well-being of farm animals than were the Catholics, even if the former were more likely to be vegetarian.[31]

As we have witnessed, a commitment to the Christian faith did not necessarily imply a commitment towards a particular stance on the treatment of animals. Orthodox Christianity views the relationship between human and animal as that of the first being the steward of the second. Humans were seen to have been given dominion over animals by God, but they were also given responsibility for their care. Jews were to sacrifice all kinds of animals to their God and were free to eat and use many animals as they saw fit, but they were also commanded to feed and care for their animals.[32] Jesus sacrificed animals, helped Peter catch a boatload of fish, and rode a donkey, but he also encouraged his followers to break Jewish laws in order to help a hurt animal. In the two competing roles of exercising dominion and being a good steward emerges the sometimes contradictory stances of Christians toward animals: the sometimes callous of treatment of animals vs the moral line taken in the argument for improved animal welfare. Indeed, the ardent Christian and anti-slavery activist, William Wilberforce, was one of the co-founders of the world's oldest anti-cruelty society in 1824.

The Renaissance era witnessed people seeking knowledge from sources outside the Catholic Church. Rene Descartes, for example, sought the self-pursuit of knowledge, and is often considered the father of Western philosophy. One particular topic of interest to Descartes was the manner and process by which nature and biology operated. He was interested in how the body of an animal worked, and whether that body was simply a machine or whether it was part of a conscious being with a soul. His conclusions were surprising.

Much like a clock that performs marvelous functions without having the ability to feel emotion, Descartes argued that animals were indeed living animals but had no capacity for feelings. Further, he argued that animals had no soul. Figure 2.1 documents how Descartes' entourage would publicly beat animals and mock those who empathized with animals.[33] While teaching a class, Descartes would nail dogs' paws to a board and dissect the animal while it was alive to study its biology.[34] Ironically, Descartes was a vegetarian, though obviously not for reasons of animal rights. He believed it was a healthier diet. Descartes is the quintessential example for demonstrating that vegetarianism is not necessarily accompanied by empathy with animals.

Though Descartes' view of the clockwork universe had enormous influence on society, there were those who challenged his view that animals could not suffer. The seventeenth century was an incubator for radical movements, many of these made up of vegetarians and animal sympathizers. A famous example is Thomas Tryon, who at age 23 became a Pythagorean, which meant he accepted Pythagoras's teaching regarding adopting a vegetarian diet. In addition to campaigning for better treatment for animals, Tryon sought better treatment for slaves, influencing the future founder of the first animal advocacy group. Moreover, it would be a mistake to assume Descartes' powerful influence over-rode other influences.[35] To the chagrin of the Protestant church, the early English still held beliefs that derived from their pagan roots: hogs could see the air, owls could deliver warnings of death, a goat could be the devil in disguise, and even today many of us pause when a black cat crosses our path. While these animals were not supposed to possess powers themselves, they were seen as portals for mystical powers, illustrating that not everyone viewed animals as a mere machine.[36]

It is often difficult to precisely pinpoint a culture's view toward animals. If the Western culture considered animals to be machines, as some did, or as sentient creatures without the ability to reason, as most did, why were farm animals put on trial for crimes? In 1379, three pigs trampled a young French boy, killing him. Not only were these three pigs put on trial for murder, but the remainder of the herd was also charged as accomplices. The three murderers and their accomplices were all condemned to death, although they were probably due for slaughter anyway! This was not a unique event; there were at least 93 cases of criminal court trials of animals between the twelfth and eighteenth centuries.[37] In 1648, the state of Massachusetts passed a measure making it unlawful for dogs to harass sheep. If the dogs could have understood that the punishment was hanging, they might have been more obedient![38]

Colonial Beasts

As the Protestant faith replaced Catholicism in England, views of livestock changed little. But British culture was one that ascribed value to people acting civilly toward livestock. Proper and successful farmers were viewed as those who provided diligent care for their animals and raised the animals in a way that was sympathetic to crop farming, to ensure a sustainable agricultural system. When the English began sailing to the American colonies they took livestock with them, not only for the food and labor but to preserve what was viewed as integral to a civilized life in the English culture. In fact, it was hoped that the American Indians would begin farming livestock, and it was thought that the mere ownership of animals would begin to temper their supposed savage tendencies. They presumably hoped for the development of a new culture among the American Indians—a culture more like that of the English.

In reality the colonial livestock brought to Maryland soon became more like the wild beasts in the surrounding woods. The land used to grow crops was littered with stumps, making animal labor infeasible. The lack of grazing material in the woods meant cows produced little milk, and so cattle were used primarily for meat, rather than for labor or their production of milk, as was the English tradition. The difficulty in providing animal feed, coupled with the unbounded forests surrounding the farm, made it more sensible to turn the animals loose in the surrounding woods to forage for their own food. Moreover, one of the most profitable crops was the labor-intensive tobacco. Farm labor was scarce and expensive. Any servant or slave a colonist could obtain was of more valuable use working in the tobacco field than in tending cattle. Sheep and goats were soon recognized as a failed import in the new land because they were too easy prey for wolves; the goats inflicted too much damage to apple trees; and the sheep's wool was too easily stripped by the forest brush. Pigs, cattle, and horses were able to withstand the threat of predators and survive in the forests. Chickens also survived by flying away from wolves and foxes. It became a spring tradition to search the woods for livestock and bring them back to the farm for some human contact, to ensure that the animals did not become completely wild.

The conditions in the New England colonies were in sharp contrast to their Maryland counterparts, developing the many cultural differences that would later lead to the outbreak of the civil war. In contrast to Maryland, the New England colonies had no lucrative cash crops to generate the means to buy what they needed, so they had to produce what they wished to consume. The only English crop that fared well in the New World was rye, so colonists relied extensively on domestic corn, some rye, and their animals. In New England, land was easier to clear and plow than in Maryland, making cattle a useful source of labor. There was a greater availability of grazing material too, so cattle produced more (and were used extensively for) milk; and despite the winters being harsh the cattle thrived. Sheep were especially valuable, as the winters were cold and the wool they provided was very useful in producing suitably warm clothing. Rather than spreading out by setting up plantations like the settlers in the southern colonies, the New England colonists lived in dense towns. Livestock animals were often kept on communally owned pasture land. So important were the oxen for farming and wealth generation in New

England that they could not be taken from a delinquent debtor. Animals used for milking and labor were kept close to the household and raised in much the same way as were their British counterparts, while animals used for meat and breeding were taken to pastures further away from the towns.

The early English culture was a meat-eating culture. England possessed the greatest number of domestic animals per capita of any European country, except the Netherlands, and this reliance on meat did not diminish amongst the English who moved out to the American colonies. The early colonists consumed large amounts of meat from both livestock and wild game. Even servants expected to eat meat regularly. Free-ranging animals became so prevalent that farmers were expected to fence in their crops to keep livestock out, rather than the other way around. Unfenced tobacco fields would be damaged by animals. Managing livestock was not just a business matter. Social norms emerged to ensure animals were protected to generate wealth to the owner and to protect damage to the wealth of neighbors. As historian Virginia De John Anderson states, "The management of domestic animals, like the suppression of sin, was too important to be left to the discretion of hard-pressed farmers. As a result, the community assumed responsibility for keeping order on farms just as it did for encouraging good behavior within farmhouses."[39] Similar sentiments can be found in the Old Testament. Exodus 21:35–36 states,

> If a person's bull injures the bull of another and it dies, they are to sell the live one and divide both the money and the dead animal equally. However, if it was known that the bull had the habit of goring, yet the owner did not keep it penned up, the owner would have to replace the other person's dead animal with a live one; the dead animal remain the property of the original owner.

Sacred Cows

It is interesting to consider how some societies might have come to deify some of their animals, forbidding their people to eat their meat. For example, there is the sacred cow of India. There is every possibility that the early inhabitants of the area that is now known as India were meat-eaters, and that they would have eaten beef as well as other meats. How might the Indian culture have evolved to worship the cow and no longer eat its meat? Historians have developed a number of plausible scenarios, such as the one that follows.

A group of pastoralists referred to as Aryans invaded the area that is now called India. The Aryans were superb producers of dairy products, including a clarified butter that would last for months in hot climates. Aryan dairy products were enthusiastically adopted by the indigenous peoples and they soon became dependent upon dairy products. As a result, the milk of a fertile cow became far more important than its meat. Early sacred Indian texts, for example, list a number of food items people ate at the time such as goat and buffalo meat, but the only beef listed as being eaten was that from barren cows. As dairy products played an increasingly prominent role in Indian society, the cow was given greater reverence for the nourishing milk it produced. The male cow was heavily relied upon for labor, and by the time it was too old to pull a plow its slaughter could only provide a minimal amount of meat that would be very lean and tough. As a result, both male and female cows grew to be more important alive than dead; this may well have led to the subsequent special meaning that was ascribed to their lives.

It would seem unnecessary for ancient Indians to deify the cow. Even if male and female cattle were worth more alive than dead, from today's point of view there seems to be no obvious reason for religions to codify what are thought to be best practices. However, consider the fact that in these times, effective agricultural practices were handed down through the generations via oral communication. Perhaps the view of the sacred cow began as a simple desire to stress to young families the importance of caring for their cattle until the cattle reached old age, and that to slaughter cattle for meat was an unwise choice in the long run. One interpretation of the existence of religious laws is that they represent a method of identifying and communicating important social norms; they are in effect codifications of best practices that serve to advance a society's interests. It is natural for societies to communicate rules for living a good life by making those rules sacrosanct. And as early as the 1600s some British writers also speculated that the act of cow worship emerged from a motivation to extract the maximum value out of each cow.[40]

Evidence from colonial America lends support to this explanation of the advent of the sacred cow. During the early 1600s, colonists in Maryland experienced great difficulty obtaining adequate food, so much so that they began slaughtering their livestock for food at such a rate that livestock herds were decimated, leading to smaller and smaller herds being left. In order to persuade the colonists to allow more of their animals to breed, colonial officials periodically issued prohibitions against the slaughter of breeding cattle. A similar story has been told of the New England region, but relating to sheep stocks; the settlers were slaughtering so many of their sheep for food that many of them then faced the cold winters without adequate woolen clothing for protection. To encourage the raising of sheep primarily for wool, the colonial governments passed measures making it illegal to slaughter sheep under 2 years old, ensuring the each sheep was sheared at least twice before it was killed for food.[41] Returning to the Indian sacred cow scenario, perhaps if these prohibitions had been made by priests, and if the rulings had been upheld over generations, it is possible that they might have grown to be viewed as religious laws.

The Aryans had no qualms about meat-eating, although they were among the first groups to identify "clean" and "unclean" meats: dogs, chickens, pigs, and camels were only to be eaten in time of famine. Alongside the Indian belief in cows being sacred beings, other cultural beliefs developed, such as the reincarnation of people, who were believed to sometimes return as animals, and the transmigration of souls. After it was accepted that a human could be reborn as an animal and an animal be reborn as a human, vegetarianism became the social norm. All animals were thereafter given greater consideration, as well as the cow. One of the Hindu chants detailing instructions for daily conduct states, "he who gives no creatures willingly the pain of confinement or death, but seeks the good of all, enjoys bliss without end."[42] (And it is interesting that this ancient chant specifically mentions *confinement* as being a source of pain, since the central debate in modern-day egg and pork production concerns whether hens and sows should be housed in cramped cages.)

When a society develops an enemy, symbols emerge to differentiate friend from foe. Vegetarianism became a symbol for Hindu Indians as they began to battle beef-eating Muslims. The sacred cow was a salient difference between the Hindu and the Muslim, and the enmity between the two peoples acted to reinforce the Hindu Indian's reverence for the cow.

Followers of the Buddhist and Jainist religions also believe in reincarnation, as these religions emerged from the same social changes that induced Hindus to become vegetarians. Their belief is that living a good life will cause one to be reborn into a higher state. All creatures—human, cow, pig, and insect—were believed to contain someone's soul: eat a pig, for example, and you may be eating your great-grandmother. Thus, Buddhists and Jainists insist on a vegetarian diet. Milk from animals or eggs from hens were acceptable fare in the view of some priests, but most Buddhist and Jainist priests today promote strict veganism. Some even go so far as to make sure there are no insects living in the fruit and vegetables they eat, and to abstain from eating garlic because harvesting it requires killing the whole plant.

A vegetarian diet is very restrictive in terms of the range of foods that can be eaten, and in previous centuries such restrictions could easily mean the difference between life and death—it is thus surprising that the vegetarian diet was ever accepted in any form. But not only was it accepted, it was also tremendously influential. The Jainist priests' arguments for vegetarianism were so influential that around 100 BC Hindu priests also began to espouse the vegetarian life (not just forbidding the consumption beef). And to this day, Hinduism, Buddhism, and Jainism, three major world religions, do not condone the eating of animal meat and at certain times of the year they even forbid the consumption of milk and eggs.

One could debate the extent to which the adoption of vegetarianism and veganism was due to a concern for the animal itself. As we have seen, one can be a vegetarian and yet possess no empathy for animals. These diets could emanate from a concern that the animal actually has a human soul, for ascetic or health reasons, or from a belief that encouraging compassion toward animals facilitates compassion toward one's fellow humans. However, many features of Jainism and Buddhism do indeed seem to require respect for the animal itself, regardless of whether the animal possesses anything resembling a human soul. Buddhism justifies vegetarianism in not only presenting the view that animals may in fact bear the souls of reincarnated relatives, but also in stressing the pain caused to the animal in being reared and slaughtered for food. It is possible that some followers of these religions simply follow the dogmatic rules of the religion, possessing little genuine care for the animal. Even within cultures maintaining a strong concern for the animals in their care, while there are some people who will work tirelessly on the animals' behalf, there are others who give little care for their welfare. Mahatma Gandhi, himself a Hindu, set up communes in which cows were kept, giving some members of the communes the task of protecting the cows. He described cow protection work as, "cattle-breeding, improvement of the stock, humane treatment of the bullocks, formation of model dairies, etc." When visiting one of the dairies at a commune, Gandhi lamented, "the so-called Hindu still cruelly belabors the poor animal and disgraces his religion."[43] Notwithstanding these exceptions, it seems self-evident that cultures which encourage compassion toward animals produce citizens with greater empathy for animals. Cultural influences are powerful, and play a significant part in forming views on the place of animals in the world.[44]

British Experiments

When the British East India Tea Company was established, trading with India, workers who traveled to India sent back detailed descriptions of a culture that lived at peace with its animals. Descriptions of the Brahmins, a priest sect within Hinduism, were especially noted, speaking of a pacifist, vegetarian, and animal-loving people. This literature written by those who traveled to India, describes the Hindu diet as one centered on the ethical treatment of animals. Readers were told about animal hospitals in India and the sacred cow. Although many vegetarian sects in Britain were motivated by health or ascetic reasons, the Hindu motivation was apparently animal welfare. British writers contrasted the seemingly peaceful, loving, vegetarian Indian culture with the meat-eating British culture, and they initiated a journey of soul-searching. One writer who ended his journey siding with the Hindu way was Thomas Tryon. Tyron was the original self-help guru. He became famous for his books on herbology, cooking, astrology, and the like. After reinterpreting the Christian Bible to confirm his new vegetarian philosophy, Tryon boldly asserted that the Brahmins led the true holy life, and that the British should follow their example and convert to vegetarianism.

Other British writers were less impressed with stories of India. Some mocked the Jainist's unwillingness to kill head lice. Allowing head lice to fester, allowing rats to reside in one's home, and making sure firewood was free of ants seemed ludicrous to many British readers. Others noted the absurdity and hypocrisy of a culture that abstained from eating meat but burned widows alive when the husband died in a practice known as *Sati*.

However, one essential difference developed between British vegetarians of the eighteenth century and those of earlier times: vegetarianism was no longer associated with Christian heretics. There were a few exceptions. For example, the writer G. K. Chesterton (1874–1936) seemed to associate vegetarianism with the ancient Manichean heresy, but such views do not appear to have been widespread. This was important, because it allowed animal rights and vegetarianism to be debated on practical, logical, and scientific grounds—instead of in terms of religious dogma (although there were some, such as William Wilberforce, who argued against animal cruelty on religious grounds). The debate about vegetarianism now centered on issues such as health, agricultural efficiency, and animal welfare.

It was a religious activist, however, who instigated the first animal welfare policies in England. Richard Martin, an Irish Member of Parliament, first presented his bill to prevent the cruel and improper treatment of cattle to the British House of Commons, and this bill was ultimately passed in 1822. Failure to enforce the legislation led Wilberforce, Martin, and the Reverend Arthur Broom to form the Royal Society for the Prevention of Cruelty to Animals in 1824, which is still in existence today. The principle objectives of the society were to ensure the law intended to protect animals from cruelty was enforced, to seek regular amendments to the anti-cruelty law, and to create and sustain an intelligent public dialog on the proper treatment of all animals.[45]

At the beginning of the twentieth century, many British citizens suffered from malnutrition caused by a combination of poor food choice and lack of food availability. Some lamented that livestock were better fed that the poor. The Vegetarian Society had been formed in 1847, and one hundred years later a number of its members suggested the society should promote a diet free from all animal products, including cheese, milk, and butter. The Vegetarian Society disagreed with this suggestion, causing those members to form their own society called the Vegan Society.

Increasing acceptance of Darwin's theory of evolution caused people to view animals as another product of nature, with no more or less right to exist than humans. One line of thought that gained attention was that of Dean Inge, an English author and Cambridge professor of divinity. After acknowledging that a process of natural evolution produced both man and animal, he stated, "if we assume that survival has a value for the brutes, no one has so great an interest in the demand for pork as the pig."[46]

It is interesting to note the similarities between the British Experiments and the contemporary farm animal welfare debate. The choice to eat or abstain from eating meat no longer has implications in relation to whether one is a heretic or Christian. The ideas that clash reflect the contrasting ideas of Western and

Eastern philosophy and religion. The British Experiments continue to this day, and have transplanted themselves west into the US and east into other European countries. Should we treat animals as our property, as in the Western tradition, or as sentient beings and even Gods, as in many Eastern traditions? Or, is there a happy medium between these two views? The modern farm animal welfare debate is an offspring of the early British Experiments.

A Complex Relationship

The human race has never really resolved the question of how to treat farm animals. We are conflicted in a myriad ways. On the one hand, farm animals provide us with nourishment, clothing, and even medicines. Human lives are enhanced by the use of animals. On the other hand, it seems clear that, like humans, animals have the ability to experience misery, and humans have a natural tendency to empathize with the feelings of others. Most humans are negatively affected if they see an animal being poorly treated, which places a limit on the extent to which people are willing to exploit animals for their own benefit.

People have resolved this conflict between seeing animals as commodities and seeing them as sentient beings by reaching various conclusions. The result of these differences is that we have widely varying practices in evidence around the world: bull-fighting, meat-eating, veganism, and cow worshipping. The differences are seen in the varying treatment given to livestock and pets: the hens that lay eggs while living in small cages with nothing to do but eat, sleep, and lay eggs; and then, conversely, the dogs and cats who often receive royal treatment, some people even being willing to quit smoking for the sake of their dog's health.[47] How will it be in the future? We have seen above that even just one or two individuals can become trend-setters, setting into motion changes that lead one society to worship animals, for example, and another to eat them. For thousands of years humans have vacillated between exploiting and respecting farm animals, and there is no reason to believe the debate about how we treat these animals will subside in the near future.

We humans have never really agreed on how to treat each other either. We rarely question the consciousness of other people because we ourselves are conscious. Yet human history is replete with slavery and racism. In a sense, it should not be surprising that we cannot decide how to treat other species when we are still struggling to decide how to treat members of our own species. In this chapter, we have looked back over human history in order to try and gain a better understanding of how human culture and their livestock are intertwined, in order to understand how the two might go on to relate to each other in the future.

It is impossible for livestock not to have a tremendous impact on our various cultures, given that they have provided for so many of our needs over the past 12,000 years. History shows that livestock have not simply been passive passengers who provided food as we humans met our destiny—animals have helped to shape our destiny. Humans and livestock rose from the wild to create modern civilization and did it together. Modern society now affords comforts previous generations only dreamed of. The question now is whether it is time that we extended some of our comforts to our farm animals as well.

In This World Together

Until about 10,000 BC, the relationship between humans and the animals that eventually became livestock was clear. Humans were predators; animals were prey. As livestock agriculture developed, human–animal relationships became less adversarial, more interdependent, and more complex. The animal still provided food and clothing, but rather than hunting animals man began caring for animals. Once an animal was fully domesticated they could not even survive outside the care of humans (with some exceptions). And yet, farm animals—cows, pigs, chickens, and sheep—are some of the most numerous animals on earth. In Darwinian terms, keeping livestock has been a spectacular success, and this success is due to the animals' ability to adapt to human needs.

Until recently livestock and humans lived in similar conditions. Both smelled nasty, were dirty, and lived with the constant threat of starvation. This is because the livestock was actually kept in the human's dwellings in many cases. Pigs and chickens were kept inside during particularly bad weather and at night. Dogs, and often pigs and chickens, were often fed with scraps that would be thrown onto the dirt floor of the house—humans and their animals literally dining together.[48] During the Renaissance most families of moderate wealth were able to avoid living under the same roof as their livestock, but their servants continued to sleep in the stables with the animals.[49] Humans and livestock lived through difficult weather conditions together. A fierce winter or crop failure was devastating to animals and humans alike. When describing the harsh winters of seventeenth-century Scandinavia, historian Reay Tannahill states, "Famine in the extreme north was sometimes so near when spring came that the cattle, skeletal from their winter diet of straw and shredded bark, had to be carried out to the pastures."[50] And if grain was scarce, there was no doubt that the animals would be those to do without.

But times have changed. Technological innovation and capitalism have steadily raised human living standards, and in some ways those of animals. In the modern developed world most humans have little fear of starvation. The eighteenth century witnessed an agricultural revolution, particularly in inventions and improvements in cropping technologies. These technologies led to an abundance of grains such as never seen before in history; this made it easier for humans to feed themselves, and made it easier to feed animals through the winter. Improved building methods and cheaper materials allowed for keeping animals in larger, cleaner barns, protected from the environment. The tractor replaced the ox and the ass. Now, all that is expected of cattle is that they eat, breed, and rest. Female sows now have private barns to raise their young, and hens have their own laying room with individual nests. Farmers developed a better understanding of the nutritional needs of their animals and how to care for the animals' health.[51] This change in the animals' living conditions does not necessarily mean that farmers now have greater empathy with their livestock—healthier pigs, for example, gain more weight which means the farmer can earn more money from them.

The modern world has attained its increased wealth by applying scientific knowledge in the workplace and persistently seeking methods to produce goods at a lower cost. Henry Ford demonstrated how a large factory, using an assembly line of many workers, could produce better cars at a lower cost. The innovator, J. P. Morgan, thought the factory method could be applied to agriculture as well. Morgan's experiments with large-scale farming were profitable. One farmer in 1926, Thomas Campbell, managed a farm of 95,000 acres.[52] This was at a time when the average American farm size was closer to 55 acres.[53] It would not be long before these production techniques were extended to the livestock production. While agricultural technologies clearly benefitted humans, their impact on animals is less clear. At no point in history have animals been given more plentiful and nutritious food, such constant access to water, and such favorable temperatures in which to live. Yet at no point in history have so many animals been confined to such small, barren environments.

The average person living in the modern-day developed world has enough money to lead a happy life. There is no longer the daily struggle to meet basic needs that existed in times gone by and the majority of us live comfortable lives. We have arrived at this point in history with the help of our livestock, but have we now forgotten about them? After our long, intertwined history with farm animals, now that we have acquired material comfort, do we leave our livestock to suffer in poor living conditions? Or can we fairly argue that farm animals do in fact live contented lives? Or do we simply not care whether they are contented or not? There are so many questions to answer in the farm animal debate, and so few have attracted the attention of objective researchers. The aim of this book is to address the most pressing of these questions, objectively and with an open mind.

CHAPTER 3

Animal Farms, Animal Activism

The Emergence of Factory Farms and Its Opposition

"...consider every flock of hens an egg factory."
Davis et al. in Livestock Enterprises, 1928

"Ever occur to you why some of us can be this much concerned with animals suffering? Because government is not. Why not? Animals don't vote."

Paul Harvey, radio broadcaster

Seeds of Animal Activism

The long history of humans and their livestock contains stories of animal abuse, but also stories of compassion. Although there were fringe groups in the seventeenth and eighteenth centuries that promoted vegetarianism and better treatment of farm animals, it was not until the nineteenth that the British experiments led to the forming of more powerful interest groups to enforce the first modern animal welfare laws. Anti-cruelty laws were introduced in England in 1800 by Sir William Pultney, but had little effect in improving animal farming practices. More effective laws were passed 20 years later by Richard Martin (nicknamed Humanity Dick). Influenced by the writings of William Wilberforce, Martin was able to bring in legislation protecting certain livestock from excessive cruelty. The first animal welfare group, the Society for the Prevention of Cruelty to Animals, was formed in 1824 to help enforce the anti-cruelty laws. The society was later renamed the Royal Society for the Prevention of Cruelty of Animals (RSPCA) when Queen Victoria expressed her approval in 1840. The efforts of the RSPCA to disseminate information made animal welfare the topic of common newspaper and editorial articles at the time. The RSPCA culture was based on Wilberforce's philosophy that the primary purpose of the society was to alter the public's moral feelings toward animals, and not just to pursue the prosecution of those who violated anti-cruelty statutes.

The United States was not immune to the changing sentiments across the Atlantic Ocean. The first US anti-cruelty legislation that made it illegal to "cruelly beat" an animal, regardless of ownership, was put in place in 1821 in Maine. Later, in 1943, a Pennsylvania man was sentenced to a one-year prison term for cruelly beating a horse, even though the state had no laws on the books prohibiting the act. In 1866, with the leadership of activist Henry Bergh, the American Society for the Prevention of Cruelty to Animals (ASPCA) was officially recognized as a group, making it the first animal advocacy group in the US. The group immediately sought to strengthen existing animal protection laws that had been largely unenforced. Local affiliates of the ASPCA soon began forming and coordinating their efforts. For example, Massachusetts and Pennsylvania established their own ASPCA charters, calling themselves the MSPCA and PSPCA, respectively. While small in number and consisting largely of women (who were less influential than men at that time), the groups were ambitious and formidable. In contrast to the advice offered by Martin on the purpose of animal welfare societies, Henry Bergh even tried acting as an arresting officer when he saw cruel acts. The ASPCA's efforts eventually led to official arrests and prosecutions. In the year following its inception, the ASPCA had convicted 66 people in 119 cases that were tried for animal cruelty.

The ASPCA largely concerned itself with the blatantly cruel acts witnessed in the busy New York streets. Road rage existed in the 1800s, but consisted of humans beating their horses. Some carriage drivers would literally work their horses to death; afterwards they would often unhitch the carriage and leave the dead animal on the street for someone else to remove. Carriage drivers who used bridles modified with nails and spikes to make the horse "prance" became ASPCA targets. The ASPCA challenged people who failed to provide adequate water to horses during races, performing vivisection, and the cruel transportation of animals; they built animal shelters and provided health care to pets and work animals. The cases the ASPCA helped prosecute included acts of stabbing horses with pitch forks and torturing dogs and cats. The exact level of such animal cruelty in the 1800s is unknown, but these acts obviously took place, and the ASPCA was the first opponent of animal cruelty of any influence in the US.

The ASPCA and its affiliates did not ignore farm animals. One of Henry Bergh's first priorities when he became the nation's foremost animal rights advocate was to force the livestock industry to utilize more humane transportation techniques. One example of cruel practice in the mid-1800s was that calves would sometimes have their legs bound and would be thrown onto railroad carts like suitcases. Other livestock animals were densely packed into railroad carts and transported for several days without food or water; when the animals arrived at their destination many would have lost 20 percent of their body weight, and in some cases the entire cart would be filled with dead animals. As depicted in Upton Sinclair's *The Jungle*, slaughtering methods also had much room for improvement.[1]

Animal advocacy groups appealed to public concern for food safety by describing the injured and putrid animals exiting railcars just prior to slaughter. These activities persuaded politicians to pursue the Twenty-Eight Hour Law, which stated that animals should be fed, watered, and rested every 28 hours. The law proved ineffective: what would qualify as a rest stop was not well-specified, there were too many exceptions, and there was no emphasis placed on enforcement.

Champions of the Twenty-Eight Hour Law were disappointed. Reaction to this "defeat" resulted in two approaches to improving farm animal welfare which continue to this day. The American Humane Association (AHA), founded in 1877, was formed to protect the well-being of both children and animals. The AHA reacted to the failure of the Twenty-Eight Hour Law by working *with* the livestock industry to address concerns. They held conferences, awarded prizes for humane railroad car designs, and provided education to the industry. The organization remains in existence today and has continued its approach of working with the livestock industry—rarely opposing the industry publicly. The AHA is seen by some as representing a more moderate view of animals in society and as taking a more pragmatic approach to improving farm animal well-being, but it has been seen by others as a *de facto* animal industry group.[2] And there was some justification for this allegation about the AHA. A federal investigation once revealed that the AHA had lied on the meatpacking industry's behalf regarding compliance with the Twenty-Eight Hour Law.[3] Nevertheless, it would be difficult to deny that AHA activities have improved the lives of farm animals, and that working *with* livestock producers can sometimes be fruitful. Today the AHA remains active, and does not appear to be a proxy for the industry. To encourage producers to adopt higher welfare standards and communicate those standards to the consumer, the AHA has created a private label called the American Humane Certified label (formerly known as the Free Farmed program). The label is designed to advertise meat and eggs from animals raised under better conditions than those on conventional farms.

The AHA is also important because it gave birth to another group that has had a tremendous impact on the animal welfare debate in recent years. In 1954, four of the AHA's officers quit the organization, frustrated by AHA's unwillingness to confront and challenge the livestock industry. After leaving the AHA, these four individuals (led by Fred Myers) formed the Humane Society of the United States (HSUS).[4] In the US, the HSUS is arguably the leading and most influential organization in farm animal welfare today.

It is tempting to think of cramped living conditions for farm animals as a recent problem, but the women's branch of the PSPCA fought the issue more than a hundred years ago. Before refrigeration, milk transportation was limited and cities such as New York and Philadelphia housed milk cows inside the city. Not surprisingly, the scarcity of space led to densely populated dairies where the manure was seldom cleaned out. Cows easily became sick and developed sores on their skin; some even claimed that some animals were so sick they could not stand without the help of belts hanging from the ceiling, although this was not substantiated. Moreover, the cows were fed garbage, a practice that was thought to make milk drinkers sick. Reporting on the practice led to the Garbage Milk Scandal, and city officials were persuaded to begin inspecting urban dairies.

Between 1866 and 1915, animal advocates in the US made the transition from being ineffective fringe groups to becoming influential and respected institutions. Still, they made little progress on key issues that interested them: eliminating vivisection, improving slaughtering methods, ensuring the use of more humane transportation and of better animal shelters, etc. Beginning in the 1950s, significant changes in livestock production occurred that would give animal advocacy groups greater reason for opposition.

Until the 1950s, the manner in which farm animals were *raised*—as opposed to how they were transported or slaughtered—gave little reason for concern, except in the case of a few exceptions, such as the Garbage Milk Scandal. Then came *factory farming*. Within a few decades nearly all egg, pork, and veal production took placed in conditions which appeared to many observers as cruel. This provided animal advocacy groups with a new mission and new ammunition.

The Factory Farm

Factory farm—do those two words make you uncomfortable? Does the idea that food may come from a business with hired labor and a mechanized, streamlined production system make eating dinner less appealing? Do you think food produced on factory farms is less safe, less tasty, less nutritious, or less humane? Before answering, consider this thought experiment. Do you prefer that your medicine be made in a modern, sophisticated factory environment, or do you prefer buying from a married couple who drive to the city once a week to sell their homemade medicine? You probably prefer medicine made in a factory, but if we asked the same question about food, you would probably say you preferred food bought from a farmers' market supplier than food from a factory farm. Why the discrepancy? Why does medicine from a factory sound appropriate but food from a factory sound unsatisfactory?

There are at least two reasons why the term *factory farming* may make you uncomfortable. The first reason is that the term is used in negative attacks by fierce critics of modern agriculture, often unfairly. Some activities conducted on large modern farms are viewed skeptically by the public, despite the fact that they are scientifically shown to be harmless or beneficial. This skepticism has provided groups opposed to modern agriculture with a tool for exploiting well-intentioned but uninformed consumers. An example is hormone use in the beef industry, which Chapter 5 will demonstrate has no harmful effects. The term "factory farming" probably makes you uncomfortable because you, like all humans, are vulnerable to propaganda. However, not all opposition to factory farming is the result of propaganda.

Factory farms also have an unappealing connotation because sentient animals are utilized there as mere "inputs." The idea of a factory per se is not unappealing, but the idea of animals in a factory is anything but pleasant. How did farm animals wind up in "factories" in the first place, and what has kept them there? We have to start by realizing that there was no "farm czar" who decided that farm animals should be raised in a factory. Nor was it a corporate conspiracy to produce profits at the expense of animals. Factory farming emerged because that method of rearing animals was deemed by the market to be superior. What do we mean by the "market?" The next section explains.

Who Decides How Farm Animals Are Raised?

There are currently few regulations regarding how farm animals should be treated. It is tempting to conclude, then, that farmers can do as they please—but this is clearly wrong. For one thing, farmers must please the customer. Asking who decides how farm animals are raised is like asking who decides what is on the internet. The answer is everybody and nobody. The internet and hog farming are the result of many individuals' actions but no one individual's intention. We are describing a capitalistic, market-based system where freedom reigns: the freedom of businesses to sell whatever they like and charge whatever price they wish, and the freedom of consumers to decide from which business they purchase and how much to purchase. For many people, a capitalist world is one that is run by a few CEOs. However, our view of capitalism and markets is quite different: a market-based system is one in which business leaders are vulnerable and the consumer reigns. In one of our favorite essays about capitalism, author Julian Gough writes, "There is nothing more powerless than a corporation."[5]

An order emerges from capitalism through the individual decisions of people as they purchase and sell goods and services. It is an order that is the result of individuals' actions, but not necessarily their intentions. No committee decided ten years ago that Wal-Mart should be the largest food retailer, but they are the largest food retailer because Wal-Mart is far more efficient than its competitors. A central force inherent in capitalism is competition. The competition for consumers forces businesses to operate efficiently: to deliver consumers the goods they desire at the lowest possible price. Wal-Mart quickly dominated K-Mart because it gave consumers what they wanted: lower prices. Whatever the evils of Wal-Mart may be, their existence has resulted in better stores with lower prices. No one is forced to work or to shop at Wal-Mart. If one dislikes Wal-Mart's business practices one can shop elsewhere. But if your criticism of Wal-Mart is that it makes too much money—then why not buy their stock, then you will also benefit. No doubt Wal-Mart hurt K-Mart, but did the automobile not also destroy the horse and carriage industry? How many of us would like to return to the days of the horse and carriage simply to protect the carriage industry? That is why Julian Gough describes the corporation as powerless. Every corporation, or any form of business, will eventually be beaten and replaced by a better competitor—providing the consumer with a better product in the process.

We live in a capitalistic society and agriculture has been subjected to the same forces that caused the makers of eight-track players and vinyl records to go out of business, allowing the introduction of the iPod. The engine of change in a capitalistic system is technological innovation. The period 1900–40 blossomed with agricultural innovations—including the all-purpose tractor, fertilizer spreader, electric fence, artificial insemination, cross-breeding of livestock, improved control of animal diseases, hybrid crops, new crop rotations, and strip cropping. It is to the individual farmers' advantage to adopt cost-reducing technologies, even if the widespread adoption of those technologies leads to greater production, which in turn lowers prices.

There were many technological innovations that led to the development of factory-farmed livestock; perhaps the most important was complete animal feed. A *complete* feed is one where the animal can obtain all its nutrient needs from each ration. Historically, animals had to live outdoors to gain access to the micro-nutrients their feed lacked. The animals obtained these other nutrient requirements through sunlight, dirt, insects, and plants. More sophisticated knowledge of nutrition later allowed for the development of livestock feed that contained all nutrients necessary for animal health. Moreover, the development of hybrid corn and soybeans coupled with improved tractors and fertilizers drastically increased the availability of animal feed. The result was that farmers could own many more animals and keep them in confined spaces without worrying about how they would be fed. Added to this, improvements in farm buildings propelled the growth of confined feeding operations. Some of the earliest exclusively indoor hog product facilities appeared in the early 1900s, but the changing technology and incentives has resulted in an outcome where space allotments went from about 32 square feet per pig in 1913 to about 8 square feet today.[6]

These technological innovations had two important consequences. First, they made each farm laborer more efficient. It took 147 hours of work to produce 100 bushels of corn in 1900, while today 100 bushels can be produced with only 3 hours of work. Similar efficiency gains occurred in the livestock sector. In 1929 it took 85 hours of work to produce 1000 pounds of broilers (chickens raised for meat). Today it takes only one hour. Chicken producers are 85 times more efficient than they were in 1929![7] Not surprisingly, this has made chicken much cheaper. Eating chicken used to be a rare treat reserved for special occasions like Sunday dinners. It is now the most widely consumed meat in the US. On average, each person in the US consumed about 28 pounds of chicken per year in 1960, compared to 85 pounds today. And it is no secret why people now consume almost three times the amount of chicken they did 50 years ago—the price of chicken has fallen over 110 percent (adjusting for inflation).[8] Similar efficiency gains have occurred in all livestock sectors. Dairy farms today only need 21 percent as many animals, 23 percent as much feed, 35 percent as much water, and 10 percent as much land as dairy farms did in 1944 to produce the same amount of milk.[9]

Second, innovation created *economies of scale*, which basically meant farmers could lower costs by producing more output. Most factories today are huge because of economies of scale. By becoming larger, the per-unit cost of production falls. When economies of scale exist, it is in *consumers'* interest for each farm to become larger. Each corn farmer should plant more acres of corn. Each cattle rancher should raise more cattle. By doing so, the price consumers pay for corn and beef falls.

With the efficiency gains fewer farmers are needed to attain the same output. The entire nation's food could be produced using fewer farm workers. No one established a plan for these changes, or necessarily realized they were needed. Market forces did all the work. In 1920, 30 percent of Americans lived on a farm. Technological innovation and economies of scale made such large numbers of farms unsustainable. The large amount of food forced food prices downward, which made agriculture unprofitable for the least efficient farmers, forcing them to pursue other occupations. Take, for example, one of our colleagues who grew up farming with his father and brother. After he graduated from college neither he nor his brother saw farming as their best career option, largely because profits in farming are difficult to make and they could make more money elsewhere. Our friend is now a banker and his brother a software programmer.

No one planned in advance for the number of farmers to fall, or for the number of bankers and software programmers to rise—after all, no one in 1950 could have forseen today's widespread use of computers and that therefore software programming might one day be a popular job! Yet it happened. Market forces made it happen. Today, only about 1 percent of the American workforce is comprised of farmers, compared to 30 percent in 1920. And while there were about seven million farms in 1935 there are only about two million farms today, and the average number of acres per farm has more than tripled over this period.[10] These changes have been even more recent and more dramatic in some livestock sectors. For example, the number of hog farms fell more than 70 percent between 1992 and 2004 while the average size of US hog operations grew from 945 to 4646 head over this same period.[11]

As the number of farmers fell, existing farmers discovered they could still make money if they enlarged their operation and specialized in fewer commodities. A typical farmer in the early twentieth century might have planted 80 acres of corn while raising 50 hogs, 20 laying hens, and 40 cows. Today, it would be hard to find any full-time farmer who produces corn, hogs, eggs, and cattle. Corn farmers now plant thousands of acres, some even tens of thousands. Modern hog farms often house over 1000 sows, and some as many as 10,000. Some modern egg-laying facilities house over a million laying hens!

No one planned for this to happen, but due to the combined forces of technological innovation, economies of scale, and the relentless competition for the consumers' patronage, many more agricultural products are produced from fewer but larger farms. Figure 3.1 shows the trends. The top graph shows that US farmers have steadily produced more animal products year after year. However, the bottom graph shows that each year the number of farms producing these commodities has fallen. The obvious implication is that the average farm size has also grown during this period: from 1950 until today, the average farm size has increased from 200 to 500 acres.

The graph also plots the number of commodities produced per farm, this being a measure of specialization. Farms are increasingly specialized in order to increase their efficiency. Increasing specialization was also a result of technological change. For example, prior to World War II, we had little knowledge of how to produce nitrogen, a key fertilizer required for plant and animal growth. In the past, nitrogen was added to the soil by planting legumes, such as clover. If a farmer had a field full of clover it made sense to raise cattle and/or hogs to eat the clover. Today, nitrogen is manufactured more cheaply in factories using natural gas, and planting clover is unnecessary for fertilizing fields. Similarly, hogs need protein for growth. Decades ago, farms would raise hogs alongside dairies, feeding the cow milk to hogs for protein. Today we know how to separate soybeans into soybean meal and soybean oil. The meal is fed to hogs for protein, and the oil is sold to consumers in the form of vegetable oil. Consequently, hogs are rarely raised alongside dairy cattle.

As another example, much of the corn fed to cattle was once indigestible. For this reason, hogs could also be raised inexpensively by letting them scavenge for cow manure. Raising hogs alongside cattle once gave the farmer more meat to sell with the same amount of corn. A livestock production published in 1928 argued, "corn salvaged from cattle droppings is clear gain. Experimental results show that for every bushel of corn fed to cattle, enough feed is recovered by swine following them to produce one to two pounds of pork."[12] Today, feedlots often process corn into a flake form, which tastes similar to the Cornflakes breakfast cereal. Flaking corn allows a larger proportion of the corn to be digested by cattle, eliminating the synergy that once existed between cattle and hog production. The corn flaking technology is just one example of a

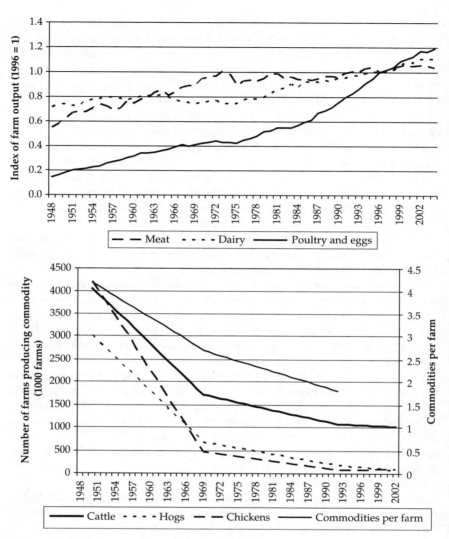

Figure 3.1 U.S. farm output and number of farms (1948–2002)

Source: Gardner, 2002.

technology that has made farms more efficient and led to them becoming more specialized.

The ultimate consequence of these changes is cheaper, higher quality food. Figure 3.2 shows the inflation-adjusted price of live cattle, live hogs, and eggs,

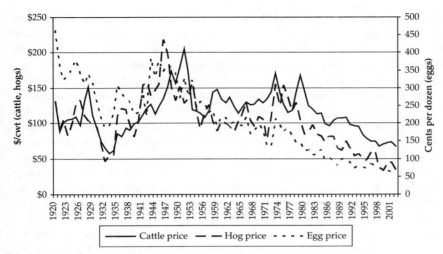

Figure 3.2 Historical farm prices for live cattle, live hogs, and eggs in 2002 prices

Notes: Cattle prices refer to live cattle (cattle ready for slaughter). Hog prices are a weighted average of barrow, gilt, and sow prices. Egg prices refer to fresh, conventional eggs. Nominal prices are converted to real 2002 prices using the standard consumer price index. Data before 1929 are from Statistical Abstracts of the US, while other data are from years reports by the National Agricultural Statistics Service. Thus, the two data series measure very similar, but not identical, goods.

given in 2002 dollars. Notice the steady decline from 1950 until today. USDA data indicate that from 1960 to 2008, inflation-adjusted retail prices for beef, pork, and poultry have each fallen on average about $0.02/pound per year, which implies that the average retail price of all meat has fallen about 69 percent over this 48-year period.[13] Virtually all economists would attribute these price declines to economies of scale and to technological changes such as improvements in animal genetics, animal nutrition, farm management practices including the use of confined feeding operations, meat processing, and meat transportation and storage. In fact, there are good reasons to believe that meat production costs have fallen even more than 69 percent over this period (see Figure 3.2).[14]

Consider a thought experiment. Suppose we outlawed the use of all technological improvements that have occurred in the meat sector since 1960. How much more expensive would meat be with such a policy? If the statistics above are to be believed, meat prices would be at least 69 percent higher than they are now. Periodically, economists can place a dollar-value on changes. Our estimates suggest that the lower meat prices due to innovation in food production benefits

Americans by $1122 per household, per year.[15] The US Census Bureau esti-mates indicate that there were about 112.36 million households in the US in 2008. This implies that the annual benefit to all US households from consolida-tion and technological change in the meat sector is at least $126.12 billion.

It is down to the individual householder to decide whether $1122 is a large or small amount of money to have saved on food costs. It is also worth asking who is the greatest beneficiary of this cost saving: food producers or food consu-mers?[16] Using a model of the interrelated beef, pork, and poultry industries, and a variety of economic assumptions, we find that the 69 percent price decline resulting from technological change generates $12.95 billion in benefits to beef, pork, and chicken producers.[17] That is a lower figure compared to consumers' benefit of $126.12 billion. The benefits of factory farming to consumers are about ten times great than the benefits accruing to producers.

In addition to the cheaper food resulting from better technologies, remark-able improvements have been made in the area of food safety,[18] food quality, and uniformity in meat products.[19] It is readily acknowledged by the public that factory farming has resulted in cheaper food, but the improved quality tends to be given little attention. Research has demonstrated that consumers prefer the chicken of today to the chicken of the past.[20] It is important to recognize these facts, because we humans have a tendency to lament the past, often experiencing unjustified nostalgia for foods of the past. John Steinbeck perhaps said it best when he said, "Even while I protest the assembly-line production of our food, our songs, our language, and eventually our souls, I know that it was a rare home that baked good bread in the old days."[21]

It is argued by some that today's low food prices actually mask other costs that society pays, and that, all things considered, our food is not cheap at all. An example is water pollution from animal manure. Raising animals requires dealing with manure, and for many livestock industries manure is stored and subsequently applied to cropland. Runoff from this manure can enter surface waters. We, however, do not agree that the environmental costs of livestock production should overshadow the price savings recently witnessed. Moreover, as we have noted, factory farming methods are highly efficient—meaning they require fewer inputs to produce each pound of food, which is of environmental benefit. We have seen no compelling evidence to suggest that the benefits of lower prices are outweighed by environmental concerns. What is of greater interest to us is whether the benefits of lower prices are outweighed by the losses in farm animal welfare.

Because so much, in our view, "foolishness" has been written about factory farms in the popular press (and by animal rights authors particularly), it is important to make our positions clear. We see no compelling evidence (and in fact we see evidence to the contrary) that food from factory farming methods is generally less safe than food produced fifty years ago. We see no compelling evidence that the environment is in any more danger from the use of factory farming methods than if other, less efficient, methods were used to produce the current amount of food needed to supply the world. We view the changes in agriculture over the past hundred years as being necessary for our current great wealth. We have many fewer farmers than we once did, but levels of unemployment are no higher than they were between fifty and a hundred years ago. As we detail more fully in the next chapter, however, the one demonstrable adverse consequence brought about by factory farming methods is a possible decline in farm animal well-being.

We began this section with a question: who decides how farm animals are raised? We conclude this section with another question: who is to "blame" for the increasing size and specialization of farms? Who is to blame for the rise of the factory farm? The consumer is—who purchased more food when prices fell, and who consistently shops at grocery stores with lower prices. Let us dispense with talk of *corporate* greed in relation to factory farms; the source of the greed, if that is what it is, is much closer to home.

Consumers and Farmers Talking

The technological innovations that occurred in the last century made farming as we know it today almost inevitable. Market forces caused producers to adopt technologies that produce food cheaply. Adoption of the technologies required less farm labor, fewer farms, and, eventually, larger farms. Hogs are raised in large confinement facilities, and they are often housed in crates so small they cannot turn around—because this is the least costly method of producing pork. Safe, inexpensive eggs also require laying hens to be so tightly crowded into each cage that they cannot fully extend their wings. Markets simply deliver what people want and people do want lower prices. Some people, perhaps, want something more, and are willing to pay more money to improve the lives of farm animals. How do people communicate these desires? How is it possible for farmers to know what consumers want?

At farmers' markets, the farmer and the consumer get to chat. Consumers who express a desire for fennel seed in sausage can express this desire directly to the farmer. If enough consumers express such preferences, the farmer will soon comply. Shoppers who have just seen appalling footage of animal cruelty at a farm can express their skepticism about animal treatment to the farmer directly, and the farmer can begin to post pictures of his farm at the market. The point is that consumers and farmers can easily and directly communicate at a farmers' market. The same level of communication cannot be carried out in the grocery store setting.

However, even in an impersonal grocery store, consumers regularly communicate with farmers, albeit indirectly. Consumer purchases, taken in aggregate, affect market prices and those changes in prices tell farmers what to produce.

That is, the consumers' voice is expressed through their purchases. For example, beef steaks taste best when they contain more intramuscular fat called marbling. Consumers can discern which steaks have more marbling either by looking at the meat directly or by relying on USDA quality grades. A steak that has been graded as "USDA Choice" will have more marbling than steaks which has been graded as "USDA Select." When Select and Choice steaks are sold at the same price most shoppers choose the Choice steak. Many people will pay a higher price for a Choice steak. This communicates to the grocery stores that shoppers prefer Choice steaks, so the grocery store offers higher prices to beef processors for steaks that can be graded as Choice. Beef processors, in turn, offer higher prices to farmers with breeds of cattle more likely to be graded as Choice, such as the Angus breed. This provides incentives for cattle producers to raise the type of cattle which consumers prefer to eat. Far more cattle producers raise Angus cattle than Brahma cattle, because meat from the former contains more marbling. Consumers have told farmers what they want through their purchases.

Communicating through markets tends to work quite well. Market forces have prompted livestock producers to respond to consumer concerns about the fat in red meat by increasing chicken production and by increasing the leanness of pork. Moreover, no meat processor wants the bad publicity (and lost business) caused by a meat recall from a food safety scare. That is, food safety is of paramount importance to consumers, and producers have responded to this concern. You may be surprised to learn that you are now much more likely to become sick from eating vegetables than meat.[22]

Until recently, most consumers have given little thought to how farm animals are treated. Grocery stores have not been bombarded by questions about animal care, and thus, there was no feedback to the farmer about how consumers want animals to be raised. If there is no feedback, farmers cannot be expected to respond to consumers' desires. The primary feedback farmers have received, until recently, was that people wanted to pay lower prices.

Consider the incentives faced by egg producers over the past century as they sought to stay in business. A hundred years ago chickens largely lived outdoors, eating insects in addition to their feed ration. The hens had access to small barns to lay eggs in nests and were allowed to roam outside. Our image of these former egg-laying conditions, however, is likely romanticized. Consider the realities. Chickens were constantly exposed to hawks swooping down in search of a meal. Chicken's sleep was often disturbed by hungry foxes, skunks, dogs, and other predators. These were real problems and it is hard to produce eggs when there are predators at large who seek to kill the chickens who lay these eggs, not to mention that it is bad for the chickens' own welfare. Some farmers began locking hens indoors to keep them safe from predators. However, buildings are expensive to build. To keep costs low the farmer had to place a large number of hens under the same roof.

However, once hens are in a cramped area they become bad tempered and begin pecking one another. Chickens have an instinct to establish a hierarchy, a "pecking order" to discover which bird is strongest. In small groups, battles to establish this hierarchy are short and rarely result in harm, but in larger groups this order is more difficult to establish. Chickens can only remember the pecking order of a flock of up to 30 birds. In larger flocks, there are continual fights. Birds injure one another, and even cannibalism can occur. Fighting chickens are bad for business and bad for the animals.

To mitigate the problem of chicken fighting farmers began to place small groups of hens (up to five) in a cage within this barn. The barn protects the birds from the weather and predators, while the cages within the barn protect the birds from each other. A number of other benefits arose from this system of egg production. For example, cages can be designed so that when an egg is laid, it rolls gently out of the cage onto a conveyor belt. Without cages, some of the eggs are laid on the floor and defecated on by other birds. Generally, consumers like eggs free of feces.[23] Being able to house thousands of birds in one barn also generates large amounts of heat, reducing heating costs during the winter.

The cage system produces cheap and safe eggs. Farmers using a cage system can sell eggs to grocery stores at a lower price. Likewise, grocery stores pass the savings onto consumers, who buy more eggs. Moreover, with safer eggs consumers get sick less often and continue to come back for more eggs. When other farmers without this cage system tried to sell their eggs, grocery stores refused to pay more than they paid the farmers with the cage system. The farms without the cage system cannot make a profit at these low prices, so they either convert to the cage system or discontinue egg production. This is, of course, how capitalism works—it forces producers to deliver to consumers the product they desire in the most efficient manner possible.

A problem now arises. Consumers begin to see pictures of farms with hens crammed into small, barren cages for their entire lives. The pictures look cruel and some consumers become angry that birds should be treated in such a manner. Farmers with cage systems are vilified. However, consumers had previously communicated to farmers through their purchases that they preferred the advantages of the cage system. As a result, the farmer had *no choice* but to use the cage system; any other system would have been unaffordable. If consumers want their eggs to come from a farm that they consider more humane, they must communicate this to the farmer. More importantly, they must communicate their willingness to pay a higher price for the product by purchasing them. Farmers will produce chicken meat and eggs in any way that appears to meet the consumers' preferences.

Currently, consumers *can* communicate to farmers whether they desire cage-free and free-range eggs. Almost every major grocery store chain now sells cage-free and organic eggs. If more people begin to buy cage-free eggs, the grocery stores will respond by asking farmers to supply more cage-free eggs. To convince farmers to switch to a less efficient production system, a higher price must be offered to the farmers. They would switch to cage-free production if they thought they could make more money by doing so. The fact that only about 5 percent of all eggs sold in the US are advertised as cage-free, organic, or free-range suggests that the current message being heard by US farmers is: *keep producing cage eggs*! Markets are dynamic, and producers have their ears open. If more people begin to buy cage-free eggs the price of cage-free eggs will begin to rise relative to conventional eggs, which will tell the producer to make a change. If these factory farmers wish to enhance their profits, you can be sure they will listen to the message consumers have sent.

Consumers also communicate with producers outside the market context via regulation. Animal activist groups are currently trying to ban the housing of laying hens in small, crowded cages (often referred to as battery cages). You will see pictures of these cages in Chapter 5. In November of 2008, Californians passed a referendum that will soon ban the use of cages in their state. Whether the ban will actually have much effect on how chickens are raised is debatable, since Californian grocery stores can easily import cheaper eggs from other states; however, consumers in California nonetheless sent a clear message to producers about the way they wanted eggs produced.

It must be said, however, that there are many who do not believe that modern farmers have "lost their humanity," as Robert Kennedy, Jr, was once heard to suggest. Trent Loos, who heard the former make this comment, is just one of these advocates of modern farming methods. Through his writing, speeches, and radio talks, Trent Loos is one of the most vehement defenders of modern farming methods. In a recent conversation with the authors of this volume, Trent expressed to us that he has made it one of his life goals to show Americans just how humane current farming methods are.

However, the point of this section is to illustrate that *if it is* the case that farmers lost their humanity (which is up for debate) then the consumers also lost their humanity. If consumers wish to employ an alternative to factory farming they must communicate this to farmers, and then they need to be willing to pay for the additional costs incurred.

A Farmer is a Corporation is a Person

Corporations make for villains, and corporations have played an increasingly important role in livestock agriculture. Historically, agricultural production took place on numerous small farms. The buyers of farm products, including bakers, millers, and other processors, were also smaller in size and greater in number. Buying and selling between numerous people is difficult and costly to coordinate, and the communication is often noisy. As a result, many of the auction markets of the past, which were used for buying and selling of farm products, began to be replaced by the setting up of private contracts. Just as there are cost advantages for a farm to become larger, there can be cost advantages to having one person or company manage the farm and the processing.

Contracts in agriculture are now the norm. In 1969, only 11 percent of agricultural products was sold under contract. Today that number is 41 percent. There are two types of contracts used: marketing and production contracts. Marketing contracts specify the quality (and sometimes the quantity) of product a farmer agrees to deliver to the processor, along with the price at which the exchange takes place. Marketing contracts allow the farmer to negotiate the sale of product while remaining a largely independent operator, while production contracts force the farmer to act as a quasi-employee of the buyer. On poultry and hog farms under production contracts, the farmer owns the land and the buildings but not the hogs or the hog feed. The contract stipulates how the hogs should be raised and how the barn should be designed. In 2005, 76 percent of all hogs produced in the US were under production contracts. For eggs and chicken meat, this number was 94 percent. Like hogs, chickens produced for meat are largely raised under production contracts.[24] Egg production is better described as a vertically integrated system; this means that the person who owns the processing plant also owns the chicken farm.

With whom do farmers engage in contractual arrangements? The primary answer is corporations. Smithfield Foods is by far the largest hog and pork producer, owning (virtually) all the hogs in North Carolina, which is the second largest hog-producing state. Seaboard Foods, another corporation, comes a distant second. A common misperception among those unfamiliar with the history of animal agriculture is that corporations, in their greedy quest for profits, entered agriculture, applied less humane methods, sold at lower prices, and drove the family farmer out of business, the family farmers being unwilling to compete by treating the animals in the new, inhumane ways being used by the corporations.

In reality, 98 percent of all farms in the US *are* family farms. Family farms are responsible for 85 percent of the value of agricultural output produced in the US.[25] It may be that the definition of a factory farm used to collect statistics is not congruent with what we envision a factory farm to be, but these statistics are nevertheless enlightening. Second, most of the large corporations in animal agriculture grew up from successful small farming operations. Wendell Murphy is a successful American pork producer from North Carolina who is largely credited with refining the current factory style of pork production. North Carolina is not well-suited for hog production. Corn, the chief ingredient in hog feed, has to be imported from distant states like Indiana, Illinois, and Iowa. North Carolina is hot and humid, and hogs cannot sweat. Successful North Carolina hog farms have to seek an advantage other than location and Wendell Murphy did so by adopting and refining a method of raising hogs inside a building and under closely controlled conditions (in addition to lobbying for advantageous environmental regulations). There is no doubt that his method is efficient. Virtually all of North Carolina adopted his production techniques, growing from a state with relatively few hogs to the second largest hog-producing state in the US.

Wendell Murphy and others like him sought to profit from the more efficient hog production system by placing more and more hogs under this system, and under their ownership. This arrangement is referred to as a *production contract*, and though production contracts were widely used in the poultry industry, Mr Murphy was the first to employ the contracts in the hog industry.[26] To support the growing number of hog farms under one business, large infrastructure investments like feed mills were needed, and no business system is better at raising investment money than that of the corporation. A corporate structure allows investors to invest money in a more risky business venture. Although an investor in a corporation might lose their initial investment they would not lose, for example, their home. Corporations, however, have added costs. When a venture makes money the profits are taxed twice, once at the corporate level and again when the profits are shared out among the share-holders as dividends. Nonetheless, entrepreneurs, like Murphy, with an improved production system, tend to utilize the corporate business structure to raise the money needed to expand their operations.

What is important to understand is that it was not just a few corporate executives who aggressively pursued the factory farm system, but the farmers themselves as well. The thousands of farmers who work under production contracts do so willingly, and often gladly. Initially, farmers had the option of raising hogs and poultry on their own using old-fashioned methods. They chose production contracts largely because these permitted access to an efficient system that generated higher profits. Numerous hog farmers in North Carolina, in discussion with the authors of this volume, have said that they make more money raising hogs under production contracts than they could as independent hog farmers. The only factors that could prevent the continued proliferation of the factory farm system are regulation, consumer backlash, or increased demand for meat and eggs from alternative systems.

The animal welfare debate is not about separating the good people from the bad or the humane from the inhumane. The debate largely involves differences in beliefs and perceptions about the extent to which animals suffer in current confinement systems and the importance of animal suffering relative to other concerns, such as lower food prices. We believe that the debate on these various issues can move forward in a constructive manner if people have access to objective, accurate information and a logical, coherent framework for thinking through the difficult trade-offs.

To move the debate forward, we must debunk the myth of the "evil factory farmer." By and large the attitudes of farmers today are no different than those of the farmers of the past. Although factory farming is a recent development, the idea is not new—a 1928 book on raising livestock encouraged readers to, "...consider every flock of hens an egg factory."[27]

Darian Ibrahim, a law professor at the University of Arizona, wrote a paper titled "Property, Profit, and the Corporate Ownership of Animals."[28] Ibrahim makes the claim that the profit motive of corporations has caused them to inflict suffering on animals that "family farmers" would not have inflicted. Ibrahim argues, "In the early 1900s, before vertical integration occurred, chicken farming was a family affair largely devoid of the profit motive." However, profits *had* to be a major concern for family farmers: after all, no business can continue without profits.

We do not agree that profits are of greater importance to farmers or the corporation managers of farm produce than they were to farmers in years past. And we find it highly unlikely that managers of corporations are any greedier than those working lower down in the corporate ranks, or any other working individual, be they self-employed, partners of a law firm, or plumbers. We would also argue that factory farming has not come about as a result of corporate ownership of farm animals. We would argue that factory farming was the end result of an evolution of practices where it was found to be the method that produced the safest animal products at the lowest cost—and until now, the farmers have heard little concern from consumers for the animals from these "factories."

Opposition to Factory Farming

We use the term *factory farming* to refer to livestock production systems where animals are confined for most or all of their lives in buildings, usually in cramped spaces that limit their ability to move and act naturally. However, we do not use the term *factory* farming to debase the method; as we have noted, there are many aspects of factory-style production that are appealing. It is the unappealing aspects of factory farming, however, that have prompted opposition and debate.

Between 1915 and 1950 animal advocacy organizations experienced a decline in activity. The movement became more conservative and institutionalized. Activities began focusing on animal shelters to accommodate the public's increasing fascination with pets, and on disseminating educational materials to the general public. Issues such as vivisection remained controversial but the debates were more tempered than in the past and gained little traction.

During this period medical and biological sciences began rapidly advancing and increased their use of animal experimentation. Government funding encouraged such research and passed a series of laws increasing the availability of laboratory animals by giving scientists access to animals in shelters. Knowing some of the shelter animals would undergo vivisection and other painful experiments, some animal advocates were outraged. The unwillingness of the American Humane Association (AHA) to actively condemn vivisection induced Christine Stevens to found the Animal Welfare Institute (AWI) in 1951. Then, in 1954 when the AHA refused to challenge the seizure of shelter animals for experiments, AHA staffer Fred Myers (who edited AHA's publication *National Humane Review*) and three others defected and founded an organization that would soon be called the Humane Society of the United States (HSUS).

The AWI and HSUS both play prominent roles in the current farm animal welfare debate. The HSUS was founded on the idea that livestock industries need to be confronted and forced to change. It is this attitude that ultimately led the HSUS to champion legislation banning the use of gestation crates, battery cages, and the like. These bans are among the most controversial issues in farm animal welfare. Although the HSUS seeks to ban certain practices, the AWI seeks to support other practices. While the AWI certainly opposes modern factory farming, they have created guidelines for an alternative method for livestock production. To understand the current animal welfare debate requires an understanding of the emergence of the HSUS and the AWI.[29]

Rattling the Cage

The daily life of farm animals was largely ignored until English activist Ruth Harrison published the book *Animal Machines* in 1964. This book detailed how farm animals such as laying hens were raised. At the time of its writing, the science behind farm animal welfare was still in its infancy; however, the book pointed out that hens were housed in small cages for their entire lives. The cages provided barely enough room for the chicken to stand and were so small that hens could not turn around without bumping into one another.

Ruth Harrison's cause was taken up by the HSUS when it hired Michael Fox in the mid-1970s. Fox sought to challenge the use of these chicken cages—both figuratively and literally. Traveling and talking to anyone willing to listen, Fox sought to make the public aware of how some farm animals were being raised. These efforts finally garnered significant publicity when Fox and his investigations into factory farming appeared in a 1980 issue of the *Smithsonian,* but farm animal welfare was still a nascent debate. Meanwhile, as Fox sought public support for changes in livestock agriculture, a recently formed group named Farm Sanctuary began rescuing farm animals and providing them with a pleasant life on rescue farms.

Fox's activities attracted the attention of livestock producers when, one year after the publication of the *Smithsonian* article, a US representative sponsored a bill to establish a House Committee to investigate factory farming. The bill was not passed, but the livestock industry saw that the issue was becoming important and began counter-attacking. The arguments lobbied back and forth in the early 1980s closely mimic those offered today. For example, in 1984 the industry stated, "[the agriculture industry] supplies consumers with a nutritious and plentiful meat supply. No one knows better than the livestock producer that sick, malnourished or suffering animals are less productive,"[30] and in 2009 they stated, "Our producers take care of their animals, and we know that an animal that isn't treated well doesn't produce."[31] The HSUS replied that an animal can exhibit high productivity in terms of egg or meat production despite the fact that some of their biological needs are not met and despite the fact that the animal is denied the opportunity to express its natural behaviors.

Through films, pamphlets, essays, and public talks, the HSUS exposed the public to (literal and figurative) images of hens, veal calves, and hogs that were permanently housed in very small cages. The pictures did the trick and were all the ammunition animal advocates needed. A picture is worth a thousand words, as they say, and the pictures of crowded hens heavily outweighed, for most people, any arguments from the producers that the cages were required for effective animal care.

However, little reform was accomplished. Much of livestock agriculture carried on as before, although the veal industry did experience a decline in demand. The Veal Calf Protection Act was introduced in 1989, and although the act did not gain much political support it produced bad publicity for the veal industry. To accompany the bill, the HSUS implemented the No Veal This Meal campaign (1982) to educate the public on how veal was produced.[32] During this period, public opinion changed and veal production became widely thought of as cruel. Veal consumption plummeted from 3.1 million veal calves being slaughtered each year in 1982 to less than one million in 2004.[33]

Another group emerged in the 1980s—People for the Ethical Treatment of Animals (PETA). The organization, co-founded by current president Ingrid Newkirk, focuses more on animal *rights* than on welfare issues. PETA's philosophy is more radical than many other animal activist organizations in that they denounce human use of animals for any purpose: food, clothing, experimentation, and even entertainment.[34] PETA engages in a host of extreme behaviors: they dress as KKK members to protest dog shows, crawl through Paris with animal traps about their feet, dump buckets of money saturated with fake blood, harass children who compete in livestock shows, and encourage ice cream manufacturers to use human breast milk instead of cow milk.[35] Although PETA's initial activities were not focused on livestock agriculture, their interests have broadened over time. Although most people (and even those involved in animal production) tend to identify PETA as the primary activist group in regard to farm animal welfare, their influence on actual livestock production practices has been relatively minor compared to that of the HSUS. It is difficult to think about animal welfare without thinking about PETA, largely because of their high-profile publicity stunts—even Newkirk admitted that, "We are complete press sluts," and that "PETA's publicity formula [is] . . . eighty percent outrage, ten percent each of celebrity and truth."[36] But if PETA represents the style of activism against livestock agriculture then HSUS represents the substance.

At the age of 13, Paul Shapiro watched a video about how farm animals were raised. Shapiro decided to become a vegetarian after seeing chickens and hogs housed in barren and cramped environments and the practices used in meat-processing facilities. A few weeks later Shapiro also learned how eggs and milk are produced and shunned these products as well, becoming a vegan. In 1995, at the age of 15, Shapiro formed an extracurricular club at his high school dedicated to reducing animal suffering, which he named Compassion over Killing (COK). The movement grew into much more than a school club. The club sought to educate people on factory farming methods, to encourage vegan diets by giving talks at workplaces and libraries, and to distribute educational materials. In a short period of time, the high school club had wide adult membership and support.

For a few years in the mid-1990s Shapiro might have been characterized as a stereotypical PETA activist. He orchestrated sit-ins at fur retailers and entrances to circuses. He even climbed onto the roof of a McDonald's to hang an animal rights banner. Shapiro was arrested at least six times, but by the late 1990s he began pondering whether his efforts were having any effect. Fur shoppers, circus goers, and hamburger eaters did not approach Shapiro and ask for more information—they avoided him. It was obvious, Shapiro thought, that a friendlier, more engaging approach was warranted. COK revamped its approach. Instead of harassing individuals entering Kentucky Fried Chicken, they began providing free samples of imitation chicken tenders made from soy.

COK became widely known for their investigations into farms, animal transportation, livestock auctions, and hatcheries. COK documentaries began to be discussed in publications such as the *Washington Post* and the *New York Times*. While working on COK projects, Shapiro became acquainted with a fellow activist by the name of Wayne Pacelle. Pacelle and Shapiro share a number of traits. One of these is boldness. Shapiro would sneak onto farms at night to capture video footage; Pacelle would go into the woods where hunters waited for their prey and make a racket to scare the prey away. Both men were leaders. Pacelle founded the first animal rights club at Yale University and Shapiro started COK. Both of them also shared a willingness to confront groups perceived to be abusing animals. When they could not achieve their objective by working with animal groups, Pacelle and Shapiro sought alternative methods for improving animal welfare, such as legislation.

Pacelle began working for HSUS in 1994, and in 2004 he was named President and CEO of HSUS. In 2002, HSUS found a renewed interest in factory farming in Florida. Members of HSUS and other organizations such as Fund for Animals and Farm Sanctuary gathered signatures for a petition to create a ballot initiative to outlaw the use of gestation crates for sows in Florida.

After receiving a sufficient number of signatures the referendum appeared on the ballot, and 55 percent of Floridians voted in favor of banning gestation crates. Pacelle played an integral role in promoting the Florida ballot initiative, and when he became the HSUS president in 2004 he called for the formation of the Factory Farming Campaign, which primarily aims to reduce the use of battery cages for laying hens, gestation crates for sows, and veal crates for calves. Subsidiary goals of the campaign include the promotion of humane eating and the banning of forced feeding of birds.[37] Pacelle enlisted Paul Shapiro to head the Factory Farming Campaign. Convinced that he could have a bigger impact on animal suffering as the head of this campaign, Shapiro left COK for the HSUS.

The campaign has witnessed many victories.[38] The biggest victories have been repeats of the Florida ballot initiative. Arizona voters decided to ban gestation and veal crates in 2006, Oregon banned gestation crates in 2007, and Colorado banned gestation and veal crates in 2008. The Oregon ban differed from Arizona and Florida in that the ban emanated from the state legislature without a ballot initiative. Colorado is particularly interesting because the Colorado Pork Producers voluntarily agreed to give up gestation crates. This was not because pork producers believed crates were inhumane but because HSUS threatened to pursue a ballot initiative in the state. Apparently, the pork producers thought they would lose a (costly) fight if the initiative went before voters and, seeking to avoid bad publicity and to gain more favorable terms, they worked with state legislators to preempt the referendum.

Proposition 2

On the heels of victories in other states the HSUS and Farm Sanctuary sought a ballot initiative that would prohibit the use of battery cages, veal crates, and gestation crates in one of the most populous US States: California. What became known as Proposition 2 (or Prop 2) specifically stated, "a person shall not tether or confine any covered animal, on a farm, for all or the majority of any day, in a manner that prevents such animal from: (a) Lying down, standing up, and fully extending his or her limbs; and (b) Turning around freely." In November 2008, 63.5 percent of Californians voted in favor of the proposition. In a sense, such a high rate of acceptance is not surprising. Who would be opposed to the seemingly innocuous issue of giving animals room to extend their limbs? The ramifications of the referendum's passage are significant, though, as much of the egg production in the state of California used a cage system in which hens could not fully extend their wings.

The battle in California was noteworthy for several reasons. First, the ballot initiative created an unusual amount of press and public attention for an agricultural issue. Second, California is a major agricultural state. It accounts for more agricultural output in dollar terms than any other state in the US (almost $32 billion in 2004)—almost twice as much as the next largest agricultural state, Texas.[39] In regard to the specific commodities affected, although California is not a major pork-producing state, it is the fifth largest producer of eggs in the US. Third, California is a trend-setting state. Regulations that pass in California often have a way of making their way to other states. As a result, the HSUS, Farm Sanctuary, and supporters of livestock agriculture such as the Farm Bureau, the United Egg Producers, and the National Pork Producers chose to "fight the fight" because of both the importance of agriculture in the state and the potential spill-over effects to other states. The intensity and importance of the debate can be witnessed by the amounts of money spent on the issues. Supporters of Prop 2 raised about $5.2 million, and opponents raised $6.9 million.[40] Finally and perhaps most importantly was the outcome of the ballot initiative. It was a landslide. Almost two-thirds of voters voted in favor of the initiative, which effectively banned the use of cages in layer production and the use of gestation crates in hog production in the state of California.

Analyzing the arguments made in favor and against Prop 2 is useful for understanding how people think about the farm animal welfare issue. It seemed clear that if Prop 2 was passed, food retailers would simply import cheaper eggs, veal, and pork from other states, these out-of-state producers being cheaper because they still employed the practices banned by Prop 2. The fear was that the only consequence of this ban would be to bankrupt Californian farmers. This is a current fear in the European Union (EU). Although the EU sets higher welfare standards for livestock than any other region, approximately 25 percent of EU meat is imported from countries outside of the EU who have lower welfare standards. The more stringent the EU standards become, the more food retailers will import meat from unregulated countries, possibly resulting in a decrease in the overall welfare of livestock. Cognizant of this possibility, the EU is attempting to require farms exporting to the EU to incorporate EU welfare standards.[41]

Were Californians aware that bans on cage-hen eggs may only alter the location in which these eggs are produced? It is unclear whether the average voter understood the consequences of the vote. It should be noted that there has been some effort to ban eggs from other states that do not comply with Prop 2 standards, but at the time of writing such efforts have thus far been unsuccessful.

HSUS and Farm Sanctuary were likely well aware that the passage of Prop 2 would have little direct effect on how animals were raised. The most logical reason they aggressively supported Prop 2 is that it represents one piece of a much larger agenda that might ultimately affect how animals are raised. Animal advocacy groups have strategically selected where to introduce and fight for ballot initiatives. State by state they have "rattled the cages" that hold farm animals. Every victory brings them closer to eliminating cages from all of livestock agriculture in the US. Moreover, fighting for passage of ballot initiatives is a convenient way to force consumers to consider where their food comes from and reconsider whether their moral beliefs are consistent with their diet.

It seems clear that animal activist groups also pursue costly legislation that has little effect on animals because it generates free publicity. When an issue gets placed on a ballot, it gets covered by local and national news. Because most people are likely to feel sympathy for animals pictured in small cages, free publicity on ballot initiatives is good publicity for the HSUS. These organizations raise large amounts of money, and Prop 2 communicates to their donors that the money is being used for the cause of animals.

By challenging the public's ideas on right and wrong treatment of animals, animal advocacy groups may have a lasting impact on social norms. It is now unfashionable to wear fur. This is in most part because people have been convinced that the fur coats are often made at the cost of cruel treatment of animals, but for some people it is also because fur coats elicit social disapprobation. When moral norms begin to shift, so do business practices. The attention farm animals have received in the press has persuaded many businesses to join with animal advocacy organizations in an attempt to preserve and promote their brand image. Numerous food providers have made announcements that they will purchase a certain percentage (or all) of their animal produce from cage-free providers. These include Bruegger's Bakery, Denny's, Safeway, Burger King, Ben & Jerry's, Trader Joe's, and Wolfgang Puck. Universities have always been an incubator for activism, and several universities (e.g., Harvard and the University of Minnesota) have adopted cage-free egg policies for food sold on campuses and others have announced efforts to increase the availability of vegetarian alternatives.[42]

Even the US government is joining the cause. Christopher Shays, a US representative from Connecticut, tried to pass a bill that would require the federal government to only purchase food from suppliers who allow animals the space to lie down, walk, and extend their limbs comfortably. The USDA purchases enormous quantities of food for school lunch programs, so such a bill would have had a tremendous impact on agriculture. Prop 2 was about more than farm animals in agriculture. The attention it drew put farm animal welfare in the spotlight and, whichever side won, the cause was likely to win more victories down the road.

Opposition to Prop 2 highlighted the fact that Prop 2 itself was likely to have little effect on farm animal welfare. For egg producers, the proposition basically stipulates that producers must use cage-free egg production as opposed to standard cage methods. Converting buildings used for cage production into cage-free facilities is expensive. Even though the farmers may receive a higher price for cage-free eggs, there is no guarantee that the price premium will compensate them for these new costs, meaning that this conversion would require egg producers to acquire more debt. It is obvious that the egg industry would oppose legislation which forces them to assume more debt for a questionable future return. Yet, there is another reason the egg industry opposed Prop 2, which was not centered on consideration of profits: this was the fact that some scientists specializing in animal welfare have argued against Prop 2 because they do not believe that cage-free egg production is in fact better for the birds.

The United Egg Producers (UEP) is a trade-group, whose membership includes a large majority of egg producers. Available to UEP members is the opportunity to raise hens under certain standards and sell their eggs with the label: "UEP Animal Welfare Certified." The welfare standards for this label are established by an independent committee of scientists, comprised of the world's most prestigious animal scientists, as well as an agricultural ethicist.

This scientific committee has established welfare standards for both cage and cage-free egg production. In doing so, they have established the opinion that both cage and cage-free egg production are humane—when implemented properly. Consequently, this committee does not agree with animal advocacy groups on Prop 2.

In the battle over Prop 2, and in other animal welfare debates, the livestock industry have asserted that their production methods are supported by scientific research, and that the production practices animal advocacy groups seek are driven by emotion and are not supported by scientific research. This is, of course, incorrect and in Chapter 5 we present numerous studies that suggest cage-free egg production provides better animal care than cage production. We argue that both private industry and animal advocacy groups have science on their side.

The second argument the agriculture industry use is the existence of what could be called a "vegan conspiracy."[43] Certainly, the leaders of HSUS and Farm Sanctuary are vegans and promote a vegan lifestyle—hence the view of some in the industry who see Prop 2 as a slippery slope toward laws that might ultimately dictate veganism. The United Egg Producers proffered the view that, "the HSUS agenda involves stopping all livestock and poultry production in the US—leaving consumers only to a vegetarian diet—as well as ending all fishing and hunting, all zoos and all human-oriented health and well-being research that uses animals."[44] Livestock industries promote the vegan conspiracy in an attempt to paint those working in animal welfare as extremists, whose views may be of questionable reliability. However, the vegan conspiracy argument is simply a distraction tactic: attacking the beliefs of the person making the argument rather than dealing with the facts of the argument itself. The answer to the question of whether farm animals should be given space to turn around should not depend on whether the person asking eats steak or salad.

A large majority of Californians voted in favor of Prop 2. Recognizing that vegetarians only comprise about 3 percent of the US population, we must conclude that Prop 2 was approved by a majority of meat-eaters. The farm animal welfare debate is driven by a sincere concern for animals, not just an obsession with veganism. Consumers believe that improving the lives of farm animals is a worthy goal, and that it is possible to extend compassion toward farm animals while retaining the option of eating them. Egg and veal production have changed little in thirty years, so why is the farm animal welfare issue only *now* being debated? There is no simple answer. One reason is that animal advocacy groups are trying harder now. Both Wayne Pacelle and Paul Shapiro have targeted factory farming methods and are persistent. They are also clever in how they achieve their objectives. It is no coincidence that most of HSUS's factory farming victories occurred when Pacelle gained greater authority within HSUS and when Shapiro assumed leadership of the factory farming campaign.

The internet has also been an important factor in the strengthening of animal advocacy support. Today, one click of the mouse can bring to your screen any number of film clips, compiled by animal advocacy groups, to demonstrate their arguments of factory farming cruelty. Being able to see gestation crates for yourself, for example, can have a powerful effect.

Changes in social norms also explain the increased prominence of the farm animal welfare movement. It was only fifty years ago that, in many parts of the US, public toilets, restaurants, and schools were segregated by race. The percentage of women employed outside the home increased from about 35 percent in the 1940s to 53 percent in the early 1980s to 60 percent today. The role of women in society has drastically changed. Issues of racism or sexism are of central importance in today's society. Views on equality have changed markedly. As such, it is not surprising to find that people are more open to discussions of equal consideration of animals than prior to, say, the 1980s.

How Radical is HSUS?

Although HSUS is a perpetual thorn in the side of the livestock industry, and is thought of as the quintessential enemy of animal agriculture, how radical is it? In one sense, its views are certainly outside those of the mainstream. The HSUS encourages a vegan diet, and Pacelle has been recorded as stating, "We have no problems with the extinction of domestic animals," and "If we could shut down all sport hunting in a moment, we would."[45]

On the other hand, some of the issues promoted by HSUS are not radical at all, and if the votes on ballot initiatives in Florida, Arizona, and California are any indication, its position on issues such as giving animals room to lie down, stand up, and fully extend their limbs are popular with the American public.

Now consider the Animal Welfare Institute (AWI). The AWI is a fierce critic of factory farming. However, the AWI dos not discourage consumption of animal food products as does the HSUS (although the current AWI president is a vegan). As such, the AWI might be considered a less radical organization than the HSUS. However, the changes proposed by AWI are much more radical than those recently pushed by HSUS. The AWI has constructed a detailed set of specifications and standards for raising animals. As an example, instead of just eliminating gestation crates, the AWI standards require that sows have individual huts for farrowing with straw, that they should be provided with twice the amount of shelter space factory farms currently provide them with, and that they be given outdoor access in addition to the shelter space. The standards are extensive; they identify all the needs of an animal and attempt to provide those needs. In our opinion, the AWI standards are among the highest animal welfare standards that exist for commercial farms today.

It is arguable that HSUS asks for only small changes because a larger request would be declined by society. They press only for marginal improvements in welfare because society only desires marginal improvements. HSUS views its role as prohibiting what it argues to be the most egregious acts of cruelty, rather than as dictating how every activity on the farm should be conducted.[46] Nonetheless, HSUS would also be in favor of livestock producers adopting the AWI standards. Equally, HSUS looks to encourage people to change their diet in order to avoid the most inhumane industries and favor those produced by more humane methods. The salient difference between the two organizations are seen in their respective objectives: the HSUS seeks for relatively small changes to be made in the set-up of factory farms, while the AWI seeks for the abolition of factory farms, calling for them to be replaced with an entirely new and, in its view, better system.

Order in the Court

The discussion thus far has concentrated on the use of referendums and legislation to influence animal welfare; however, the judicial system is at the center of the debate as well. One particularly interesting development is taking place in New Jersey. The New Jersey legislature passed a law in 1996 requiring certain "standards for the humane raising, keeping, care, treatment, marketing, and sale of domestic livestock." However, the regulations exempted animal practices that were considered "routine" practices before the regulations were passed. A number of animal advocacy organizations (e.g., HSUS, AWI, Farm Sanctuary) challenged the exemption and the challenge was heard by the New Jersey Supreme Court. Both sides claimed victory in the final ruling: the Supreme Court decided that a practice cannot be considered humane simply because it has been used in the past; however, it also ruled that the use of gestation crates, veal crates, and the transportation of sick animals were allowable practices.[47]

There are few federal laws aimed specifically at the well-being of farm animals. At the federal level, the Humane Slaughter Act prohibits needless suffering of animals. Critics argue that the Act is ineffective and enforcement is non-existent. The Act also exempts poultry, for historical reasons not considering birds as animals. Another example is the Twenty-Eight Hour Law, which requires animals be given at least one period of rest and nourishment when they are being transported over periods lasting longer than 28 hours. However, the maximum penalty for violating the law is only $500. And those are the only laws at the federal level pertinent to farm animal welfare (there are some important laws regarding the slaughter of horses, but for the purposes of this book and relating to modern-day animals, we have placed horses under the category of pets).

The laws most relevant to farm animal welfare are state-level, anti-cruelty laws, which apply to all animals, not just farm animals. Moreover, as in New Jersey, most anti-cruelty laws exempt practices that are considered to be customary farming practices. Prosecuting farmers using such laws has proved difficult, as was illustrated in the HBO documentary *Death on a Factory Farm*. Although an undercover activist working with the Humane Farming Association (HFA) had documented numerous examples of cruelty (e.g., euthanizing sows by hanging them with a chain, failing to feed sick sows, killing piglets with a hammer, throwing pigs by their back feet, and leaving dead sows in pens to be cannibalized by other sows), the farmer was only convicted of one charge in relation to how piglets were taken from their mothers to be weaned.[48] The only penalty was a $250 fine and a one-year probation for "improperly carrying or transporting animals." Although other practices used on the farm might also be considered cruel by others—such as the use of gestation crates where the sow cannot turn around or comfortably lie down—under law, such practices are not considered cruel. The Ohio anti-cruelty laws specifically state that no person shall, "Keep animals other than cattle, poultry or fowl, swine, sheep, or goats in an enclosure without wholesome exercise and change of air, nor or [sic] feed cows on food that produces impure or unwholesome milk." As a result, while it is illegal to keep a dog or a cat in a small cage for all of its life, farmers are free to use such practices with livestock.

Such considerations emphasize the importance of the New Jersey case. If courts should rule that anti-cruelty legislation must apply in equal manner for farm animals as for companion animals, the egg and pork industries would have to undergo a total alteration. If the average egg, veal, or pork farm had to comply with Ohio's requirement that animals receive "wholesome exercise," they would have to completely change their production system or go out of business.

Back to Britain

It was the British Experiments (see previous chapter) that gave birth to the modern animal advocacy debate, so it is not surprising that Britain and much of Europe have been more active in regulating how farm animals are raised. As early as 1976 the member states of the Council of Europe signed the European Convention for the Protection of Animals Kept for Farming Purposes, which required that farm animals be provided with care in a manner "appropriate to their physiological and ethological needs." The European Union subsequently adopted a number of laws in the 1990s specifying minimum standards in veal production and for laying hens. Minimum standards that dictate space requirements were passed in the early 2000s.

The EU has also passed laws on slaughter and transport. Perhaps most interesting and radical is an EU law requiring that all operators involved in farm animal production be given training on animal welfare.[49] The European Union has banned the use of veal crates and has decided that battery cages will be prohibited after the year 2012. Some individual European countries have adopted more stringent regulations. For example, the gestation crate is banned in the UK, and Switzerland prohibits battery cage production and gestation crates, and requires hogs to have straw or litter.

There is a salient difference between animal welfare activity in the US and in Europe: the debate within the US is over *whether* animal practices should be regulated, whereas that debate has been settled in Europe, leaving only the debate of *how* animal welfare should be regulated. In Europe almost every aspect of a farm animal's life is dictated to some degree by existing or pending legislation. For a list of specific examples, see Table 3.1, which shows five welfare regulations and each country's position on those regulations. That salient difference between the US and Europe, mentioned previously, is evident by the resounding list of "no's" for the US and "yes's" for Europe.

Animal Rights Terrorists

Colin Blakemore is a renowned physiology professor at Oxford University. In addition to his scientific accomplishments, as part of one of his experiments he sewed together the eyelids of kittens. However, Blakemore is no monster: he avoids eating meat to a large extent because he does not approve of factory farms and is devoted to his family cat. Blakemore conducted experiments on cats to understand the causes and nature of blindness. His research on kittens helped lead to a cure for a form of child blindness called amblyopia.

Blakemore is fortunate to still be alive. Barry Horne is an animal rights extremist in Britain who was sentence to 18 years in prison for fire-bombing stores that sold fur coats. Horne began a series of hunger strikes aimed at forcing the British government to cease support for research involving animal testing. Horne's hunger strike in 1998 placed Colin Blakemore's life in danger. The demands issued by Horne included an end to all vivisection by 2002. An extremist animal rights group called the Animal Liberation Front issued a threat that if Horne died from his hunger strike ten animal researchers who

Table 3.1 Animal Welfare Regulation Activity across Countries

Country	Density and housing limits	Banned gestation crates	Banned battery cages	Banned beak trimming	Banned castration without anesthesia
United States	no	no	no	no	no
United Kingdom	yes	yes	By 2012	By 2011	no
Sweden	yes	yes	yes	yes	no
Netherlands	yes	yes	By 2012	no	no
Denmark	yes	By 2013	By 2012	no	no
Germany	yes	By 2013	yes	no	no
Finland	yes	yes	yes	yes	no
Switzerland	yes	no	yes	no	yes
Norway	yes	yes	By 2012	yes	yes

Source: Adapted from Matheny and Leahy (2007).

conducted vivisection at that time would be assassinated; Colin Blakemore was among those listed. After some discussions between Horne and the British government, Horne agreed to end his hunger strike in return for a Royal Commission on animal testing.[50]

Extremist animal rights groups continue to threaten researchers. A group called the Animal Liberation Brigade set fire to a car owned by neuroscience professor, David Jentsch, as he slept in his house. Gene Block, the Chancellor of the University of California at Los Angeles, has dealt with vandalism, threats, and protestors not just outside his medical research lab but also at his home. The protestors made it known to all of Block's neighbors that he is a murderer and posted his name and address on the web, encouraging others to do him harm.[51] Universities and companies now go to great lengths to keep the location of the research facilities and the names of the researchers a secret, as extremist groups will readily threaten both the facilities and the researchers.

Sometimes the extremists' protests have gone wrong. Extremists once tried to destroy a researcher's car, only to mistakenly destroy the car of someone else.[52] In 2008, members of the Animal Liberation Front vandalized a chicken farm in Finland. In the process, they also broke the ventilation system, causing 5000 of the chickens to suffocate.

In Britain, animal rights activists have physically assaulted people. Groups like the Animal Liberation Front and the Primate Freedom Project have threatened to assassinate researchers who perform vivisection. One group member openly said, "I don't think you'd have to kill—assassinate—too many vivisectors...before you would see a marked decrease in the amount of vivisection going on. And I think for five lives, ten lives, fifteen human lives, we could save a million, two million, ten million non-human lives." Before you think that these groups are far from traditional animal advocacy groups, note that the President of PETA, Ingrid Newkirk has stated that, "I will be the last person to condemn ALF [Animal Liberation Front]...no one has been hurt, they have stopped the hurting of animals."[53]

This book pays much attention to the efforts of animal advocacy groups, but in this category we are not including animal rights terrorists. Groups such as the HSUS and Farm Sanctuary do not condone violence. The HSUS may have leaders that, from the perspective of many, are unusual, but their proposals for change are not extreme—nor are their tactics.

Animal Rights and Animal Welfare

Until the 1970s, efforts to improve the lives of animals can largely be considered part of the farm animal welfare movement. The object was to improve the lives of farm animals, not alter their social status. The 1960s witnessed a number of social movements centered on the "rights" of certain people within society. The women's liberation movement sought to provide women with equal pay and equal freedom to raise a family while pursuing their career. The movement sought not only to change legislation but also to alter the way we think about women in society. Alongside this change in women's rights, the Civil Rights Act of 1964 opened all schools to black children and gave African Americans the right to eat in any restaurant. Women are equal to men; black people are equal to white people—these were the dictums passed down from the 1960s. In this time of change people also began to ponder the rights of animals.

During his clinical training in the 1960s, Richard Ryder witnessed a number of experiments carried out on monkeys and decided to change the psychology profession. After forming friendships with three like-minded philosophy students, Ryder began to question the role of animals in society. To consider women inferior is sexist; to consider black people inferior is racist: what word, then, would describe the attitude that views animals as inferior? *Speciesism* was Ryder's answer, a term that is still used today. With the aid of other students the team edited a series of essays titled *Animals, Men, and Morals*, which set to logically argue against the killing of animals for any human purpose, and called for extending liberty and equality for animals. The group was located in England, and the English public was relatively uninterested in the collection of essays. One of the group's members, Peter Singer, hoped to find an American outlet for the book before the message was forgotten, and he wrote a review for the *New York Review of Books*. It was Singer's review that is said to have introduced the term *animal liberation* to the world, and landed Singer a job teaching a course called "Animal Liberation" at New York University (the first such course ever offered). When Singer published *Animal Liberation* in 1975, some say the animal rights movement took hold.

Animal Liberation was written for the masses. It is well-researched and easy to read. *Animal Liberation* prompted dialogue among animal advocates, and encouraged others to consider the issue of animal rights. One prominent example is Tom Regan, a former professor at North Carolina State University. Regan's 1983 book entitled, *The Case for Animal Rights*, gave the idea of inalienable rights for animals a more prominent place, and the idea of "rights" attracted a following.[54]

The term *animal rights* did not begin with the work of Singer and Regan. During the 1960s groups like the Fund for Animals (formerly run by Wayne Pacelle) began injecting the term into their literature and campaigns. In 1972, the National Catholic Society for Animal Welfare changed its name to Society for Animal Rights, using the civil rights movement as a justification for the name change.

What exactly are animal *rights* or animal *liberation*? The answer to this question can be seen by contrasting such terms with the concept of promoting animal welfare, which relates only to the well-being of animals. Improved animal welfare can be achieved without altering the status of animals as being property or altering the end-use of animals. The animal rights movement is best described as a movement that seeks to improve the welfare of animals while promoting their status within society. Animal rights advocates do not necessarily suggest that all animals should be "set free" to fend for themselves. Peter Singer introduced the world to the term *animal liberation*, but within Singer's philosophy one could ethically allow for animal ownership and animal consumption (something Singer himself, however, eschews). Some animal rights proponents see nothing immoral about raising animals for food or companionship, so long as they are well cared for, but there are others who vehemently challenge the current role of animals in society; for example, they oppose the keeping of animals as pets. Demonstrating the extreme view we have, for example, PETA's argument that, "it would have been in the animals' best interest if the institution of 'pet keeping' ... never existed."[55, 56]

There are some who wish to eliminate the classification of animals as legal property and to give animals some of the legal rights currently enjoyed by humans. An example is the Spanish parliament, who officially declared that Great Apes are entitled to the rights of life, liberty, and protection from torture. But what does this actually mean? It certainly does not mean that the animals are *liberated*: apes are not set free on the streets of Madrid. The intended result is likely to be that the declaration (which is expected to become law) should prohibit the use of apes in circuses or for filming. Apes could just as easily been protected from such activities without an official declaration of rights, so it is not exactly clear what is achieved by conveyance of rights. Only time will tell whether there might be unintended consequences to providing animals with such rights.

There is a distinction between advocates of animal rights and advocates of animal welfare. Those who tend to advocate animal rights are more likely to eschew the use of animals for food, research, entertainment, or any other goods such as shoes and clothes. Animal welfare advocates, by contrast, are concerned with how animals are treated but are not adverse to eating meat or some (but not all) animal experiments. Animal welfare and animal rights groups often work together towards similar goals but they do not always agree, as we have seen.

Animal rights groups claim that animals have certain inalienable rights, but arguing for animal liberation does not mean arguing that animals should be set free. In the context of farm animals, *animal liberation* would be best described as *animal extinction*, as farm animals would dwindle to very small numbers if they were given the right not to be used as food.[57] Farm animals need us. Most could not survive in the wild, and those that could, would survive only in small numbers. Regardless of how many rights are conveyed to animals, their well-being depends on how humans decide to treat them. Even wild animals that are supposedly liberated from humans are affected by our choices. Deer populations must often be thinned for their own benefit, and the ability of a wild animal population to flourish depends on the extent to which we develop wild lands. Even if we turned the animals loose into the wild, humans would still have to make decisions that related to the animals' welfare.

Animals, especially farm animals, can never really be liberated. Farm animals owe their existence to us and their well-being depends on our actions. In our assessment, the real issue is how we choose to treat farm animals. We can see farm animals as being a means to an end—useful only for the meat, milk, and eggs they provide—or we can treat them starting from the point of view of them as being sentient beings with natural instincts, who happened to be raised for human consumption, but who must also be treated with compassion. Going a step further, we could decide that the category of "farm animals" should no longer exist, and allow this form of relationship between man and animals to be relegated to history. Our choice will depend on the view we decide to take regarding the ability of farm animals to feel pain and suffering, and our ethical beliefs about how animal well-being and needs should be weighed in relation to our own well-being and needs. We now take up these topics in the following chapters.

CHAPTER 4

Animal Qualia

Investigating Animal Sentience

> *"Love the animals: God gave them the rudiments of thought and an untroubled joy. Do not trouble it, do not torment them, do not take their joy from them, do not go against God's purpose. Man, do not exalt yourself above the animals: they are sinless."*
>
> Fyodor Dostoevsky in *The Brothers Karamozov* (1879)

Are farm animals conscious? Can they feel pain? If the answers to these questions are "no," this book is unnecessary, as is the entire farm animal welfare debate. People today differ in their beliefs about animal sentience, as have humans throughout history. Recall that Descartes believed animals to be machines, but the Dalai Lama believes animals have sentience like humans. Thus, rather than simply assuming that animals are indeed sentient, we have decided to take up the argument for and against their being sentient in this chapter, arguing from the point of view of fact and logic, rather than religion or culture.

Farming Aliens

The culture in which we are raised has a powerful impact on our ideas of right and wrong. Sometimes these prejudices are passed down through the generation. Such prejudices are be beneficial, but others may serve no useful purpose, and may create unnecessary and undesirable strife between groups.

The consumption of meat, dairy, and eggs in our Western society is part of a longstanding tradition. This tradition owes its existence to the fact that consumption of animal foods helped past generations to thrive. However, people today can easily obtain their nutrients from non-animal sources. Hence, now is a good time to set our cultural prejudices aside and ask ourselves whether the consumption of animal foods, being no longer necessary, is ethical. Let us engage in a thought experiment which will help clarify our ethical beliefs by analyzing the decision to consume an animal food product, by replacing the subject of animals with a new subject—that of an alien life-form.

Imangine that somehow, the human race has captured an alien life-form. This alien species can be easily held and bred in captivity. Some product of the alien can be consumed as food, whether it be an egg-like substance the alien produces or whether the alien must be killed and consumed directly. Not only does the alien meat provide tasty food, but it is safe to eat and represents a healthy food source (when consumed in moderation and alongside other more conventional food). Is it ethical to breed and raise this alien solely to provide humans with food? Probably, your decision will depend on the extent to which the alien resembles humans and displays humans' ability to think and feel emotion.

Taking the analogy a step further, let us consider that there are two disparate types of potential aliens. The first is a very simple form of alien life, one that resembles earthly bacteria or yeast. Besides the fact that it contains DNA and reproduces, it does not seem very much like a living thing. It has no nervous system, no body parts that resemble animals, and certainly no brain. There is little about this alien that makes you think twice before eating it. (Just as vegans have no qualms about baking bread that toasts and kills the living yeast within, so they would presumably have no problems with eating this alien life either.)

Conversely, suppose the alien resembled the fictional characters in the film *Star Wars*. They are obviously not human, but they have eyes, ears, noses, arms, legs and appear to be vertebrates. Their face seems to smile when they are happy, and they appear to exhibit fear and sadness. They seem to communicate to one another in a sophisticated language. Though you may have no proof, there seems little doubt that the alien can think and feel. The prominence of these human qualities suggest that humans should pause before killing these alien life-forms. Only with a sense of guilt could it be thrown alive into a pot of boiling water (as is done so frequently with crabs and lobsters).

Before raising the alien for food, society would insist on scientific experiments discerning what kind of living creature this was, whether it contained any kind of nervous system or brain that could register pain, and whether the alien was conscious of its own identity. Society would be interested in more than just whether the alien could feel. We would be curious as to whether the animal was just sentient, or whether it was capable of complex thought and could contemplate its own existence and mortality. We would be curious about whether this alien shared our susceptibility to pain, fear, and depression. We would be curious as to whether it was sentient and/or self-contemplative.

Sentience and Self-Contemplation

Imagine playing an outfielder in a softball game. The opposing team's batter hits a ball high and in your direction. Just before getting in position to catch the ball your eyes are distracted by the bright midday sun. Because of the sun's obstruction the ball lands not in your glove but directly on your nose. At first you are stunned, and then the throbbing pain arrives. Nevertheless you scramble to find the ball to throw out the runner rounding the bases. After throwing the ball you fall to the ground holding your bloody nose. All of this activity is the result of your deliberate will, and within range of your perceptions.

However, this sequence of events unfolds without a need for you to engage in self-dialogue and self-contemplation. You did not deliberately think to yourself, "This hurts—really, really, hurts—and I want to fall down and cuddle my nose, but right now it is far more important that I pick up the ball and throw it." There was no internal dialogue debating the advantages and disadvantages of throwing the ball versus tending to your pain. This sequence entailed sentience but not self-contemplation; the decision to keep playing despite the pain seemed to make itself.

On the contrary, once you have thrown the ball, you begin to think about your injury and all the consequences of the injury. Now an internal dialogue takes place: "Wow, I haven't hurt this bad in a long time. I knew I shouldn't have played softball today. I really needed to work. I have gotten behind on my bills. My wife is going to kill me if there is an expensive hospital bill. This is all my fault." This dialogue makes explicit references to the self. There is a profound difference between the events. The first event regarded emotion that seemed to happen instinctively, while the second set of thoughts required explicit communication between the human and himself. The first dialogue describes *sentience*, and the second dialogue describes *self-contemplation*.

It is a plausible, or at the very least interesting, thought that farm animals may possess sentience but not self-contemplation. The strike of fear is intense in man and chicken alike, but only the man holds a deliberate, internal debate about whether the fear is justified. Perhaps animals possess sentience, which entails a form of consciousness, but self-contemplation, a higher form of consciousness, is in the sole domain of humans. This view of consciousness, where animals can experience a flood of emotions despite their ability to rationally deliberate upon those emotions, is a view appealing to some scientists.[1]

This chapter looks at whether animals experience and are aware of emotion. The exploration requires an understanding of what it means to be conscious. We must first recognize, however, that there will always be a metaphorical canyon separating human and animal intelligence. Rene Descartes, for example, concluded that, "there are no men so dull and stupid, not even idiots, as to be incapable of joining together different words, and thereby constructing a declaration by which to make their thoughts understood; and that on the other hand, there is no other animal, however perfect or happily circumstanced, which can do the like."

However, it is important not to deny a cow's ability to feel pain just because they cannot write a poem. We know human infants can feel pain despite their inability to communicate their thoughts. We cannot rule out a form of consciousness in which a being is sentient even if they are rarely or never self-contemplative. For this reason we seek to apply both the attributes of consciousness *and* sentience. Our goal is to evaluate the logical and scientific evidence for animal sentience and consciousness, and we also consider the likelihood that animals experience pain and pleasure even if they cannot write, say, a poem.

Qualia

Matter is made up of atoms, and atoms are made up of protons, electrons, and neutrons. Light is a series of photons. Literature is a sundry of letters, arranged by a lyrical maestro. Music is a mystical conglomerate of air waves. Everything is constructed from units of something, including consciousness. The private experience of consciousness, that undeniable feeling of identity; memories, fears, and hopes both mundane and salient—all of these conscious experiences are made off *qualia*. Qualia refer to subjective conscious experiences; they can be thought of as units of consciousness. Although most of us know we have something like qualia, some philosophers have searched frantically for proof of such subjective experiences. Some argue that qualia do not actually exist, whereas others believe proof of such phenomena will be discovered in time. In our thought experiment regarding whether the alien discussed above possessed conscious emotions, we were asking whether the alien possessed qualia.

The Philosopher's Zombie

No doubt you are a busy person. Would it not be nice if you had a clone to do all the tasks you hate: mow the lawn, sit in committee meetings, change the baby's diaper? Perhaps you might even be willing to pay a scientist friend to develop a robot that looks and acts *exactly* like you. If you argue with your spouse about whose turn it is to do dishes, the robot will argue with your spouse. If you are afraid of clowns, the robot is too. When you stop to marvel at Magnolia tree blossoms, the robot does the same. The robot might even be able to become so familiar with your life that it might gossip with your friends and even try to manipulate you. The scientist's work is a true success: your friends and family cannot distinguish you from the robot.

This clone is referred to as the *Philosopher's Zombie*. This zombie is a thought experiment that all philosophers of consciousness are well aware of and have frequently debated. The debate is not whether this zombie can be built on practical grounds. Instead, the debate concerns whether, even with the most advanced science, a zombie who is absent of consciousness could be identical to you. If this zombie could be absolutely identical to you, then consciousness serves no purpose, and is simply a superfluous trait of human nature that is ultimately unnecessary.

Think about it. If consciousness serve no function, it would be possible to build a clone that possessed no consciousness but that is your absolute identical match. A perfect human clone cannot be built without a heart, because hearts serve a critical function. A robot could be built with an artificial heart that acted in many ways the same, but it could not be identical to the biological heart. The same logic extends to a sense of consciousness. The thought experiment is important, because if the Philosopher's Zombie exists in the imagination alone, then consciousness has an influence on human behavior; consciousness has a *function*. If it has a function we can search for this function and in the process better understand whether animals need this function, and hence whether they possess consciousness.

That humans have consciousness suggests that it has a beneficial function, otherwise it would not have emerged and remained with us over the ages. Though some parts of the body may serve little or no function, these parts are the exceptions rather than the rule. The fitness or functionality of consciousness provides us a framework by which we can logically discuss the concept of qualia, and use that framework to discuss whether non-humans possess qualia. In what follows we attempt to further define consciousness, understand its role, and study the function of consciousness to help determine whether it is held by animals such as pigs and chickens.

Meet Your Consciousness

The consciousness resides, of course, within the complex brain. The brain works by sending electrical pulses from one neuron to the next through synapses. With 100 billion neurons, and each neuron connecting to 1000 other neurons on average, there are 100 trillion neural connections in your brain.

The electrical pulses process information and produce decisions.[2] Consciousness has no relationship to the part of your brain that controls breathing and digestion. Reflexes and snap judgments are likewise conducted without seeking input from what you generally think of as "you"—your rational consciousness. As professors, we are well aware of the research showing that students form an impression of a teacher's effectiveness largely based on initial snap judgments, rather than based on the teacher's actual effectiveness. Likewise, food purchases are not just determined by conscious decisions about expected taste and price. They are also profoundly influenced by subconscious reactions to seemingly irrelevant attributes, such as color and package shape. These subconscious reactions often serve us well. For example, Vic Braden is a man who can almost perfectly predict when a tennis player will double fault, even though he does not know how he does it.[3]

The conscious and subconscious act independently as well as together. Consider an agricultural example. Baby chicks are separated by gender because only females can lay eggs. The problem is that, for most people, male and female baby chicks look identical. Only trained *sexers* can look at chicks and determine which are males and females, but they often have no idea *how* they do it. Chicken sexing is an exclusively subconscious activity. Because sexers are unsure of how they know separate sexes, when they teach their trade to others they simply go through a slide show showing a chick on each slide and announcing whether it is male or female. There is no other way to teach the trade other than watching pictures of "male, female, female, male, female, male, male..." Teachers hope that the student's subconscious picks up on the differences between the males and females, but not all students develop the skill. The point is that both the subconscious and conscious minds exhibit a profound influence on our choices. To understand consciousness is to understand how the conscious works in tandem with the subconscious, and where consciousness begins.

To see how the conscious and subconscious affect decisions, try the following experiment. Figure 4.1 asks a series of questions. Take a sheet of paper and cover the figure, and then slide the sheet of paper down as you read each question. Answer the questions as quickly as you can. The first question asks whether there are any Os in a black background. If you are like most people you are able to detect the black background almost instantaneously. You did not have to "think" about it; the answer came immediately. Your brain did not utilize your consciousness to answer the question. The answer came as quickly and mysteriously as your instinct to shut your eyelids when a fly comes close to your eye. There is an algorithm hard-wired into your brain which can detect stark discrepancies in color, and can make distinctions without the assistance of conscious reasoning. The second question asks if there are any Xs in the collection of Os. Again, you probably noticed the lone X immediately without really thinking.

Now attempt questions 3 and 4—which ask you to identify whether there is an X in a black background. Most people cannot automatically detect the presence or absence of an X in a black background without first finding the letter X and then asking, "Is the X in a black background?" There is a salient difference between the first two and the last two questions. The difference is that answering the last two questions requires conscious, deliberate thought. The transition from the first two questions to the last two questions illustrates the conditions where your consciousness does and does not exist. You have now met your consciousness. This example does not imply that you did not possess sentience when answering the first two questions. Most readers likely felt a

First, take a sheet of paper and cover everything below this caption. Then slowly pull the paper down and read the question. Underneath the question will be a series of 12 letters. Pull the paper down until you see the letters and answer the question as quickly as possible.

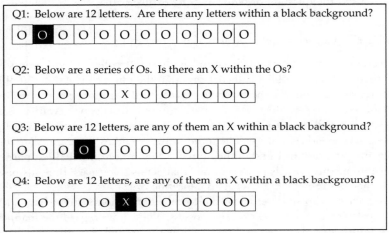

Figure 4.1 Meeting your consciousness

sense of anxiety or curiosity. Something different happens in the last two questions though, and that difference is a vivid illustration of the singular self that constitutes each individual human being.

The explanation for the difference in speed of response lies in the *computational theory of the mind*, which posits that the brain (or the mind) is essentially a computer, an information-processing system. Within the processing system are numerous informational processing routines, much like computer programs, except they run on neural networks instead of silicon chips. Some routines run independently (e.g., your heart beats regardless of what you do) and others run simultaneously. One routine in your brain allows for quick identification of discrepancies in color and another allows for quick identification in discrepancies of shapes (allowing you to differentiate Xs and Os almost instantaneously). Fast judgments in these areas are necessary for you to respond to changes in the environment, such as a stoplight changing from red to green or a lion stalking you through tall grass. Changes in color and changes in shape happen all the time, so your brain devotes many resources to detecting such changes using independent and fast routines.

There are so many different combinations of color and shape in the world that unique routines cannot be efficiently established in the mind for detecting simultaneous changes in color and shape. Thus, when the mind must identify shape and color at the same time, two independent routines (one for shape, one for color) in the mind report to a "judge," who takes both sets of information into account to yield a verdict. That "judge" is your consciousness. Scientists study two features of consciousness: one is mostly understood, but the other is as mysterious as the question of why anything exists at all.

One Face of Consciousness: Information Processing

The computational theory of the mind is the most coherent and empirically valid theory of the mind. This theory says that the mind can be thought of as a computer, and although some psychologists contend it is more complicated than this, the computer analogy is adequate for our discussion. Animal brains evolved over millions of years, developing differing complexities that helped species survive and reproduce. The mind of the bat allows it to translate sound waves into pictures; the mind of the human allows it to translate light particles/waves into pictures. Birds possess minds that act as a compass; humans possess minds that can remember or forget the location of car keys.

Researchers who study efficient biological information systems have identified four laws that all *efficient systems* possess. These laws regard the manner in which information is stored and retrieved by the animal. The first two have to do with probabilities. The first says that the greater the number of times a certain piece of information is requested by the mind the more it will be readily available. The second says that more recent information queries are given greater priority than historical queries. Suppose an actress named Katy is currently starring in a play called *The Merchant of Venice*, and twenty years ago she starred in the play *To Kill a Mockingbird*. The second law says Katy is more likely to remember the lines of *The Merchant of Venice*, but the first law states that the greater the number of times she starred in *To Kill a Mockingbird*, the more lines from that play she will remember. Wild animals do not have to know lines from a play, but they do have to remember the locations of water holes and which plants are poisonous. It is obvious why these two laws would benefit the survival prospects of an animal.

The third law suggests that experience is "colored" by emotion. All animals are driven by goals which help them survive, prosper, and reproduce. Information from past experiences needs to be stored in a manner that determines whether certain actions help or impede achievement of these goals. Eating a poisonous plant results in the emotion of sickness and fear, helping to make sure that the next time the human sees this plant they will keep their distance. Eating food provides us with great pleasure, which is nature's way of reminding us that food is necessary for nutrition and survival.

The fourth law is the most relevant for this chapter. Psychologist Steven Pinker describes this fourth law as, "the funneling of control to an executive process: something we experience as the self, the will, the 'I'."[4] An executive director is needed to coordinate bodily functions in pursuit of a goal. If the eyes did not focus on the object toward which the feet were walking, bad things would happen. For a predator to increase its chances of catching prey it must make sure that all feet are running towards the prey, that the eyes are focused on the prey, that the teeth or hand with the spear are performing the right functions to kill the prey, and so on. The subconscious part of the mind will act to increase the predator's heart rate as they attack the prey, but these subconscious actions are in harmony with the other actions the body is undertaking. The consciousness decides *what* to do, and the subconscious portion of the mind decides *how* the heart should beat to help achieve that action. Furthermore, as was shown previously, the conscious is often needed to reconcile and combine different information signals provided by different mental routines.[5]

Another Face of Consciousness: Sentience

Consciousness as a single authority for information processing is largely accepted by psychologists, but it is not what we typically think of as "consciousness." The idea of being self-aware largely regards an almost magical, undeniable awareness that the *I* feels myriad emotions from pain to boredom to ecstasy. The unfortunate man who cries to the heavens, "Why me?" epitomizes the sentient perception of the self. The concept of a soul flows from consciousness. Consciousness is undeniable, but personal. It is one of the few concepts that is unverifiable yet rarely questioned. This alternative fact of consciousness leaves scientists baffled. The importance of this statement bears repeating: there is no scientific evidence or reasoning to explain why or how the biological functioning of the mind leads to the awareness of emotions and the self.[6] We do not even know the proper questions to begin the exploration.

Many animal species work in groups. Consider, for example, the wolf pack. On a hunt each wolf watches what the other wolves are doing and modifies its own behavior to increase the probability of a kill. Wolves stalking from separate locations often converge on prey from multiple sides. A wolf that sees an unusually large gap between two dogs may choose to fill the hole. Communication is not necessary for group cooperation. Once the prey knows it is being pursued, one wolf might bark. Barking helps signal location. The wolves closest to the prey might bark the loudest so that the other wolves know where to go. For each dog to know its optimal location in the hunt, it must know the position of the other animals, and that is achieved by barking.

Researchers have also noted that the beautiful patterns that emerge in bird swarms (for example, the V-shaped formation flown by ducks), arise, not from explicit communication, but from each individual bird following a few simple rules like "stay within a certain distance of the next closest bird." Such simple coordination is so effective that scientists have recently applied observations on animal swarming behavior to develop computer algorithms (referred to as particle swarm algorithms) to help solve complicated optimization problems.

These examples illustrate that group behavior does not depend on explicit communication, but such communication can certainly help. Communication is advantageous for a variety of reasons. Birds sing to communicate their territories. Male deer roar at each other during mating season. The depth of the roar helps communicate the size and strength of the deer, and hence who would likely win a fight. Roaring helps male deer determine superiority without a physical fight, which could very well spell death. This type of communication, however, is rudimentary.

More sophisticated communication can be seen in the animal kingdom. Rodents can communicate which types of foods are nutritious and which types are poisonous, and they even pass this information down from one generation to the next. No wonder humans have such a hard time eliminating mice infestations.[7] Vervet monkeys use subtle grunts to communicate many different types of information. The human can communicate practical pieces of information as well as abstract philosophy. Heightened communication skills and the intense awareness of the self have resulted in a species that enjoys hearing stories, reading novels, watching television, and going to the movies.

Early humans, no doubt, told stories of their hunts. Perhaps the clan would laugh when a young hunter described how he attempted to spear a wild pig, only to have the pig turn on him and chase him up a tree. The clan was probably entertained by the story but its plot also served a purpose. From that story other humans learned that when hunting pigs the predator may become prey. Evolution endowed animals with an intense love for food. Food is essential. By programming the animal to enjoy eating, its survival chances are enhanced. Similarly, humans are programmed to love stories because they help them learn from the conflicts of others.

Humans love stories because of our heightened sense of the self. If it were not for the ability to put ourselves in the place of story characters, movie theaters might not exist. Have you ever jumped in fear watching a movie? Have you cried watching a film, feeling as if it were you that were betrayed? The intimate understanding of the self also allows us to predict how other people will react in situations. By understanding what makes us angry we can predict what makes other people angry, and therefore abstain from performing such actions. This ability to project the self onto others provides us a conceptual framework for social interaction that reinforces social bonds and encourages cooperation. Almost 200 years ago, Adam Smith, the moral philosopher and father of economics, stated, "As we have no immediate experience of what other men feel, we can form no idea of the manner in which they are affected, but by conceiving what we ourselves should feel in the like situation."[8]

The understanding of self, therefore, facilitates an understanding of others, which helps species that live in groups to prosper. It seems to make sense, then, that evolution might bestow self-consciousness on species that live in social groups and that cooperate to meet their basic needs. At a minimum, this constitutes a reasonable hypothesis. Consequently, one indication of consciousness is cooperation through complex communication.

Searching for Consciousness, Searching for Sentience

Much about sentience and consciousness, which hereafter are considered the same thing, is unknown. We are convinced that we are sentient. There are enough similarities between people that we make the leap from knowing *we* are sentient to believing that *other people* are also sentient. Indeed, this is hardly a leap but a logical inference. What about our alien that exhibited human-like qualities? Is the alien sentient? What about a chicken? Inherent in these sorts of questions is a tendency to conceptualize consciousness as an "either/or" phenomenon: either an animal is consciousness or it is not. A more proper view, however, is to think of consciousness existing on a spectrum, where animals are seen to possess varying degrees of consciousness. The closer the probability of consciousness approaches 100 percent, the more important the topic of their welfare becomes.

The computational theory provides a background from which we can articulate a few rules that might identify whether a being has consciousness. If an animal appears to have an executive director controlling the routines of the mind, they are likely to possess consciousness and sentience. When animals appear to exhibit emotions which, given our human experiences suggest they truly feel the emotions, they are likely to possess consciousness and sentience. Emanating from these propositions are rules for discerning consciousness and sentience.

An animal is more likely to possess consciousness and sentience if:

1. The animal possesses a brain and nervous system resembling human systems.
2. The animal appears to exhibit intelligence.
3. The animal appears to have an awareness of its own suffering and joy.
4. The animal appears to make rational, calculated trade-offs.
5. The animal lives in social groups who utilize sophisticated communication.

The idea of intellect as consciousness is useful because it indicates the presence of an executive director within the mind that filters through all the competing subconscious processing routines to arrive at a final decision. The executive director must reconcile competing emotions for a single action. Suppose you are considering purchasing a case of beer. There is one emotion in your mind signaling that you do indeed crave that drink. However, there is another emotion worried that the price is too high, and a third emotion worried that if you drink the beer you may have trouble performing at work tomorrow. These competing signals must be transformed into a single decision. The more signals and the more decisions an animal must make, the greater the need for an executive director to filter and reconcile these signals.

Santino is a chimpanzee who lives in a German zoo. Santino rather dislikes being on display for humans. In the early morning before visitors arrive, Santino piles rocks in one specific spot. Later, when visitors arrive, Santino begins throwing rocks toward the visitors. When Santino is unable to scare away the humans (because there is a wide body of water separating visitors from animals) he becomes infuriated.[9] It is hard to argue that an animal which displays this type of intentional planning is neither intelligent nor sentient.

The presence of a consciousness is also related to the ability of animals to make rational trade-offs. In humans, for example, we tend to consume less of a good as the price rises; this behavior reflects an attempt to balance a desire for consuming the good in question with a desire for other things that money can buy. If livestock also appear to make these sorts of rational decisions, we can be more confident that they are conscious beings. Moreover, the fact that animals make trade-offs implies that they have *preferences*—that they desire some things more than others. If an animal knows it prefers outcome A to outcome B, this is suggestive of consciousness, as it implies a sort of self-awareness where the animal knows what *it*, the *self*, wants. Indeed, some scientists accept the existence of preferences as evidence of consciousness partly because it can be *measured*, whereas many other features of consciousness cannot.[10] For example, Colin Spencer argues, "Immediately, if we admit that an organism may have preferences, we must admit that it also has consciousness, i.e., an awareness of options which necessitate a choice."[11]

Interestingly, economists have conducted numerous experiments with animals which indicate that animals have preferences and make rational trade-offs in the face of changing incentives. In the book, *Economic Choice Theory: An Experimental Analysis of Animal Behavior*, economists John Kagel, Ray Battalio, and Leonard Green show that the economic theories that have been developed to understand human trade-offs and preferences apply equally well in explaining the behavior of animals—mainly rats and pigeons.[12] For example, people generally demand less of a good as the price increases, people desire more of a good as their income increases, and they supply more labor when the reward of labor rises. Interestingly, in cases where humans fail to make decisions consistent with the rational theories of choice (these typically involve particular choices between lotteries), animals do too—in much the same way. The existence of preferences and the ability to make rational trade-offs is not at all unique to humans, and more interestingly, the artifacts of irrationality are not terribly different between humans and animals.

Evidence of rational thought is one marker of consciousness, but so is evidence of feelings. It is not naive to suppose that an animal which appears to exhibit an emotion actually does experience that emotion. Is this not one of the reasons we assume our fellow humans share our sense of the self? It would seem absurd not to count the appearance of emotion as *some indication* of sentience.

Some theories describe consciousness evolving alongside sophisticated communication. Communication between species at a sophisticated level requires some sort of internal dialogue accompanying the dialog with others. For example, if you decide to write a review of this book you cannot communicate your thoughts to others without first formally arranging a personal set of thoughts, parsing concepts in your own mind, and arranging your own set of logic. The review requires you to make comments such as "I think the authors...," and "...doesn't make sense to me," and "the authors fail to consider..." Perhaps you will even critique our discussion of consciousness in your mind, only to realize that this internal dialogue is exactly the mechanism we were trying to describe.

A sophisticated species needs to consider hypothetical settings. Suppose that a primitive species wants to improve a hunting method. They might draw a map in the sand, use sticks to represent a spear being thrown at a mammoth, and then point to where a canyon exists. The idea being formed is that mammoths, for example, can be more easily killed by running them off a cliff. Thinking hypothetically requires *self*-thought and *self*-communication, which indicates a sophisticated consciousness. Communication encourages and perhaps requires some sort of consciousness. Any type of communication, however rudimentary, between animal species that appears to change under alternative environments is indicative of a being with the potential for self-thought.

The greater degree to which an animal possesses the five aforementioned traits the greater the chance an animal is sentient and perhaps even self-aware. In one of the classic books on animal rights, *The Case for Animal Rights*, Tom Regan outlines what he calls the Cumulative Argument for Animal Consciousness. Regan's argument is partly as follows. If an animal possesses the five aforementioned traits, their possession of the traits is best explained by consciousness emerging in both humans and animals as they evolved. Any alternative theory that posits humans are conscious but animals are not is (a) less consistent with the world we observe, (b) more complicated in that it explains the same phenomenon with more assumptions, and (c) most likely contrived, developed specifically to give the result that humans, but not animals, are conscious beings. Given these considerations, Regan argues that the degree of concordance with the five traits is indicative of conscious.

This is the general framework we will employ to assess the consciousness of farm animals. It is not an infallible procedure, but it represents a systematic way to rationally consider the question of farm animal consciousness. We are aware of no better process for attempting to identify consciousness. When Regan sketched out his Cumulative Argument for Animal Consciousness he stated, "The preceding does not constitute a strict proof of animal consciousness, and it is unclear what shape such a proof could take." In similar spirit, biologist Christof Koch states, "there is no accepted theory of consciousness, no principled theory that would tell us which systems, organic or artificial, are conscious and why."[13] Thus, if any uncertainties remain after we finish considering the evidence on farm animal consciousness, we can take heart, because the experts feel likewise.

Assessing Farm Animal Consciousness and Sentience

The previous section outlined a series of rules by which we can assess whether animals are conscious. In this section, each rule is taken individually and is evaluated in light of the research on farm animals.

Do farm animals have a biological structure similar to the nervous system and brain of humans, which would suggest a similar ability to transmit and feel pain?

All vertebrates have a biological makeup similar to humans. There are significant differences on the outside, of course, but vertebrates share the same biological building blocks including circulatory, immune, and nervous systems, and brain structure.[14] Pigs resemble humans to such a large extent that the heart valves from pigs can be transplanted into humans, and some believe that humans will begin receiving kidney and heart transplants from pigs in the near future. We are unaware of any scientific evidence suggesting that the basic biological makeup of farm animals would prevent them from possessing sentience. The influential Brambell Report of 1965, commissioned in the UK to address nascent concerns about factory farms, concluded that, "all mammals may be presumed to have the same nervous apparatus which in humans mediates pain. Animals suffer pain in the same way as humans."[15] Subsequent research findings have only reinforced this conclusion. Even chickens have a brain structure similar to humans, as they use the left and right side of the brain for different purposes.[16]

The fact that farm animals and humans have very similar nervous systems and brain structures is important. In a recent discussion with a farmer in the livestock industry, the farmer dismissed animal welfare concerns as silly. His argument was that plants also respond to injury by sending out electrical signals,[17] and thus, showing concern for animal welfare is akin to showing compassion for plants. From this logic, the farmer sought to demonstrate the futility of the farm animal welfare movement, as humans must eat, and if plants and animals have each been shown to feel pain, what can we do about it? What will we be left with to eat? However, the farmer's logic is faulty. Plants are not equivalent to animals: they contain no nervous system or brain structure that is even remotely similar to those of vertebrates. There is no evidence to support the argument that plants feel actual pain, but evidence abounds for farm animals. It is very unlikely, however, that this farmer believed his own argument. To our knowledge, only the disciples of Scientology have given serious consideration to the sentience of plants.[18]

Are Farm Animals Intelligent?

A researcher once pumped cigarette smoke into a rat cage in an effort to study the effect of second-hand smoke on health. Not surprisingly, the rats were bothered by the smoke. What is surprising, however, is that the rats picked up their own feces and used it to block the tunnel pumping in the smoke. The experiment failed because the mice out-witted the researcher. If this is not evidence of intelligence and reasoning, what is?[19] Many experiments have been conducted with animals and the general consensus is that many animals are able to think rationally, solve problems, and sometimes communicate solutions to problems.

Pigeons and dolphins can recognize their reflection in a mirror. Rats can count, and will count if it helps them navigate a maze. Birds have a remarkable memory for where they store seeds, even when researchers manipulate the birds' environment to make the discovery process harder. Some birds use sticks and stones as tools to obtain food. Apes can engage in thinking games with one another, and will try to manipulate each other by guessing how the other animal is thinking. Apes relay information from one generation to the next— information such as how to extract termites using sticks and to wash sweet potatoes in the sea.[20]

Birds, including chickens, can predict the trajectory of an object when the object disappears behind another object. Chickens are capable of abstract thought. Even if part of an object is hidden, a chicken recognizes the object as a whole. Chickens can count, conduct geometrical calculations, and they enjoy mental stimulation.[21]

Chickens are certainly less intelligent than apes, and perhaps less intelligent than mice. However, their small heads contain efficient thinking machines. Chickens are social animals that normally live in flocks of up to thirty birds. They know the precise pecking order in a thirty-member flock and the identities of all thirty members. They learn from their environment. If an object passes over them, such as a chick, they will duck and freeze as if a hawk were swooping down. But once they observe that the object is benign, if it is a regularly occurring event, they learn to ignore the object, while still looking out for other overhead objects.[22]

Chickens can associate objects and activities, much like Pavlov's dog salivating at the sound of a bell. Chickens can be trained to perform certain actions, such as pecking at a plate in return for rewards, and they learn better in groups where they can learn from fellow birds. A bird's behavior is affected by its environment when young. Chickens raised in small cages with no enrichment demonstrate greater stress, are more reluctant to explore, and do not display some of the normal behaviors of their counterparts who have been raised in an enriched environment.[23, 24] Unlike human infants, baby chicks will search for an object they have seen being hidden behind a screen.[25] These birds are smart enough that they experience boredom. Even when food is plentiful chickens prefer to work for food.[26] The question, then, is whether they also have the intellectual capacity for emotional pain.

As is commonly recognized, pigs are as smart as dogs. Like dogs, pigs can recognize their name and can be taught to respond to commands such as those for fetching an object or moving an object from one place to another. Pigs can manipulate a joystick to play simple video games, and are unique in that they understand the rules of games after only playing once. Researchers have found that when pigs are given a food reward in exchange for solving a problem, they are persistent in solving the problem for the love of the game and for the love of food—not *just* for the food. When they watch other pigs, they are also able to predict what that pig thinks and sees.

Pigs' intelligence is illustrated nicely by the following, a common occurrence witnessed on hog farms. Some farms have automatic feeders which respond to computer chips placed in collars on each pig. When the pig approaches a feeding stall, the collar communicates with a computer whether the pig has been fed yet. If the pig has not eaten, the stall automatically drops feed. Pigs have come to understand that it is the collar that causes feed to drop, not the pigs themselves. If a collar is found on the floor a pig will pick it and take it into the stall to receive an additional feed allotment. This is not something that has happened as a one-off coincidence, but it is something that has been repeatedly observed, making the observer confident that the pig must be thinking, "If I pick up this collar and take it into the stall, I will get feed."[27]

Hog producers who raise hogs in confinement facilities would have no awareness of the animal's intelligence, because the above sorts of behavior are only witnessed on farms where the hogs are allowed out of their pens, and where they are treated humanely.

To illustrate with a personal example, Bailey once showed a pig for the first time in college. He had to train the pig to respond to his commands so that he could lead it around the show arena. After Bailey chose his pig he needed to take the pig away from its barren concrete floor to a larger place where he could work with the pig alone. When Bailey asked the farm manager where he should take her, the manager replied, "anywhere."

"Anywhere?" Bailey responded, confused.

"Yes, right out front there is fine." The front of the barn was a simple grassy area without a fence.

"Won't the pig run away?" Bailey just assumed that if the pig was allowed out into the open for the first time in its life it would take off for freedom.

"No, the pig will actually probably try to get back into the barn."

Bailey did not really believe the manager, and suspected the manager was playing some kind of practical joke on him, and that Bailey would end up spending all day trying to catch this pig. Bailey first led the pig out of the cramped, concrete pen where the pig lived with five of its peers. The pig did not want to leave. Once Bailey got the pig out of the pen and to the barn exit, the pig froze. Directly ahead of the pig was an open, grassy area. No fences. Nothing to hold it back. For the first time in its life it was free to dig in the dirt, run, to do anything it wanted to. But the pig froze, it was scared to walk on grass. The pig's entire life had been spent on a concrete floor. Whatever this green thing was in front of it the pig wanted nothing to do with it. Bailey finally convinced the pig to walk on the grass, and just like the farm manager said, the pig kept trying to get back into the barn. It wanted to get back to the barren, cramped concrete pad because it was the only environment it had known and this new experience frightened it.

However, this example does not tell the whole story of what pigs prefer in terms of living conditions. If you were to take the typical hog who has experienced a more natural, freer life which has included pasture access and if you were to give that hog an opportunity to explore and root in a new area, you had better stand back because it might run over you in its keenness to get to that new area! Pigs who are not raised in an open, enriched environment will fear this environment, but this does not prove that all pigs fear an open, enriched environment, and so to provide them with it does not equate to providing them with higher welfare conditions.

Bailey would remember this event years later when he watched the movie *Instinct*, starring Anthony Hopkins. The movie studies gorillas that have lived a natural life versus those raised in cages. The caged gorillas, when adults, do not even think of leaving the cage when the door is open wide. The concept of life outside the cage does not exist. Though the gorilla exhibits no desire to leave the cage it would have been a much happier ape if raised outside the cage, in a natural habitat. The similarities to the plot of *Instinct* and the story of modern hogs are compelling.

The problem is that the typical hog farm removes baby pigs from their mothers at an unnaturally young age, and places them in a crowded, unenriched environment. As a result these pigs do not mature mentally. Their arrested mental development and lack of environmental enrichment causes them to experience great stress when encountering a new environment. Anyone who has entered a nursery (an enclosed barn where recently weaned piglets are kept in small groups) will attest that any sudden move by a human causes panic throughout the whole barn. Such pigs also perform very poorly when asked to solve problems compared to hogs reared in a more natural environment. Piglets allowed to nurse for many weeks and who live in an enriched environment demonstrate higher intelligence and confront new environments with more bravery. This is an important point, because livestock producers commonly complain that animal welfare activists and consumers generally do not really know about hog well-being. *Paradoxically, it may very well be that hog farmers themselves are the least aware of the animals' mental capabilities.*

Cattle are not as intelligent as pigs, and often appear little more than grass-eating machines. Research has shown, however, that there is more complexity within the head of a cow than a first glance would suggest. Cows can not only solve simple problems but they become excited when a solution is found. Cows can be trained to perform simple feats, such as pushing a lever for food, and they can read certain signs. Cows are especially adept at remembering directions and geographic locations, and at recognizing their peers. After years of absence a cow can still recognize up to 50 cattle and ten human faces. In the hunt for food, cows democratically elect herd leaders who are older and more skilled at finding food and water.[28]

We do not expect chickens, cattle, and hogs to score well on an IQ test; however, this does not mean they do not possess any intelligence. Farm animals exhibit a social intelligence in that they can remember and identify group members, along with showing a capacity for communication. Farm animals respond to changes in their environment by learning new things. In our assessment, farm animals do have a moderate intellectual capacity, and one that appears to have the potential for experiencing certain types of emotions, including suffering.

Do Farm Animals Make Rational Trade-offs?

Assessing whether farm animals have the capacity to make consistent decisions when they are faced with conflicting emotions would suggest the presence of an *executive director*—an unambiguous sign of consciousness. Making rational trade-offs does not have to be the result of the presence of a consciousness—we could easily write a computer program that could perform this feat. However, asking whether farm animals make rational trade-offs provides one more piece of evidence in the assessment of sentience. The answer to this question also provides evidence on what animals like and the intensity of that desire—issues of importance when determining the well-being of farm animals.

Considerable evidence supports the notion that pigs and chickens make rational trade-offs. Moreover, the preferences they display help us categorize animal desires. To assess these issues, chickens and pigs are trained to perform acts before receiving rewards. Chickens must squeeze through a hole or pigs must push a panel to receive a reward. The "price" of the reward is altered by decreasing the size of the hole or increasing the number of times a panel is pushed before a reward is given. Animal preferences can be studied by identifying the maximum price they will pay for different rewards, or by determining how responsive animals are to changes in the price; the measure of this sensitivity is "elasticity," which is a term replete in economic studies. Research shows that pigs like to socialize with other pigs; however, they will pay a much higher price for food than for socialization. Hens will pay a very high price for access to nests. The ability to dust-bathe and scratch in the dirt is highly valued by the chicken, as the maximum price they will pay is larger than the respective price for a larger, though barren, space allotment. When chickens are injured they can even learn which foods contain pain relievers.

In short, animals have preferences, and they change their behavior rationally in response to different incentives. When pigs are given the opportunity to socialize with another pig by pressing a lever with their nose, the researcher can measure the value of socialization by the maximum number of presses the pig will exert. Chickens do the same. Chickens and hogs are not just reacting blindly, but rationing their energy in a way to make them happy, just as we ration our paychecks in a manner that best suits us.

Economists have developed a reliable set of tools to measure human preferences for goods and services. These tools generally assume that humans make rational trade-offs and have well-defined preferences, meaning they know what they like best and tend to make decisions that achieve these desires, while prudently minding their limited time and energy. Economists and animal scientists have found that the tools used to measure human preferences work just as well for chickens and hogs, which provides us with one more piece of evidence in favor of the notion that farm animals are sentient.[29]

Do Farm Animals Experience Emotions?

Almost anyone who has spent time with animals will attest to their apparent ability to experience emotions. Cats and dog owners often believe they can tell how their pets are feeling. Likewise, their pets seem able to perceive the moods of their owners. Dairy cows who have recently calved are clearly agitated and frustrated if their newborn calf is taken from them. Sows grow bored and frustrated when confined to tight cages, exhibiting abnormal behaviors like repeatedly biting the cage bars or scraping the floor with their hoofs. Animals who are being castrating show every sign of experiencing genuine pain. It is hard to deny that animals feel pain, enjoy eating, and experience fear.

Scientific studies have been done which provide evidence of animal emotion. Cows produce 10 percent less milk when a person who frightens them is present. Cattle respond to positive treatment from humans, becoming less fearful and more approachable, for example, after being petted by a human.[30] Calves separated from their mothers experience frustration and have elevated levels of stress hormones, and the evidence furthermore indicates that it is their mothers they are missing and not just the mothers' milk.[31]

Recent studies suggest that one of the reasons human children from poorer families tend to remain poor as they grow up is that the stress they experienced in childhood has impaired their mental development. Their intelligence levels seem to be lower than they might otherwise have been due to their childhood stress. Negative emotions experienced as a human child permanently impair the ability to function in life.[32] Similarly, research shows that birds, pigs, and cattle that experience greater stress in their youth experience greater difficulties in solving problems, in adjusting to new settings, and they have greater welfare problems in general.[33]

Hens will go to great lengths to reach a nest to lay an egg; they become fixated with finding a nest. Chickens and hogs become agitated when placed in cramped groups, often expressing their frustration by hurting other animals. As demonstrated above, some animals are known to be capable of making rational trade-offs, and both hens and hogs are willing to work, making trade-offs, in order to socialize with other animals of the same species. This indicates that they have a preference for being in social groups.

There have been some studies made that show indication that some animals are capable of an aesthetic appreciation, which is arguably one of the most sophisticated forms of emotional processing. Studies have been made where pigeons have demonstrated an ability to predict which paintings human children would judge as *good* and *bad*.[34] This suggests that they have similar aesthetic preferences to humans, and that they perhaps have an appreciation of beauty.

When given different foods and exposed to different temperatures, rats experience the same range of emotion as humans, causing researchers to conclude, "the physiological and behavioral parts of the mechanisms in ourselves and those in other species are so similar that the leap of analogy we would have to make to assume that conscious experiences are also similar is reduced to the barest minimum."[35] Early experiments on dogs revealed the "learned helplessness" concept that now forms the basis of modern psychological thought on depression. When dogs were tortured they fell into a shell of despair, believing nothing they could do would change their lot, just as depressed humans tend to believe only misery awaits them in the future.[36] It is likely then that hogs, pigs, and cattle experience similar "helplessness" as well when mistreated.

Objective measurements confirm the emotion hypothesis. Cortisol, the hormone commonly known as the stress hormone, has some positive effects, such as boosting immunity and lowering sensitivity to pain, but its presence in the body is generally a symptom of stress. Higher cortisol levels can be beneficial in emergencies or life-threatening situations, but if experienced over long periods of time they lead to compromised health. For example, pregnant human females with higher cortisol levels are more likely to experience a miscarriage. The cortisol does not necessarily cause the miscarriage, but it does indicate the presence of stress and other physical problems.

Given that cortisol is an indicator of stress, we now have an objective measure of whether an animal experiences stress and thus emotion. Table 4.1 shows the cortisol levels in sheep that received conventional, minor surgeries, such as having tails or testicles removed, as compared to a control group that did not undergo any surgeries. Results indicate that 24 hours after the surgery, cortisol levels are significantly higher for sheep which had had surgery, compared to the control group which had not. Moreover, cortisol levels were seen to increase with the number of surgeries performed, as would be predicted in a sentient animal. Cortisol levels can even be measured in response to different living environments. Pigs housed inside buildings, living exclusively on small concrete lots and in close proximity to many other pigs, with severely constricted space allotments experience stress. Table 4.2 demonstrates this. The higher the stocking density, the higher the cortisol level. Similar research has shown that calves also experience heightened cortisol levels when housed in small crates where they cannot turn around or groom themselves.[37] This should come as no surprise—think of the stress many people experience in a crowded elevator. Because of the objectivity of the measure, cortisol level comparisons are used extensively in farm animal welfare research.

Of course, it is possible that the hormonal response of farm animals mimics that of humans, while their minds do not. Animals may release stress hormones but fail to actually "feel" stress. There is no proof that hormone changes actually translate into conscious feelings of emotion in the animal. Yet, we have no iron-

Table 4.1 Cortisol Levels and Surgical Procedures Performed in Sheep without Anesthetic

Surgical procedure	Cortisol (n mol 1^{-1})
None (control group)	87
Tail docking	136
Castration	171
Mulesing and tail dock	187[*]
Mulesing, tail dock, and castration	232[*]

Note: Mulesing is the surgical removal of wrinkled folds of skin, done to prevent maggot and fly infestation.

[*] Denotes that animals experience high cortisol levels even after 24 hours.

Source: Broom and Johnson, 1993.

Table 4.2 Cortisol Levels for Pigs at Different Stocking Densities (Floor Area per Pig in Square Meters)

	High density—0.51 m^2 / pig	Low density—1.52 m^2 / pig
Male pigs	158.9	87.7
Female pigs	107.1	90

Notes: Numbers denote plasma cortisol levels after ACTH challenge test, in ng ml^{-1}.
Source: Broom and Johnson, 1993.

clad proof that humans experience conscious emotions either. True, humans can tell researchers that they experience these emotions, but even pigs have been taught to communicate their anxiety to researchers.[38] What we can say is that animals act *as if* they have feelings of emotion, both physically and biologically. Cows, pigs, chickens, and all other farm animals act exactly as one would expect of a sentient being.

Do Farm Animals Prefer to Live in Social Groups and Communicate?

Chickens, hogs, cows, sheep, and goats all prefer to live in groups. Chickens and hogs prefer group sizes of 30 or less. Much like the large buffalo herds that once roamed America's prairies, cattle are comfortable in very large herds, but they also form smaller, stable cliques within the herd.

Social animals generally possess sophisticated language. We all know animals posses some verbalization skills that they use to communicate with each other. Animals also communicate information to one another in a code that researchers cannot crack. No matter how hard they try, when it comes to wild ruminants and birds, researchers are unable to predict which male will be selected by the female, or which animals will ultimately be the strongest and healthiest. The females still manage to select those males that turn out to be the strongest and healthiest. These animals are able to communicate something to members of their own species about their physical fitness in a way that researchers who have studied them for years cannot fully identify.

Baby chicks are born with many instincts, but much of their behavior is learned from their mothers. Mother hens use a system of pecks, scratches, and chirps to tell babies what is acceptable to eat. Roosters help find food for hens and their offspring. Roosters also help hens build nests and they coordinate much of this through oral communication. Chickens will make alarm calls when a predator is spotted, but interestingly enough, only when other chickens are around. Chickens are social animals that know how and when to communicate with peers. Chickens communicate using up to thirty types of vocalizations.[39]

Pigs' grunts and squeals can take on many different meanings. Hogs also communicate through body scents and motions. Able to remember and recognize up to 30 of their peers, pigs prefer to live in small herds and often sleep very close to one another even if the additional warmth is unnecessary. Hogs develop complex social structures. A pig moved into a new group will experience stress and often injury as it adjusts to its new peers and finds its status in the group's existing hierarchy. To prevent fighting among groups of hogs it is often better to either keep a pig within a stable group or to keep it isolated. This complex social structure is indicative of a higher level of sentience.[40]

Cows are also social creatures, although their social behaviors appear to be less complex than that of pigs. When placed in a pasture, cattle naturally cluster into groups, as opposed to spreading themselves uniformly throughout the pasture. Within a large herd, smaller subsets of cattle tend to form. Herds often elect leaders that find food and water. Fighting tends to be less of a problem for cattle compared to chickens and pigs. New members can be introduced into herd-groups with little adverse consequence for the new members. Cows generally, though not always, get along well with one another, and while they will establish a hierarchy, it usually involves little stress or injury for any of the members of the herd. The rare exception includes bulls fighting for dominance, or in feedlots, where cows are placed into small pens. Frequently, a cow will begin "riding" other cows, acting-out the motions of a bull mating. These "riders" can cause significant stress and injury to the cows that they ride, and must be separated from the group. Cows communicate vocally as well as through grooming, odors, body language, and the like.[41]

There is an important difference between communication and language. A French speaker and a Russian speaker can communicate certain pieces of information without knowing each others' language: a smile, a frown, and a fist all signify something. Equally, pigs' grunts and a chickens' clucking relay some sort of information—but do they form a language? Language is an intricate tool which can be used for expressing virtually any thought. Many attempts have been made to teach apes how to communicate with us using sign language, objects, or keyboards. And while they have been seen to learn some features of language and to be able to communicate simple phrases, their communication seems to be largely the result of repetition rather than an actual understanding of language itself. Steven Pinker, an expert on language, concludes that, "fundamentally, deep down, chimps just don't get it." The basic conclusion of most scientists is that only humans use language.[42] However, it is undeniable that pigs and chickens can and do communicate with other members of their social group on a daily basis, and this is an indicator of the levels of awareness they possess.

Even though animals may not have language, their communication levels are in some cases surprisingly complex. Take, for example, vampire bats, which have a sophisticated system of reciprocity and altruism in their dealings with other vampire bats. The success of a vampire bat's hunt is largely due to chance.

A lucky hunt yields a bat more blood than they need and an unlucky hunt leaves the bat hungry. To help smooth out the variation in their hunting success, the bats that experienced a lucky hunt share blood with the less successful bats. The hungry bats remember who shared with them and in most cases they return the favor when they later encounter a successful hunt. However, the bats that share their food will remember a bat that did not return the favor, and will never share with them again.[43] Bats do not just help other bats because they care about them, but because these acts of reciprocity are integral to the success of the species.

However, this sort of reciprocity is not found amongst farm animals, who spend much of their time competing with each other, constantly seeking to take food from others and establish superiority in the group hierarchy. Farm animals communicate to breed, to eat, and to avoid being eaten. Farm animals can be gentle and caring with their young, but they can also be cruel and negligent. In summary, farm animals do communicate and they live as social animals, even if their communication is not always an indicator of good social behavior within their groups.

Just because farm animals are naturally social animals does not necessarily imply that they *should* live in groups. Good things occur in natural groups, such as chickens helping each other forage for food. Yet, as any livestock producer can attest, animals will hurt and even kill each other within their natural groups. If mother sows are not kept in crates when nursing, some will crush and even cannibalize their young. This is also important, for example, when we compare cage to cage-free egg production—sometimes farmers find that cage-free methods of rearing chickens find the chickens killing one another.

Are Farm Animals Sentient?

Determining whether farm animals are sentient is a key question in the farm animal welfare debate, with obviously important implications for the ethics of the way we raise animals for food. Are animals sentient? They have moderate intelligence, but nothing as impressive as dolphins or apes, and certainly nothing on the level of humans. Farm animals have preferences and make rational trade-offs, suggesting the presence of an executive director in the mind which filters conflicting emotions in the pursuit of well-defined goals. Farm animals seem to possess rudimentary communication skills and live in moderately complex social cultures, given the living conditions in which this is possible. These factors provide some evidence to support the notion that farm animals are sentient. In our assessment, however, it is the display of emotions that provide the greatest support for animal sentience. Farm animals react to injuries exactly as a sentient being would be expected to, and, equally, they exhibit an ability to take pleasure in certain activities. Furthermore, the biological systems of farm animals respond to stressful environments the same way human systems respond—by increasing levels of the hormone cortisol.

Marian Dawkins is a biologist fascinated with the concept of animal consciousness. In her magnificent book *Through Our Eyes Only?* she takes a thorough and objective look at the possibility of animal sentience. She comes to the conclusion that animal consciousness is possible, and maybe probable, as she asserts, "We now know that these three attributes—complexity, thinking and minding about the world—are also present in other species. The conclusion that they, too, are consciously aware is therefore compelling."[44]

It seems that most people agree with Dr Dawkins. When we conducted a telephone poll with a random sample of US citizens, we found that only about 12 percent did not believe that animals have roughly the same ability to feel pain and discomfort as humans (see Table 4.3).

Nevertheless, not everyone does believe that animals are conscious beings. In fact, there are some people who do not even believe *humans* are conscious. A particularly interesting view is that of Julian Jaynes, who believes that many of our ancestors, including Abraham and Homer, were unconscious. Consciousness, Dr James suggests, emerged about 1000 BC along with more symbolic language and writing.[45] It is not necessarily absurd to conclude that animals are devoid of consciousness and are thus unable to perceive pain and suffering. When a farm animal gives apparent screams of agony when they are hurt, does that mean that they have the same sense of "self" as a human? As we have seen earlier in this book, there are many who argue that they do not.

The question of animal sentience and consciousness is, and will probably always remain, an open question. Even if one feels strongly that animals can or cannot feel pain, any objective reading of the evidence must admit there is *some* chance of being wrong either way. However, this uncertainty need not mean indecision.

Table 4.3 Result of a Nationwide Telephone Survey of over 1000 Respondents

Question posed to respondent: Do you agree with the following statement? Farm animals have roughly the same ability to feel pain and discomfort as humans.

Answer	Percent of respondents
Strongly agree	57%
Agree	24%
Neither agree nor disagree	5%
Disagree	7%
Strongly disagree	5%
Don't know	3%

Dealing with Uncertainty

There are many reasons to support the idea of animal sentience, and some reasons to cast doubt on the idea. As humans, we often try to avoid uncertainty and possess an inherent desire to decide on an answer absolutely before deciding how to act. Nature abhors a vacuum and humans abhor uncertainty. However, pretending to be certain when there is little reason to be so is not an approach we embrace. We prefer the truth of uncertainty to the comfort of certainty. On the subject of animal consciousness we must accept the possibility that animals may or may not be conscious, sentient beings. Fortunately, the economic toolbox kit is replete with tools to deal with the uncertainties.

What is the most you are willing to pay for a lottery ticket when there is a 99 percent chance of winning nothing, but a 1 percent chance of winning $100,000? Some simple algebra indicates that the *expected* or *average* winnings equals (0.99) ($0) + (0.01) ($100,000) = $1000. If you purchased many, many such lottery tickets, on average each would yield $1000, even though most tickets will win nothing. If the lottery ticket price is $1, logic suggests the ticket is a good purchase. With a price of $1 and an expected winnings of $1000, the *expected* profits are $999. Of course, you will almost certainly not actually win the $1000; 99 percent of the time you will lose your $1 but 1 percent of the time you could win $100,000. But there is the chance that, if you played the game over and over, your winnings would average $999. Thus, we can think rationally and reasonably about whether we are willing to buy a lottery ticket, *even though the outcome is uncertain*. Similarly, it is possible to think rationally and reasonably about animal sentience even though the truth is not known.

Just as the fact that each lottery ticket gives us some chance of winning money, making the expected value of the ticket positive, the fact that there is some probability animals are sentient suggests the *expected* animal sentience is positive and that animals do have some level of sentience. As there is a fair possibility that animals can feel pain it seems reasonable that this pain should be given *some* consideration. Even if one believes animals are most likely unconscious machines that just appear to feel pain, so long as there is some probability they are sentient they should be treated at the very least as partially sentient creatures.

Expected Level of Sentience = (P)(*animals are sentient*) + (*1*-P)(*animals are machines*)

As we have already argued, sentience can then be seen on a continuum scale between sentient and non-sentient. In the equation above, *P* is the probability that an animal is sentient and *Expected Level of Sentience* is an overall assessment of animal sentience, accounting for the uncertainty. Sentience, then, can be expressed in degrees. Arguments about whether animal sentience exists are probably best understood in terms of *expected animal sentience*.

Rational thinking would suggest that there is some expected level of animal pain and suffering in farm animals. Just because animals are sentient and can feel pain does not necessarily imply that anything should be done to improve their lives. No doubt your neighbor is sentient, but not everyone would necessarily agree that any suffering on your neighbor's part imposes a moral obligation on you. Whether and how animal welfare is to be balanced with that of humans is another question, and is the realm of moral philosophy. Before we turn to these issues it is important to delve more deeply into the daily living conditions of modern-day farm animals. Now that we have concluded farm animals have an expected level of sentience that is at the very least "some sentience", we must consider the environment in which they are raised to assess whether they experience suffering.

CHAPTER 5

Raising the Animal

The Life of Birds, Pigs, and Cows

"Put the fowls in coops so small that they cannot turn around."

A. H. Baker in *Livestock: A Cyclopedia* (1913)

"The hen home should be a place of comfort, safety, contentment, cheerfulness, and happiness. Given these, the hen responds."

Davis, et al. in *Livestock Enterprises* (1928)

Measuring Animal Welfare—Indicators of Well-Being

As professors, we frequently find ourselves in the position of providing advice to students who are conflicted between two job offers. Many of these dilemmas take the following form. They can take a job near their hometown and be close to family, or they can take a significantly better-paid job out-of-state. As far as job choice goes, *closeness to home* and *level of salary* are often two *indicators* of our students' well-being.

Imagine a situation where one of our students is confronted with a choice between two otherwise identical jobs: one that pays $30,000/year in her hometown or another that pays $60,000/year that is 500 miles from home. It is unclear which job the student will prefer because each job performs well in one indicator of well-being and poorly on the other. To help the student assess which job will make them the happiest we often ask the importance they place on living close to home and the importance of making a large salary. By assessing the relative importance of these two indicators to the student, and by observing the difference in the salaries and distances between the two jobs, students can often walk away with a better idea of which job is best.

When asked whether a hen is better off in a cage or cage-free system animal scientists often respond in much the same way we respond to our students. Farms typically have two basic options for raising eggs, which results in the *egg dilemma* depicted in Figure 5.1. Almost all egg-laying chickens in the US are raised using the cage system shown on the left of the figure. A few birds are placed in a small, barren cage for virtually their entire life. An alternative is the cage-free system shown on the right. The cage-free system is better than the cage system in that the birds have more freedom to move, lay eggs in nests, scratch in the dirt and dust-bathe, and have more space. But the cage-free system is not superior in every respect. A drawback of the cage-free system is the very large flock size. In such a large flock hens cannot establish a pecking order because the flock size tends to exceed the maximum of thirty, which chickens prefer. Consequently, when placed in large flocks, hens have a tendency to engage in bouts of aggression. Birds in a cage-free system thus experience higher injury and higher mortality rates than those in cage systems.

Is the cage or cage-free system superior? Just as our students have to weigh the relative advantages and disadvantages of different jobs, so too must we weigh the advantages and disadvantages of cage-free egg production. In both cases, there are *indicators* of well-being. For hens these indicators are space allotments per hen, nest availability, rates of injury and mortality, and the like. The desirability of any system depends on the provision of these indicators and the importance of the indicators to the birds' overall well-being. This is the *egg dilemma*, and there is no easy answer.

If we are to talk about animal well-being it is important that we should be able to measure well-being and describe how and why an animal is said to experience high or low levels of well-being. Can animal well-being be measured? Yes, but not perfectly. Think about whether a co-worker is considered to be happy: what are the signals to observe? If they frequently smile, seem active with friends, and seldom complain, we might conclude that the co-worker is happy, but if they

Figure 5.1 Cage (left) versus cage-free (right) eggs

Note: Permission granted by United Egg Producers.

frown a lot, are isolated, and frequently complain, you might conclude the opposite. Most people would agree on the indicators of happiness and sadness. What is difficult, however, is when the signals conflict. For example, what are we to make of someone who rarely smiles but appears to actively engage in life?

The same quandary exists for assessing the well-being of farm animals. Experts generally agree on the indicators of animal well-being, but when conflicting signals arise, experts disagree on how to reconcile these signals to assess the overall level of welfare. The bottom line is that we possess indicators of animal well-being levels but we have no method of aggregating those indicators into a single well-being measurement that is agreed upon by all scientists. We can develop overall subjective welfare measures by deciding, based on the evidence, how important space allotment versus nest availability is in determining overall well-being of hens, but experts have legitimate disagreements on such issues, just as two different people might legitimately disagree on the relative importance of a high salary versus living close to home. In this section these indicators of animal well-being will be discussed, along with a description of attempts to aggregate them into a single animal welfare measurement.

The Five Freedoms

The impetus for the contemporary farm animal welfare debate stemmed from the book *Animal Machines*, by Ruth Harrison. *Animal Machines* allowed individuals with no agricultural background to see for the first time how modern farms produce eggs and other animal food products. Many did not like what they read. In response to public concerns, the United Kingdom formed a committee to investigate these farms. In 1965, the results of the investigation were documented in the Brambell Report, which officially scrutinized certain production practices, such as confining hens into cramped cages. But it also highlighted animal *needs* and animal *urges* as issues to be considered when designing livestock production systems.[1] For example, when discussing the advantages and drawbacks of housing animals in cramped cages, the report states, "The degree to which the behavioral urges of the animal are frustrated under the particular conditions of the confinement, must be a major consideration in determining its acceptability or otherwise."[2] Although the Brambell Report made a number of specific recommendations for each livestock type (many of which are still not implemented in the US today), perhaps its most lasting impact was to encourage policymakers, society, and farmers to take into account animal suffering and needs. As a result, a number of voluntary codes were established in the UK prescribing how animals should be raised. The codes were issued by Britain's Farm Animal Welfare Council, which in 1979 modified the codes to suggest that all farm animals should be provided the Five Freedoms delineated in Figure 5.2.[3]

1. Freedom from thirst, hunger, or malnutrition.
2. Appropriate comfort and shelter.
3. Prevention, or rapid diagnosis and treatment, of injury and disease.
4. Freedom to display most normal patterns of behaviour.
5. Freedom from fear.

Figure 5.2 The five freedoms

Despite numerous reports, scientific studies, and deep thought on farm animal welfare since 1979, the Five Freedoms remain the best succinct summary of the criteria entailed in establishing high welfare. The freedoms are intentionally vague. No description of animal needs can be succinct without also being vague. Moreover, by focusing on animal *outcomes* rather than specific production practices the Five Freedoms are relevant regardless of the technologies available to farmers.

The Five Freedoms provide a means of assessing whether an animal experiences a high level of well-being in a particular production system. Production systems could be designed to provide animals with each of the freedoms listed in Figure 5.2, but producing eggs and pork in an affordable manner often requires sacrificing one or more of the Five Freedoms. The key is to identify which of the freedoms are most important for animal well-being and which are associated with the highest costs.

Although the Five Freedoms provide a simple and holistic view of animal welfare, they are often not specific enough to establish whether one system is preferred to another—especially when the farmer is constrained by monetary concerns. For example, the egg dilemma requires assessing whether greater bird freedom is worth greater rates of injury. That is not an easy determination. Both could be provided, but probably at a financial cost greater than most egg consumers would be willing to pay. To assess whether animals are better off in one system versus another, animal scientists seek *indicators* of animal well-being and observe how indicators vary from one production system to another. In what follows, we discuss a number of well-being indicators.

How to Tell a Happy Hog: Health and Profitability

When confronted with accusations of animal suffering livestock producers routinely reply that they would be unable to make money from animals that suffer. For example, in a statement arguing against the need for animal welfare legislation, Scott Dewald, Vice-President of the Oklahoma Cattlemen's Association, argued, "Our producers take care of their animals, and we know that an animal that isn't treated well doesn't produce."[4] There is certainly some truth in the statement. Many studies have shown a close relationship between health and happiness, even for humans. Happy people live longer than unhappy people, even if both are afflicted by the same diseases. Unhappy people are also more likely to contract sickness and take longer to recover from disease and injuries than happy people. In "how to" books on human happiness psychologists routinely argue that exercise and maintaining a healthy body are important keys for happiness.[5]

One indicator of animal well-being is animal health and productivity. A farm animal that experiences significant stress will grow at a slower pace and will be less likely to reproduce. Like humans, animals that are depressed are more likely to become sick and take longer to recover from sickness. Farmers have an incentive to ensure their animals are at least somewhat happy because it improves their bottom-line figures.

The first of the Five Freedoms, freedom from thirst, hunger, or malnutrition, is not a concern on the typical farm, except very rarely. Profitable farmers ensure that animals have ample access to food and water. They would do so even if they cared nothing for the animal. The same can generally be said for treatment of injury and disease. Large livestock operations often employ full-time veterinarians to ensure adequate health. The farms that provide the worst health care are often small farms, because size limits the ability to pay for veterinary service and such producers have less expertise with animal health care. Although, it must be said that some small farms with particularly attentive owners can take better care of their animals than those where there are more animals for each worker.

When attempting to measure an animal's well-being our first instinct is to try to measure emotions. Measuring animal emotions is fraught with difficulty, but we can measure whether animals in a particular setting are reproducing well and growing rapidly. "What can't be measured can't be managed," an animal scientist argues, and for this reason some scientists contend that we should place greater focus on "performance" indicators of animal well-being. The argument is that, "an animal is in a poor state of welfare only when physiological systems are disturbed to the point that either reproduction or survival is impaired."[6]

If animal welfare could be inferred directly from the productivity and profitability of the farm the egg dilemma would be no dilemma at all because birds are healthier and more productive in a cage system. Whereas hens in cage systems produce up to 270 eggs a year, cage-free hens produce only 259 per year. Caged hens live until 115 weeks of age while cage-free hens only live until 80 weeks. In addition, mortality in a cage facility is only 3 percent while the figure in cage-free facilities is around 8 percent.[7] We are simplifying the story somewhat. Cage and cage-free system use different breeds of birds and those genetics alone explain some of the productivity differences. Also, we are ignoring a few issues. For example, birds in a cage-free system have stronger bones because they exercise more. Nonetheless, birds in the cage system are more productive than birds in a cage-free system. If farm productivity served as the sole proxy for animal welfare, the cage system would appear to be a more humane egg production system. Unfortunately, the relationship between animal productivity and welfare is not that simple.

Productivity is indeed a useful indicator of animal welfare, but only when applied to a *single* animal, rather than a group of animals or the entire farm. As the egg dilemma illustrates, farmers often have an incentive to trade some *individual* animal well-being for increased *group* performance. To say that an individual animal's performance is high when well-being is high is *not* the same thing as saying that a farm's total output is high when each individual animal's well-being is high. The reason is that when farms are constrained by land and building size, and machinery, it often makes economic sense to use more less-happy animals than fewer happy animals.

Animal health and farm profitability play an important role in establishing animal welfare guidelines in the US. The United Egg Producers (UEP) Certified is a voluntary program in which producers can receive a UEP Animal Welfare Certified label on their products if certain conditions are met. As of 2008, about 80 percent of all US eggs are produced under this label.[8] The UEP certification standards are constructed by an independent panel of scientists. These standards state that each hen should be afforded a minimum of 67 square inches per bird. Producers who do not abide by these standards typically provide each bird with around 48 square inches of space. As Figure 5.3 shows, these are tiny space allotments. The figure shows four hens: a box has been drawn around each hen to represent the amount of space a typical hen requires to stand comfortably (about 75 square inches). The black square on each side of the

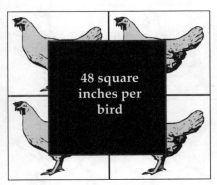

Hens require 75 square inches of space to stand comfortably. In cage systems with 48 sq. in. per bird, these four bids must be crammed into an area the size of the black square shown above.

By increasing the per bird space allotment to 67 square inches, the United Egg Producers come much closer to meeting the birds' natural space needs.

Figure 5.3 Comparison of two space allotments

figure represents the space that is provided for the four birds under the two different standards. The left diagram illustrates the 48 square inch space allotment: the four birds would have to be literally crammed into the cage. However, the diagram on the right of the figure shows the 67 square inches space allotment: the black box almost covers all the hens, illustrating that 67 square inches per bird comes much closer to providing enough space for the birds to stand comfortably.

Data show that increasing space from 48 to 67 square inches leads to increased egg production on a per-hen basis. Each hen produces more eggs, and the mortality rates are lower.[9] Here is a case where using productivity as an indicator of welfare equates to improving the well-being of the birds. While moving from 48 to 67 square inches per bird may be an improvement, the argument that birds in such cages are "happy" is not convincing. As Figure 5.4 shows, if the birds were to attempt to flap their wings they would have almost no room to do so. Why not give the hens more room? One reason is that when the hens are given more than 67 square inches they become more aggressive, plucking each others' feathers, which reduces bird health and welfare. From the prospective of productivity and profits, the 67 square inch space allotment does, however, seem about optimal, which is why the EU producers tout the cages as providing "optimal hen welfare."[10]

Maximizing profits is not the same thing as maximizing animal well-being. There are situations when a producer can sacrifice some animal well-being and still achieve increased profitability. Consider a hypothetical example. Over the

While the 67 sq. in. allotment provides adequate space for standing ...

... birds need 303 sq. in. to flap their wings comfortably. If all four birds tried to flap their wings simultaneously each birds would have to do so within the confines of the black box.

Figure 5.4 A 67-square-inch-per-bird cage meets some space needs but not others

course of a year a typical laying hen produces, on average, a little less than one egg per day (generally about 270 eggs over a 365 day period). Suppose a farmer has a single barn and must make a choice between two production systems. In System A the barn will contain 1000 laying hens and each will be laying 270 eggs per year for a total of 270,000 eggs per year. In System B, the farmer can crowd 1500 hens into the barn, but because of the decreased space, each individual hen's well-being falls. Associated with the decline in well-being is a decline in productivity, say, down from 270 to only 230 eggs per year.

It is easy to see that System B will generate many more eggs per year for the farmer than System A ($230 \times 1500 = 345,000$ versus $270 \times 1000 = 270,000$). Clearly, this farmer can make more money by sacrificing some hen well-being. The profit motive can work counter some of the Five Freedoms, and profit-maximizing outcomes are not perfectly correlated with those that lead to optimal animal well-being.

It needs to be stressed that, while per-hen productivity rises as the per-hen space allotment is increased from 48 to 67 square inches, total egg production within a barn falls because there are less hens in the building. The egg industry has aggressively forced through the measure ensuring farmers provide their hens with the increased space allotments of 67 square inches per hen, and it has done so at the expense of farm productivity. These more generous space allotments are provided to birds due to a real effort to improve animal welfare, and/or to protect the image of eggs and the egg industry.

Consider the relationships shown in Figure 5.5. The top graph shows how hen productivity and well-being change as the space per hen is increased. Hens are more productive on a per-hen basis *and* their well-being is higher when given more space. Note, however that while the two curves are positively correlated, they are not identical. In particular, there are ranges over which increasing space per hen would increase well-being substantially but would only marginally increase egg production. Does the argument put forward by the livestock industry that well-being and productivity are correlated imply that farmers will choose to maximize well-being? The answer is clearly, "no." The bottom chart in Figure 5.5 shows outcomes at the barn level: it illustrates total egg production from the entire barn, assuming a producer has a fixed barn space. If farmers are forced to increase space per hen and if they have a fixed barn size, then the number of hens in the barn must, logically speaking, fall.

This induces a trade-off for the farmer. Starting from a very small space per hen, where hens are crammed as tightly as they can be inside the barn, animal welfare is seriously compromised and the hens will lay very few eggs. However, as space per hen increases, the increase in animal welfare is significant. This welfare enhancement causes each bird to produce more eggs, and the increase in the per-bird laying rate is greater than the decrease in egg production stemming from a lower hen population, and output for the barn rises. However, as the space per hen continually increases there comes a point where bird welfare

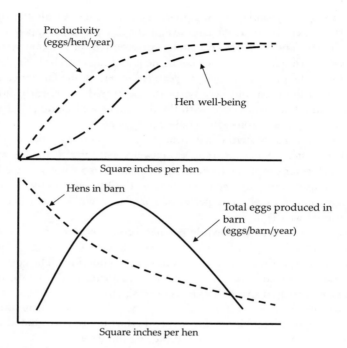

Figure 5.5 Productivity, profitability, and welfare

increases, but not by much. In this case, the increase in the per-bird laying rate is not enough to counter the impact of a lower hen population, and total output for the barn begins to fall.

One can take this thought experiment to the extreme and ask what the barn output would be if the barn only contained only five hens. The welfare of the hens would be very high, but total output from the five hens would be incredibly low. The farmer must make a trade-off between productivity per hen and number of hens, and as shown in the bottom portion of Figure 5.5, there will be a particular space allotment that will maximize the total number of eggs produced in the barn. The key point of Figure 5.5 is that the space allotment chosen by farmers does *not* maximize animal well-being, or even individual animal productivity. When farmers are constrained by land, labor, barn size, or even availability of capital, the economically optimal stocking density will be more crowded than would suit the animals.

Furthermore, not all scientists agree that animal welfare for the individual animal can be measured by animal productivity alone. In fact, it is our impression that most scientists disagree with the notion that productivity is a sufficient indicator of animal welfare. Farm animals are specifically bred to reproduce and

Many witnesses have represented to us that the growth rate of an animal for meat or the egg production of a laying hen are the only reliable objective measures of their welfare. It is claimed that animal suffering of any kind is reflected by a corresponding fall in productivity. The argument is that in the absence of any scientific method of evaluating whether an animal is suffering, its continued productivity should be taken as decisive evidence that it is not. *This is an oversimplified and incomplete view and we reject it.*

Figure 5.6 Does profitability ensure high welfare?

Source: Brambell Report, 1965, pp. 10–11, emphasis added.

grow fast and can do so under even the worst of conditions. Some, perhaps most, methods of enhancing welfare can actually work against profitability. Laying hens raised in a cage-free facility are allowed to walk around a rather large space and thus they walk more than they would if raised in a cage. The ability to walk no doubt improves the bird's welfare (temporarily leaving aside mortality and injury issues from larger flock sizes), but also the additional exercise diverts some nutrients from producing eggs into the energy needed for walking. In this case the animal welfare improvement is accompanied by a decline in productivity. Finally, it needs to be stated that there are cases where animal health is compromised without a corresponding compromise in productivity. Remember that farm animals are a commodity and they are treated well to the extent that it is profitable. On farms with large numbers of animals facing meager profit margins it can seem economically necessary to withhold care for a single animal. For example, rather than fixing a pregnant sow's broken leg, a producer may wait till the piglets are born and then euthanize the sow.[11] Providing the care needed to reduce what would be significant suffering to one sow is not profitable, and thus is not provided.

The weight of the evidence indicates that animal well-being cannot be measured by the productivity and profitability of the animal *alone*. As far back as the Brambell Report, most experts have rejected the view that animal productivity is the sole indicator of well-being (see Figure 5.6). Productivity is one indicator of animal well-being but not the only one. There are other factors besides productivity that can be measured and thus managed— if not perhaps in the same way.

How to Tell a Happy Hog: Stress Hormone Levels

In the preceding section, we argued that the productivity and physical health of the animal is an imperfect indicator of individual animal well-being, and even if individual productivity perfectly measured individual animal welfare, it would be a mistake to infer individual animal well-being based on the good productivity of a barn or a system.

We have already seen in Chapter 4 that sheep receiving minor surgeries, such as tail docking or castration, without anesthetic, tend to exhibit higher levels of the stress hormone cortisol. Cortisol levels can be measured by taking a sample of saliva from the animal, and because cortisol stays in the animal for a long period, the sample can be taken hours after the event hypothesized to cause stress. If impaired welfare leads to stress and this stress can be easily measured from cortisol, we have what appears to be an objective measurement of animal well-being.

Stress hormone levels are indeed routinely used as an indicator of animal welfare levels. Most animal welfare studies use a number of different welfare indicators, and stress hormone levels are almost always included in this list. However, there are instances when the level of stress hormones conflicts with our expectations about animal welfare, and the usefulness of hormone measurements becomes suspect.

Where cortisol is not a useful indicator of well-being is in veal production. Veal calves are typically tethered in tiny crates so small they cannot even lie down comfortably. Because of consumer backlash the veal industry is currently transitioning to a group-pen system. Are the veal calves happier in group-pens? Surprisingly, cortisol measurements would indicate the calves are worse off. When veal calves are first placed in crates, cortisol levels are higher than those in group-pens, as one might suspect. However, weeks later when the calves have had time to adapt, calves in crates actually have lower cortisol levels than those in group-pens. Researchers suspect this finding may, however, be a result of the measurement process. For cortisol measures to be taken from calves in pens the calves must be physically caught by humans, a process which is stressful. Calves in crates do not need to be chased to obtain cortisol measurements. Because we are almost certain that calves in pens have higher levels of well-being than calves in crates, and because cortisol levels do not confirm our expectations, it is thought that cortisol is not a sufficient indicator of animal welfare in all settings.

Consider another example. Experiments have shown that horses exhibit higher levels of cortisol when they are transferred to new environments but not when they are denied food and water.[12] Horses certainly must suffer without food and water. Thus, again, because inferences from cortisol do not confirm our notions we again question its validity in this instance. In summary, cortisol measurements are useful for measuring animal welfare in some cases, but in others it is of questionable value; other indicators as well as the use of common sense are needed to validate it.

How to Tell a Happy Hog: Behavioral Indicators

Animals that are stressed, bored, or discontented have a tendency to behave in uncharacteristic ways, and studies have shown that experienced researchers and farmers can detect an animals' state of well-being through observation.[13] Although there are behaviors that tend to be correlated with high levels of well-being, researchers often tend to focus on those abnormal behaviors indicative of low levels of well-being. An animal exhibiting abnormal behaviors is thought by many experts to be suffering. Although it is common for hogs to fight to establish social order within a herd, when pigs are crammed into small pens they have been known to bite off each others' tails—that is abnormal. Animals can be abnormally aggressive in response to poor living conditions, but they also exhibit other types of behaviors. A *stereotypy* is an abnormal behavior in which an animal repeats an action over and over again, with little variation and for no obvious purpose. Sows housed in gestation crates are known to continually bite cage bars and slide their mouths back and forth, back and forth. Such behaviors are thought to arise from stress or boredom and the inability to satisfy normal behavioral needs. For another example, chickens raised in cages are denied the ability to scratch in the dirt and dust-bathe. Probably just for the need to do something, they will sometimes pick at the feathers of other hens, causing injury.

In determining the well-being of farm animals researchers utilize a variety of behavioral indicators. When studying the well-being of sows, researchers might observe the following indicators of levels of well-being: time spend standing, pacing, bar-biting, attempting escape, attempting to turn, vocalizations, number of times weight is shifted from one foot to the other, number of slips, and so on.

How to Tell a Happy Hog: Animal Preferences

Presumably, animals know how to make themselves happy. If given the choice to sleep on straw or concrete most hogs will choose straw, and in so choosing a hog has revealed themselves to be happier on straw than concrete. However, when the weather is hot they might prefer the concrete, which is cooler than straw. We have previously discussed experiments designed to measure animal preferences. In these experiments researchers observe how hard animals are willing to work for one outcome (e.g., getting food) over another (e.g., socialization with other animals). Such research suggests that pigs are willing to work harder for food than for socialization. Similar experiments have been conducted to determine animal preferences for different types of flooring and living arrangements. These studies show that if the door to a sow's gestation crate is opened the sow will leave. However, when allowed to enter and leave the gestation crate voluntarily, sows remain in the gestation crate about 80 percent of the time.[14]

Does this mean sows are well off in gestation crates? Not necessarily. The results show that sows prefer the gestation crate *to the alternative presented*. In these studies the alternative was a small, barren, concrete floor. Thus, it might be argued that the sow was given a choice between two evils. It would be like arguing a prisoner likes being in his cell because, when given the choice between his cell or the prison yard where fights routinely break out, he chooses the cell. It seems obvious that, given the chance between his prison cell and freedom, he would choose freedom.

It would be more interesting to study a sow's choice between a gestation crate and an open area with straw for rooting and access to the outside. More research is needed on sow preferences between gestation crates and an enriched environment, though there seems little doubt over which option the sow would choose. Some studies have shown that sows like turning around even for no specific reason, something they are unable to do in gestation crates.[15]

Experiments have been conducted with dairy cows and chickens to determine which types of floors they prefer. In studies, dairy cows are shown to prefer soft rubber mats over concrete; hens are shown to prefer to stand on wire with a small mesh size because it provides more total surface area to support their feet. Other preference studies generate results that coincide with our intuitive notions of what animals prefer: hens prefer large to small cages, enclosures with litter to scratch instead of barren wire floor, and eating with familiar birds over strange birds. Although such outcomes seem obvious, it is important to determine whether human intuition coincides with animals' revealed preferences, and perhaps more importantly, there are situations in which it is unclear exactly which outcome an animal might prefer. Experiments with animals can be designed to determine not only what the animal likes, but whether they like one thing more than another. Chickens have been given the choice of a large cage with a barren floor or a small cage with litter to scratch, and they prefer the latter. It appears that scratching material is more important than cage size, which is useful information for designing enriched cages.[16]

Studies show that hens strongly prefer laying eggs in nests as opposed to on a barren wire floor. When the hen sees a nest but must overcome an obstacle to reach the nest it will put in considerable work to reach the nest. One might wonder, though, whether a hen suffers from not having a nest if she has never seen a nest. Studies have shown that hens strongly desire a nest to lay their first egg, which casts doubt on the out-of-sight, out-of-mind hypothesis.

Nevertheless, animal preferences are influenced by their previous experiences. Think for yourself whether you suffered from not having a high-definition television in the 1990s. Consider again the barren hen cage, and then ask yourself whether a hen, who has known nothing but this cage, really suffers. Research has shown, for example, that the well-being of sows in farrowing crates depends on the type of systems in which they were previously housed. Sows raised in gestation crates seem to get along better in farrowing crates than sows raised in open pens.[17] Thus, it is important to understand that animals' preferences and choices are, at least to some extent, conditioned by their own earlier life experiences.

How to Tell a Happy Hog: Intuition

There are some obvious similarities between humans and farm animals. It is important not to take anthropomorphism too far. However, given the biological similarities of livestock and humans, it is equally important that we make use of these similarities and ask how we would feel in certain settings, and project those feelings onto the animal. Because a bullet to the chest would hurt us, we suppose it would do the same to animals. Because we would suffer living in a barren wire cage, we suppose a chicken or hog would as well.

One can over- or under-anthropomorphize. It would be invalid to believe a cow would enjoy reading *East of Eden* simply because we do. It would also be invalid to assume animals do not feel pain when injured simply because they are not humans. Making inferences about animal well-being based on our own intuition often seems sensible. After all, even animal scientists must rely on intuition and common sense. Previously we discussed using cortisol levels to measure stress. We argued that cortisol measurements are sometimes of limited use in determining the welfare effects of one production system over another, because the measurements do not always conform to scientists' intuitions. Here, scientists' intuitions lead them to believe that there must be another factor (the need to catch the animals) that must lead to increased stress levels in open pen systems. Likewise, we have found that the "average" consumers' beliefs about animal welfare and expert assessments are highly correlated.[18]

Measuring Animal Welfare—Putting It All Together

Animal well-being can be measured using health, productivity, stress hormone levels, behavior, and preference indicators, along with intuition and other factors. One of the primary reasons for measuring animal well-being is to determine the farm systems that provide the highest levels of animal welfare. Ranking farm systems requires observing animal welfare indicators for each system and reconciling conflicting indicators to arrive at an overall level of animal care. Rarely is the decision unambiguous. Even people with identical training and experience will disagree on matters such as whether behavioral indicators are more important in inferring welfare than productivity indicators. If the desire is to think about farm animal well-being in a useful context, ambiguity is unavoidable. How do we attempt to reconcile competing beliefs and rank animal production systems? In what follows, we discuss three approaches: expert assessment, non-expert assessment, and mathematical models.

Expert Assessments

We often turn to experts for help when making complex decisions. In the case of animal well-being, we often turn to scientists who conduct research and review the literature to arrive at an educated assessment of animal well-being. Many decisions in the farm animal welfare debate are made in this manner. For example, the United Egg Producers (UEP) commissioned a panel of animal welfare experts and asked the experts to provide them with recommendations. As discussed previously, one recommendation was to increase space per hen in cages from 48 to 67 square inches in cage systems. However, this group of experts did not recommend provisions for dust-bathing, scratching in the dirt, and nests for laying. The reasoning was that such provisions require raising birds in larger flocks with higher mortality rates. In these experts' opinion the increased mortality rate did not justify the provision of nests. There might have been cost considerations as well.

The problem with expert opinion is that experts are people like anyone else, and deciding whether hens are better off in a cage or cage-free is difficult—especially when the issue of cost comes into play. Experts are not free of bias and *all* experts face incentives to arrive at conclusions that may differ from certain facts. The style in which scientists write and the institutions they construct tend to create a facade of objectivity that does not really exist.

In our interviews with many animal welfare experts in the US, it appears that they tend to favor the cage system because of the lower mortality rates and protection with which cages provide the hens. Animal welfare experts in Europe disagree, believing most every system is better than a cage system.[19] Such differences in opinion *cannot* be result of differences in information—all experts have access to the same body of information in published scientific studies. One difference may be that agricultural scientists in the US work more closely with the livestock industry than their European counterparts. For this reason, agricultural colleges may attract the types of people most likely to support industry practices. These different cultures result in different scientific assessments in ways that cannot be easily reconciled by scientific measurements.

When judging farm systems, welfare indicators often conflict to such a degree that scientists are unwilling to deem one system better than another. For example, it might seem obvious that a gestation crate system in which sows live their entire lives in cages too small to turn around leads to a suboptimal level of welfare. Yet, a 14-member task force of the American Veterinary Medical Association concluded, "no one system is clearly better than others under all conditions and according to all criteria of animal welfare."[20] The conclusions of this report are especially telling and speak quite openly to the limits of using scientific experts to decide between production systems. The team indicated,

There is no scientific way, for example, to say how much freedom of move-ment is equal to how much freedom from aggression or how many scratches are equal to how much frustration. In such cases, science can identify problems and find solutions but cannot calculate and compare overall welfare in very different systems.

This is a bold statement for a group of scientific experts. After reviewing more than 1500 pages of peer-reviewed scientific papers, the team basically said, "we don't know." This is not to say that the report failed to draw any conclusions or that expert opinion should be ignored, but the basic point is that the usefulness of expert opinion is limited. Ultimately, you must rely on your own judgment.

Non-Expert Opinion: Consumers

It might seem odd to argue that one should take into account the opinions of non-experts. After all, what does the average person know about the well-being of farm animals? However, we must recognize that it is the average person who is buying the eggs and bacon—and thus when questions of cost are involved it seems prudent to ask what consumers think about the well-being of farm animals. Ultimately it is the consumer who pays for the raising of animal welfare standards. But even more importantly, there is a large body of research showing that aggregating the opinions of a diverse set of people is often more reliable than relying on the advice of a few experts.[21]

What does the average person believe is required for farm animals to experience good levels of well-being? We administered a nationwide telephone survey to a random sample of over 1000 Americans to address this question. We asked people which of the nine farm practices shown in Table 5.1 are most important in determining the well-being of a farm animal. Each participant was asked a question such as, "Is it more important that farm animals be allowed to exercise outdoors, or that farm animals be provided with shelter at a comfort-able temperature?" These choices, along with a statistical model, were used to calculate the relative importance of each farm practice on a scale which sums to 100, where a higher number indicates that the consumer places greater emphasis on the practice.

As shown in Table 5.1, people believe that it is most important for animals to receive ample food and water—a belief incidentally, which coincides with expert opinion.[22] That receiving ample food and water received a score of 38.43 percent implies that 38.43 percent of Americans would be predicted to say that this was the most important factor in determining farm animal well-being. The importance scores can also be interpreted as relative measures of importance. The score for outdoor exercise (8.01%) is roughly twice that of shelter at a comfortable temperature (4.43%), which indicates that twice as many people

Table 5.1 Importance of Livestock Production Practices as Perceived by 1007 US Consumers

Production practices refers to farm animal...	Importance score (higher score indicates greater importance)
Receiving ample food and water	38.43%
Receiving treatment for injury and disease	29.05%
Being allowed to exhibit natural behaviors	8.01%
Being allowed to exercise outdoors	7.95%
Being protected from harm by other animals	5.90%
Being provided shelter with a comfortable temperature	4.43%
Being allowed to socialize with other animals	2.76%
Being raised in a way to keep prices low	1.75%
Being provided with comfortable bedding	1.72%

believe outdoor exercise is more important than adequate shelter than those who feel shelter is more important than outdoor exercise. Thus, for the average American outdoor exercise is twice as important as adequate shelter.

The survey results suggest, not surprisingly, that food, water, and health care are the most important factors for ensuring animal well-being. Outdoor exercise and allowing animals to behave naturally are believed to be more important than protecting animals from injuring one another, giving them shelter, giving them access to socialization opportunity, and providing them with comfortable bedding. Expert opinion tends to suggest that outdoor access is of relatively low importance for farm animal welfare.[23] It is evident from Table 5.1 that consumers disagree. This is important because farms designated by scientists to have high welfare levels may not be the farms desired by consumers. Some consumers may view any product claiming to provide high standards of care to the animals of dubious value if they discover it never allows for animals to go outdoors. Scientists and consumers may disagree, but we must remember who is paying the money—as the saying goes, "the consumer is always right."

As consumers exert influence, they should be reminded of their limited knowledge. A failure to understand the unique psychology of the animal as distinct from that of humans can lead consumers to make incorrect inferences. Consumers may look at a farm picture and feel disgusted at the conditions in which the animals live, when in reality the animals may be receiving high standards of care. For example, in a series of personal consumer interviews conducted by British researchers, consumers learned how (broiler) chickens were raised, and they were not impressed with the farm conditions. However, once they learned that the mortality rate was only 5 percent, they were shocked. In surprise at the low mortality rate, one of the consumers stated, "I'd have expected it to be around 30 or 40%, what with all the chickens being so close together in such a warm atmosphere."[24] Despite the understandable ignorance of the consumer, our research suggests that consumer beliefs about animal welfare and expert assessments of animal welfare are highly correlated. One big difference, however, is that our research shows that most consumers drastically underestimate the extent to which egg and pork production occurs in cage systems. They believe hogs and laying hens have far greater access to the outdoors and room to walk than occurs in practice, and when consumers learn how animals are currently raised most become more concerned than they once were.

Mathematical Welfare Models

Regardless of whether one wishes to use input from experts or consumers, some mechanism must exist for aggregating the input in a systematic fashion to decide whether one system is preferred to another. Some people try to avoid the difficult choice. The American Veterinary Medical Association Task Force, to which we referred previously, simply said that "Because the advantages and disadvantages of housing systems are qualitatively different, there is no simple or objective way to rank systems for overall welfare."[25] However, a farmer must make a choice when deciding how to utilize his barn. Proposition 2 on the ballot in California did not have an "I don't know" option; citizens had to vote in favor or against banning cage-egg facilities. The reality is that people make choices everyday which require an *overall* assessment of the well-being of animals in different systems.

Mathematical models represent one convenient way to synthesize any assessments. Such models take information on a farm system's attributes (e.g., space per hen, flock size, availability of nests) along with information on the relative importance of the attributes (information similar to that given in Table 5.1) as inputs, and outputs a single number indicating the overall level of welfare. The scores are often normalized such that the worst system receives a score of zero and the best system receives a score of ten. One particular model called the FOWEL model has been used to compare the welfare of hens in 19 different farm systems. We utilize this model later in the book and discuss it there in more detail, but a cursory description is provided here. In the FOWEL model, the cage system receives a score of zero, a barn system (barn with hens uncaged on the floor) receives a score of 5.9, and an organic system scores 7.8. The absolute level of the scores is not particularly important. A score of zero for the cage system does not necessarily mean assuming that with this system the hens suffer (although they may), but any system with a score greater than zero would be deemed better than the cage system. According to the FOWEL model the cage system, the one currently used in almost all farms in the US, was the worst of the 19 systems considered.[26]

A similar model named SOWEL has been developed to assess the well-being of sows in various farm systems.[27] The developers of this model showed that the model's well-being scores were highly correlated with expert opinion.[28] After comparing seven systems, the model determined that the typical US cage system, which will be discussed in greater detail below, is associated with low levels of sow welfare, receiving a score of only 0.66 (the only system lower was a system in which sows were tethered—receiving a score of 0.00). Systems such as "family pen" system received a score of 8.31 and a pasture system with electronic feeders received a score of 9.89.

The FOWEL and SOWEL models can be thought of as models that aggregate opinions and scientific findings from all animal welfare studies that have been conducted. The models are constructed based on a *representative expert* view drawing on published research that expresses "scientific statements" about what makes laying hens and sows happy. The models are useful in that they aggregate the findings of all scientific studies to provide unambiguous welfare assessments. The mathematical structure can also impose a form of rationality that personal opinions sometimes do not. Moreover, the model structure is such that it can easily be modified if one disagrees with a particular expert assessment of, say, the importance of space relative to the importance of outdoor access.

Animal Agriculture: A Few Generalities

Below we will take a look at modern livestock farms. For those who have spent little time on a farm, a few generalities need mentioning first. There are a few unpleasant factors common to *all* livestock farms of today and in the past. Understanding these factors is necessary for those who desire an accurate, complete knowledge of how farm animals are really raised.

Some level of suffering and premature death on farms is normal and unavoidable. *One cannot farm without becoming somewhat desensitized to this suffering and death.* Some will die young. Some will become ill. Some mothers experience intense pain giving birth and some kill their offspring. Some animals will be injured and will be killed by their fellow herd or flock mates. Some animals will be drowned in floods, burned in fires, or freeze in winter. Moreover, the farmer must be able to live with the fact that some of this suffering could have been avoided, but only at significant cost. Allowing animals to periodically suffer because it is unprofitable to prevent that suffering is an unpleasant but mandatory task of the farmer. For these reasons livestock producers can sometimes appear insensitive to the average person who owns animals but simply as pets.

Virtually all farms are *smelly and plagued with flies*. Farms are generally not the natural environments we might picture as best suiting farm animals. Agriculture is the engineering of biology to suit human needs, and this process requires keeping animals at a higher density than would be seen in a natural habitat.

There will be cases where *some people treat farm animals inhumanely*. This is a fact well established by undercover animal rights groups and pictures of these malpractices can often be readily accessed on such websites as YouTube.com. Let us think about this fact in the proper context though. There are cases where human parents abuse their children but we do not consequently call for an end of procreation of the human race. Equally, abuse on farms should not be taken as final justification for the ending of all animal agriculture.

Similarly, *it is impossible to raise animals for food without some form of temporary pain, and you must sometimes inflict this pain with your own hands*. Animals need to be castrated, dehorned, branded, and have other minor surgeries. Such temporary pain is often required to produce longer term benefits. Just as we allow our children to receive painful immunizations, in a similar manner all livestock suffer some forms of temporary pain. The analogy between immunizations and the castration and dehorning is probably an unfair comparison. It takes a level of desensitization to pull the testicles out of a newborn calf with your hand, and toughness to chop off a calf's horns in the same way you trim hedges. All of this must be done knowing that anesthetics would have lessened the pain but are too expensive, and these procedures are sometimes performed hundreds of times in a single day. Dehorning and castration are nonetheless a necessary part of working on a cattle ranch or a pig farm.

Animal Industrialization—Eggs

Historical Egg Production

Historically, the same type of chicken was utilized for both egg and meat production, but today different chicken breeds are employed for each end use. In the past, chicken and egg production were typically a small part of the overall farm operation, and was usually managed by the women and children. A typical farm in the 1940s might have a flock of 50 to 300 hens.

Seventy years ago hens on the typical farm were provided shelter with at least four square feet (2,304 square inches) per hen. The shelter contained perches and nests with bedding. Eggs were gathered by hand. The barns were designed to allow fresh air and sunlight to provide Vitamin D, which was a problem before modern feed technologies. Except in winter, the hens were allowed outdoors to exercise and forage for feed. Because the science of nutrition was in its early stages, the grain rations given to hens did not meet all their nutritional requirements. Consequently, outside access was critical for hen health, as hens searched the ground for the vitamins and minerals their feed lacked. Egg production was seasonal. After a year of laying eggs hens would cease laying eggs for several months during a molting period, after which laying resumed. This molting was not artificially induced and the hen was not treated any differently because she was molting. Ample space and small flock sizes meant that aggression between birds was uncommon and beak trimming was unnecessary.

Chicken life in the 1940s was not a paradise. Poor nutritional and medical knowledge often resulted in sick and diseased hens. Moreover, chickens were often killed by predators and were exposed to extreme temperatures in the summers and winters. In other respects, however, chickens of yesteryear lived more spacious and varied lives. Do not think that these farmers necessarily gave chickens more space and outdoor access out of a heightened concern for the birds though. Choices about hen living conditions, like most other farm activities, were driven by the profit-motive.

During the first half of the twentieth century, scientific experiments were conducted to improve hen productivity. It was known that some hens laid more eggs than others. It would have been desirable to breed only the better producing hens, but identifying superior hens in a flock was difficult. In an attempt to identify higher performing hens, trap nests were designed. The trap consisted of a cage containing a nest. When the bird entered the nest to lay her egg, a trap door would shut allowing the farmer to monitor egg output from each bird.

While initially designed to identify genetic differences between birds, farmers soon learned that with more advanced feed rations, the hen could be kept in the cage permanently. Farmers learned that the cage did not need bedding; chickens seemed to get by just fine on wire floors. Moreover, it was soon learned that cage floors could be slanted so that eggs would roll out onto a conveyor belt, which eliminated the need for hand gathering. Farmers found that several hens could be housed in a single cage without a large effect on production. Many of the hens would fight and peck one another, so farmers trimmed their beaks so that the pecking would inflict little injury. Better feeds contained all the Vitamin D the bird needed, so sunlight was unnecessary. Farmers soon learned to manipulate the indoor lighting (leaving the barn lights on well after sunset during the winter), to increase egg production. Because chicken feed now met all the animals' nutritional requirements outdoor access became superfluous. The modern cage system was born.

Although much of the egg production which occurred in the 1940s took place on diversified farms, some families specialized in egg production. Specialized farmers raised a different chicken breed (the white leghorn), which was adept at egg production, at the expense of meat production. Chickens on specialized farms were kept in slightly smaller spaces and given less room to roam outside as compared to more diversified farms, but they still provided about 4 square feet of shelter space per hen and hens were given outdoor access during the warmer months. Again, larger space requirements and outdoor access were the norm because profits depended on it—available technology would not permit constant indoor housing.[29]

Egg Farms Now

The advent of the battery cage marked a significant change in how hens were raised. The life of the hen changed dramatically, and so did production. In the early 1930s the most productive egg farms only produced about 153 eggs per hen per year.[30] Today farmers produce more than 250 eggs per hen per year.[31] That is a remarkable achievement, the benefits of which are passed almost entirely to consumers in the form of lower egg prices. Figure 5.7 shows egg prices paid to farmers from 1857 until today. The industrialization of egg production began in the 1940s and steadily progressed over time. The decrease in egg prices over this time is truly amazing. Prices in 1913 were over 9.5 times higher than they are today. Prices in 1943 were over 6.5 times higher than they are today. If animal welfare has suffered from the industrialization of eggs, it is the consumer who has benefited from their plight.

A modern egg producer buys pullets (female chicks) from a breeder specializing in the production of high-performing white leghorn chicks. Some producers utilize brown birds which lay brown eggs, but it costs more to produce eggs from brown birds because the brown birds are less productive.[32] The market share for brown cage eggs is thus small. Male chicks born at the breeders are killed almost as soon as they are hatched because it is unprofitable to raise birds of this breed for meat. In the past, male chicks were often thrown into trash bags to suffocate, but modern farms usually place the male chicks in a grinder that kills them instantly. Researchers have recently identified the genes which control gender in chickens, which should permit the advancement of technology to significantly alter the sex birth ratios.[33] In the future, we may witness pullet operations in which a very high percentage of female chickens are born, negating the need to kill so many male chicks.

Pullets are raised in a brooder house until about 17 weeks of age, after which they are moved to the hen house where they are kept permanently in a cage with four or five other hens. Hen houses are large metal buildings containing from 100,000 to one million hens in cages stacked up to six rows high. Modern hen houses are alive with conveyor belts. One system of belts brings chicken feed, another carries away manure, and yet another transports eggs to a cleaning house. Despite many rumors to the contrary, the feed mechanically transported to egg-laying hens is *not* supplemented with growth hormones, nor is the hens' food supplemented with antibiotics.

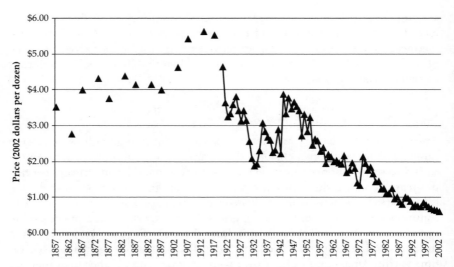

Figure 5.7 Inflation-adjusted egg prices paid to farmers from 1857 to 2002

Notes: The data from 1920 and before are taken from Michell (1935), data between 1920 and 1928 are from Statistical Abstracts of the US and the subsequent data are from NASS yearly reports. The three data series do not refer to the exact same goods or use the same sampling methodology, but no one identical data series covers the entire time period. Prices are deflated using the standard consumer price index.

Manure falls through the wire cage and is collected and removed by conveyor belts. Although some older systems poorly manage manure, with the feces of hens from upper layers falling on hens below, most modern facilities ensure that hens are kept free of manure. Large fans manage air flow, and heating and cooling is provided to ensure the birds are maintained at a comfortable temperature. As might be expected, the barn is smelly because of high levels of ammonia. While the high ammonia levels are uncomfortable for most human visitors, the birds perform well.

As described previously, typical cage systems provide 67 square inches per bird (76 square inches if the bird is a brown bird raised for brown egg production). Although there is room for the bird to move and walk about the pen, conditions are cramped. For example, a hen cannot fully extend its wings. The birds are kept in small groups of four to six hens to a cage, which allows better monitoring of individual birds, permits better health treatment, and reduces aggression between birds. Pictures of the cage system are shown in the top row of Figure 5.8. The pictures shown represent a well-managed farm with high cleanliness standards.

Approximately one-third of all premature hen deaths can be attributed to aggression between hens, but beak trimming greatly reduces the effects of such aggression. As a result, virtually all operations in the US trim beaks to reduce the injuries and mortality. Beak trimming occurs shortly after hatching, and entails removing one-third to one-half of the hen's beak. Although beak trimming reduces aggression between birds—which promotes animal welfare—the beak contains high populations of nerve fibers and studies have shown that beak trimming may lead to chronic pain. The younger the age at which beak trimming occurs, the less likely such pain will occur, which is why most beak trimming occurs when the bird is less than 10 days old.[34]

Hens do not receive much exercise. This makes their bones brittle and easily broken. Hens begin laying eggs at 17 weeks of age and are removed from the flock when productivity falls at about 115 weeks (2.2 years) of age. During her life, a hen will lay about 500 eggs. Most of the nutrients consumed by a hen are put to work in egg production. Consequently, there is very little meat to harvest at the end of the hen's life. As a result, spent hens are used mainly in the production of pet food.

Approximately 95 percent of all US eggs are produced under the cage system. Animal advocate groups dislike the cage system for a variety of reasons. The space allocations are an obvious drawback, limiting a bird's range of motion; the simple act of turning around often requires a hen to bump into other birds. Moreover, birds have a natural desire to dust-bathe and lay eggs in private nests, both of which are unavailable in a cage system. In a cage, hens are unable to perform these natural functions and this frustrates them.

The cage does provide some benefits for hens. The wire floors allow manure to pass through, keeping the hen free of excrement. Such a system results in cleaner eggs and healthier hens. Small group sizes and beak trimming keep injuries and mortality low. The cages make it easier to identify and treat sick hens. Being housed permanently indoors implies continual protection from the elements and predators.

Molting

No discussion of hen well-being would be complete without mentioning the topic of molting. After reaching adulthood hens naturally go through a molting stage where they lose and then grow new feathers. Interestingly, a hen's egg-laying productivity surges following molting. Farmers learned that when adults become less productive, they can be induced to molt, boosting their subsequent productivity. Molting can be induced by withholding food and water. Clearly, such a practice results in substantial reductions in hen well-being, and as a result, animal advocacy groups and some animal scientists have been vocal in calling for an end to starvation-induced molting.

Cage system—67 sq. in. per white bird (76 sq. in. for brown birds), barren cage, small group size

Barn system (brown birds)—200 sq. in. per bird floor space, perches, litter, sawdust, nests, large group size (with or without free-range facility)

Aviary system (brown birds)—144 sq. in. floor space per hen, 240 sq. in. floor and tier space per hen, litter, sawdust, nests, multiple tiers, large group size (with or without free-range facility)

Enriched cage system—116 sq. in. per [white] bird, small group size, perches, nest area, dust-bathing ares

Figure 5.8 Illustration of egg production systems

In recent years, the egg industry has largely abandoned the practice of forced molting through starvation. Some producers continue to induce molting, but by adding more fiber to hens' diets. Other producers find that molting is not profitable; it is cheaper for them to harvest and replace the birds when they become unproductive. Approximately 80 percent of all eggs in the US are produced under the United Egg Producers Certified standards, which does not permit molting by starvation. Molting by starvation is largely a practice of the past. Although the practice may still occur on a few farms, it is no longer a widespread practice, nor is it is unique to cage production.[35] Regardless of your views on animal agriculture, one must applaud the United Egg Producers for their aggressive elimination of molting by starvation.

Cage-Free Egg Production

The cage system has advantages and disadvantages. Although it is understandably difficult to imagine hens being "happy" in a cage system, the other alternatives should be considered before judgment is made. Almost all grocery stores now carry eggs produced by cage-free or organic production practices, along with other labels such as Certified Humane or Animal Welfare Approved. These alternative egg products sell at higher prices than conventional eggs, and presumably provide a better life for the hens. Let us explore the alternatives to cage production to see if this is the case.

The three most prominent alternatives to the cage system are shown in Figure 5.8. The barn system is one in which a large flock of birds roams freely inside a barn. Food and water are continuously provided at various locations in the barn, perches are available, sawdust is provided for scratching and dust-bathing, and hens lay eggs in nests. The nests are usually made out of a rubber material that allows the eggs to be removed using conveyor belts, but the nest curtain provides the privacy hens desire when laying. The barn system provides hens more room to move and exercise, and as a result they have stronger bones. Some barns have slatted floors that allow the manure to fall down to a conveyor belt for removal. Others barn systems cover the entire floor with sawdust, and the hens live on top of their manure (though hen manure is generally dry). The sawdust is removed and replaced as new flocks are brought into the barn. In the barn system, hens have an average of 200 square inches of floor space per bird, which is a large increase over the 67 square inches provided in the cage system.

The aviary system is similar to the barn system except that multiple tiers are added to the barn to allow hens to fly or walk up to elevated platforms. The added platforms allow hens to escape bully hens more easily than in a barn system. The choice between an aviary or barn system depends on the preferences of the producer. Aviary systems are more expensive to build but some producers believe they reduce labor and management costs. Hens in aviary systems have less *floor* space per hen than in a barn system, but they might have more overall space when the room on the elevated platforms is included. Beak trimming is a common practice in both barn and aviary systems. Both the barn and the aviary systems are considered cage-free systems, but a majority of cage-free eggs are produced in barn systems.

Many consumers are willing to pay more for brown-colored eggs and consumers often associate cage-free eggs with brown eggs.[36] Our analysis of retail egg prices indicates that cage-free, brown eggs sell at a premium of about $1.74/dozen over white cage eggs. However, $0.72 of this premium is a result of the difference in eggs being brown instead of white. Stated differently, about 42 percent ((0.72/1.74)*100) of the overall premium normally observed for cage-free eggs in the grocery store is a result of the eggs being brown versus white, *not* a result of cage-free versus cage.[37] Although consumers seem to prefer brown eggs the reason most eggs are white is that brown hens, which lay brown eggs, are more costly and less productive than white hens, which lay white eggs. For example, for every pound of feed a white hen eats a brown hen will eat 1.11 pounds. Brown hens are less productive, producing 10–20 fewer eggs each year than white hens. Brown hens do not live as long as white hens either; brown birds tend to become unproductive at 80 weeks of age compared to 115 weeks for white hens. Despite the disadvantages of brown hens, they are better suited to cage-free production systems. Brown birds are calmer in large flocks and around humans, making their disposition better suited to the cage-free system. It is for these reasons that production of brown eggs becomes more economical in cage-free systems.[38]

Which system is better for hen welfare: the cage or the cage-free system? There is no easy answer. The cage-free system provides greater space allotments and allows hens to engage in natural behaviors. However, cage-free systems often contain flocks of 30,000 hens or more, and in such large flocks hens cannot establish a pecking order. Feral hens tend to roost in flocks of six to thirty birds, suggesting a much smaller natural flock size than is present in cage-free systems.[39] When hens cannot remember the pecking order they continually fight. Sometimes the spats are minor and the submissive birds run or fly away, but other times the increased prevalence of fighting results in feather pecking, injuries, cannibalism, and higher mortality rates. We have interviewed several egg producers who use both cage and cage-free systems, and all of them state that hens were better off in the cage system where they were protected from hen aggression. These producers report much higher rates of injury, cannibalism, and death on cage-free farms. When we asked these producers about their perception of animal welfare in the two systems, they universally favored the cage system, asserting that animal welfare could not be high in a cage-free system where twice as many birds die.

In a 2008 referendum, Californians banned the use of cage systems, requiring egg production to take place in cage-free systems by 2015. In response, an egg farmer who raises hens in both cage and cage-free systems wrote a letter to the editor of the newspaper *Feedstuffs*, which is reproduced in Figure 5.9. This letter expresses well the sentiments we have heard from egg producers in the US. It illustrates general perceptions amongst farmers relating to the different farm systems, but it is not our intention that it should be seen as a final argument for or against any one method.

Egg producers who utilize cage and cage-free methods also report more health problems in cage-free systems. In large flocks of thousands of birds, if a sick bird is spotted it can be difficult to treat, the bird being difficult to catch; and if the bird knows it is being chased, it will begin acting healthy to avoid detection. One farmer that we interviewed utilizing cage and cage-free systems told us that his employees would not eat eggs from the cage-free system because they felt the eggs were more likely to come from sick birds. That is the perception of one farmer. Another different farmer had different thoughts, saying that catching sick birds is not difficult, and that his employees showed no preference between consuming cage or cage-free eggs.

There is evidence that Salmonella is more prevalent in cage compared to cage-free systems,[40] but there is also evidence of the opposite result.[41] In cage-free systems, 2–10 percent of the eggs are not laid in nests but on the ground where bird feces collects. Some farms clean and sell these eggs while others discard them. The overall safety of eggs from each system is still being assessed by scientists and no firm conclusion has yet been reached. At this point, it seems to us that the safety of cage and cage-free eggs are similar and safe enough, and that food safety should not be a concern when deciding which egg product to purchase.

The cage-free buildings are two to three times more expensive to build than buildings utilized for the cage system, on a per-bird basis. The brown birds used

---

Dear Editor:

The chickens of California are the latest victims of Paul Shapiro and the Humane Society of the United States' Factory Farming Campaign.

It is cruel to subject chickens to the less humane and substandard conditions the so-called animal rights people want.

Hens die two to three times as fast under the cage-free conditions championed by the so-called animal rights people. This is a proven fact, and gathered data show this.

What looks good to humans does not work as well for chickens. Safety pens (cages) protect hens from killing one another, eating their own feces, big temperature changes and diseases caused by lack of protection other systems fail to provide.

We quit using the systems wanted by the so-called animal rights people when we found that cages worked better for the hens. This was 40-50 years ago. We recently tried these systems again, and they still don't work as well for hens.

Only those who are clueless, those who have been lied to or those who are misguided about hen welfare want to ban cages. Knowledgeable and caring people want to keep hens in cages until a better system is found, because having hens in cages is the right thing to do for the chickens.

Figure 5.9 Editorial from a farmer criticizing cage-free systems

Printed in *Feedstuffs*, January 12, 2008; p. 8. It is important to note that this farmer raises eggs in both cage and cage-free systems. The name of the person and the farm is withheld in this book.

in cage-free systems are less efficient. More labor is required in cage-free systems. Hen mortality is higher in cage-free systems. All of these factors make cage-free systems more costly.

Many farmers and animal scientists in the US believe the cage system to provide higher welfare. While a barn system allows hens to walk around and behave naturally, the higher prevalence of injury and higher mortality rates counteract those benefits. Most European experts agree that mortality rates are higher in cage-free systems, but in assessing overall well-being they tend to place less importance on mortality than many US scientists, and more emphasis on allowing hens to exhibit natural behaviors. As a result of these differences in beliefs, the European *LayWel Report*, which compared animal welfare under different egg production decisions, concluded the cage system is the *only* system that *cannot* provide satisfactory welfare for the laying hen.[42]

Much of the differences in opinion can be explained by different perceptions of the importance and magnitude of the mortality rates. Based on the available evidence, our estimate of the mortality rate in cage systems is about 3 percent. By contrast, we estimate mortality rates of 7 percent in cage-free systems and 9 percent in free-range systems (this cage-free system with outdoor access will be discussed below). Organic systems have higher mortality rates of about 13 percent because of feed restrictions. Organic producers cannot supplement animal feed with "unnatural" synthetic (man-made) amino acids. Another obstacle is the fact that a farmer cannot treat a sick animal with antibiotics and then sell the animal for organic food. This causes some farmers to deny antibiotics to sick animals. As a result, hens suffer. A number of animal scientists in the US believe organic production is cruel to hens for this reason.[43]

Comparing mortality rates across production systems, however, is a bit like comparing apples to oranges. The reason is that different kinds of birds with different genetics are used in cage and cage-free systems, and these differences in bird genetics (rather than differences in the systems) might explain much of the difference in mortality rates. There is good reason to believe that if hens of the same genetics, age, and beak trimming status are placed in a cage and cage-free system, the mortality rates would be more comparable.[44] But, different kinds of birds *are* used in different systems. Although the choice of bird breed is a primary cause of higher mortality rates in cage-free systems and not the housing system itself, higher mortality rates still matter. One cannot just put white hens in cage-free systems and expect consumer demand and costs of production to remain the same. Consumers expect and desire cage-free eggs to be brown. Thus, our assessment is that differences in mortality are a legitimate factor that should be considered when comparing welfare of cage and cage-free systems as they are used in practice.

In the end, the decision of whether hen welfare is higher under a cage or cage-free system boils down to a belief about the relative importance of mortality rates and opportunities for natural behaviors. Are behavioral factors more important (as the authors of the LayWel report argue) or are productivity and health factors more important (as many US animal scientists argue)? Farming experts and scientists disagree on this matter, which suggests that science alone cannot answer this question.

Enriched Cages

The enriched cage, also referred to as a furnished cage, was designed to avoid criticisms of both the cage system (e.g., inability to perform natural behaviors) and the cage-free system (e.g., large flocks and higher mortality). Shown in the bottom row of Figure 5.8, the enriched cage provides more space, a small perch, a pan for dust-bathing, and a private nest for egg laying. Group sizes are kept smaller to reduce fighting. There are some experts who believe that the enriched cage system provides the highest level of animal welfare among the commercially viable production systems, and many animal scientists in the US prefer enriched cages over cage-free production systems because of the higher levels of animal welfare.

There are others, however, who disagree. Although the enriched cage possesses many features that promote animal well-being, it has its drawbacks. The cage is still small, and as illustrated in Figure 5.8 hens defecate in the dust-bathing tray. The Compassion in World Farming organization, for example, prefers cage-free to enriched cage systems because they argue that competition for nest boxes in the barren cage environment reduces the hens' ability to use the box, and in some cases up to 35 percent of the eggs are laid outside the nest.[45] Others argue that the perches are so low to the ground that the hens perceive them as an alternative type of floor instead of as an actual perch.

Free-Range

Free-range is a ubiquitous term describing a wide diversity of production systems. In the US a carton of eggs can be legally labeled as "free-range," according to US Department of Agriculture standards, if the chicken is raised with *access* to the outdoors. The organic labeling standards for eggs in the US also only indicate that *access* to outdoors must be provided. That is, a chicken never actually has to *go* outdoors to be considered a free-range hen according to labeling guidelines, and the outdoor area to which chickens are provided access is not regulated; there are no size requirements, no requirements for grass or water, and no requirements for shelter or predator protection. Thus, it is perhaps not unsurprising that there is a wide diversity of farming systems in the US that sell free-range eggs. European standards are stricter. For eggs in the European Union to have a free-range label hens must have continuous daytime access to outdoor runs, which must be covered with vegetation and meet certain space requirements.[46]

As a result of the relatively loose labeling requirements in the US, most organic and free-range eggs come from systems very similar to the open barn or aviary systems, with a door on one end of the barn leading outdoors to a small, covered dry lot. Such systems arguably provide very little improvement in animal well-being as compared to the open barn or aviary systems. The birds may only seldom venture outside, as the outdoor pen is usually so barren that there is nothing to attract the birds.

There are some free-range systems which operate in the true spirit implied by the "free-range" term. Most of these operations are small in size and mostly cater to localized, niche clientele—often farmers' markets. We have visited several such farms that have flocks of a few hundred hens. These farms use buildings or houses which are mobile, and are moved around in a pasture as the need arises. The hens are locked inside at night to protect them from predators. In these systems hens are allowed free access to the outside during the day, where they forage for food (the hens are also given a normal corn feed ration and are provided water). These birds have a very diverse diet—eating bugs, grass, and whatever else is growing in the field (see Figure 5.10).

Hens in such systems live a relatively natural life. Our visits reveal what outwardly appear to be happy hens. The hens will venture into the pasture on their own quite boldly. We have seen a rooster catch a bug and present it as a gift to a hen. With a natural environment come natural predators, though. Hawks and other predatory birds are the biggest obstacle to animal welfare on such farms, and this is the reason the European Union free-range regulations require that the outdoor runs be covered. One farmer we visited told us that out of a flock of 250 hens, in one year he lost 50 to hawks (a predator mortality rate of 20%!). This farmer said that he never had to decide what to do with spent (i.e., no longer productive) hens because the hawks kill the chickens before they reach this unproductive age.

The free-range system, even in the most pastoral conditions described above, present some of the same drawbacks as the cage-free method. Undoubtedly, free-range hens live something very close to a natural life. But is it a *good* life?

Figure 5.10 Illustration of free-range egg production systems

The hens have the ability to walk, forage, dust-bathe, roost on perches, and lay eggs in individual nests with straw. Yet, as we have seen, predator problems are no minor issue. Indeed, researchers have measured the level of stress hormones in caged and free-range hens, and have found that free-range hens are as stressed as their caged counterparts. The researchers attribute the finding to a higher incidence of parasites in free-range hens, and the hens' constant fear of predator attack.[47] This said, we should bear in mind that stress hormone measurements are not always reliable predictors of overall stress in an animal, as cortisol levels could have been raised simply as a result of the birds being chased and captured for the tests to be carried out (see Chapter 4).

For high hen welfare, outdoor access must be accompanied by predator protection, such as tall wire fences and some sort of roof, similar to the bird netting at zoos. In these systems, hens will soon learn there is no threat of predators, and stress levels should fall. Under the right conditions, outdoor access can improve hen welfare. However, perhaps we place undue importance on outdoor access. In the FOWEL model, for example, outdoor access is ranked as only the nineteenth most important characteristic (out of 25 total characteristics) affecting hen well-being.

The desirability of any free-range system depends crucially on predator protection and the indoor housing facilities provided. For this reason, we do not consider free-range a separate system, but as a characteristic that could be added to the barn or aviary system. This component of the free-range system would include a rather large, fenced, and covered area that grows grass during the warmer months, has a solid roof on some part of the area, and is fortified against predators.

Another reason for defining free-range as an optional component of the cage-free egg farm has to do with costs. The free-range farm depicted in Figure 5.10 has dramatically higher costs of production. This farm allowed the authors to study their accounting data, and we determined that they lose a considerable sum of money in their egg enterprise. Unless they find a creative way to charge higher prices or lower their costs they stand no chance of making money. In fact, they would probably have to charge more than $6.00 per dozen to break even on their egg production enterprise. Although the free-range farm in Figure 5.10 is interesting, the economics of the enterprise cause us to eliminate the system as a viable alternative. Free-range would then be defined as an optional outside access component to conventional cage-free systems. That is, we consider the possibility of a cage-free facility with outdoor access, which we call a cage-free with free-range system. A farm such as the one in Figure 5.10 is not studied, though perhaps in the future it may become a viable alternative.

How to Make a Hen Happy

Of the four egg production systems shown in Figure 5.8, which is best one for the hen? The advantages and disadvantages of these systems have been articulated, and we now seek to synthesize what we have learned. The egg industry certainly prefers the cage system. The United Egg Producers claim that the cage system with 67 square inch per bird space *"optimizes"* hen welfare, and it is commonly asserted that "it is not possible to house hens in a cage-free system at the higher safety and welfare standards available in cage systems."[48] Our conversations with animal scientists in the US suggest that many of them believe the cage system to be humane, to be superior to cage-free systems, but believe the enriched-cage system to be the ideal system, as long as consumers are willing to pay higher prices.

Other experts believe hen well-being is higher in cage-free systems as compared to both enriched and traditional cages. The LayWel Project is a European consortium of researchers who have been comparing alternative egg housing systems since 2002. After carefully scrutinizing a variety of systems akin to those shown in Figure 5.8, the researchers came to the conclusion that,

> With the exception of conventional cages, we conclude that all systems have the potential to provide satisfactory welfare for laying hens...Conventional cages do not allow hens to fulfill behavior priorities, preferences, and needs for nesting, perching, foraging and dust-bathing in particular. The severe spatial restriction also leads to disuse osteoporosis. We believe these disadvantages outweigh the advantages of reduced parasitism, good hygiene, and simpler management.[49]

FOWEL is a mathematical model which takes information from numerous animal welfare studies, and aggregates the findings to construct a single number representing the welfare-level of the system. The model can be thought of as an aggregator and synthesizer of scientific assessments. The authors of the model put it to work comparing 19 different farm systems, including four systems similar to those in Figure 5.8. On a scale of 0 (worst) to 10 (best), the FOWEL model gave the cage system a score of 0.0, the enriched-cage system a score of 2.3, the aviary system a score of 5.8, barn system a score of 5.9, and a the barn system with free-range received a score of 6.3. No production system currently in large-scale commercial operation received a score greater than 8.0.[50] Another, similar model reports that enriched and cage-free systems provide a similar quality of welfare to the birds.[51]

Now we weigh in. We believe the outcome of the LayWel Report and the results of the FOWEL model are more reliable than the other sources. These sources use the methods to arrive at conclusions that are (1) transparent; (2) logical; and (3) based on the results of many scientific studies. This is not to say that other studies are not transparent, logical, or based on other studies, but that they do not do so to the extent of the FOWEL model. As economists, to compare the costs and benefits of egg systems that utilize lower and higher welfare standards, we require the ability to cite research documenting why one system is superior to another. The FOWEL model and LayWel Report are the only studies considering a wide variety of systems that we can cite with confidence. Thus, we conclude that cage-free production and enriched-cage production are more humane than cage egg production. Perhaps consumers will disagree; we shall see. We are in no way arguing that the farm systems that maximize hen welfare *should* be the system implemented—this involves a determination of costs and consumer demand, which are the topics that we take up later in the book.

Animal Industrialization—Chicken Meat

Historical Broiler Production

Before the 1950s, much of the chicken meat produced came from birds that were also used for egg production. Whereas female chickens were retained for egg laying, male chicks were raised for meat. The cockerels were typically separated from the pullets at about 9 weeks of age. Before this separation they would have been raised in a brooder house, which can be thought of as a nursery. At separation, the cockerels were then placed in a house, cage, or small pen to be fed for one to two more weeks, at which point they weighed 1.5–2.5 lbs and were harvested. Even on these older farms, the final "finishing" phase took place in a dark room to reduce fighting and minimize movement.[52] Expert advice on fattening the chickens in 1913 urged farmers to, "Put the fowls in coops so small that they cannot turn around."[53] The chickens were typically harvested when they were about the size of today's Cornish hens (2 lbs), and were often cut in half and broiled, which is why they are referred to today as *broilers*.

Although many farms raised birds for both egg and meat production, some farms specialized. In the more specialized meat-producing farms, particular breeds of chickens emerged that were particularly suited to meat production. The modern breed of broiler used today has its roots in the Chicken-of-Tomorrow contests, held on state and national levels in the late 1940s. The contest was designed to encourage chicken farmers to breed birds with especially large breasts and high rates of feed conversion. To participate in the contest, producers would ship the fertilized eggs to a central location at which they would be incubated, hatched, and the chicks raised under uniform conditions. Researchers monitored the chickens closely and documented weight gain, mortality rate, and the size of the chicken carcass.[54] The contest finals were accompanied by a parade, dances, a rodeo, and even a crowned queen. This contest sparked the beginning of the modern broiler breeds.

Broiler Farms Now

It has been suggested that the broiler industry started when, in 1932, Wilmer Steele, a Delaware woman, purchased 500 chicks and sold them when they reached a weight of 2 lbs. The novelty of Mrs Steele's business model was that she purchased chicks for the *exclusive* purpose of meat production.[55] It would be 20 years before others widely adopted Mrs Steele's business model, but when the idea eventually took, it spread to encompass most all of the broiler industry.

Today's broiler chicken is twice as efficient as its 1940 counterpart; modern broiler breeds produce twice as much meat for the same amount of feed. In 1925, it took 4.7 lbs of feed for a broiler to gain one pound, but today it takes only 1.95 lbs of feed. Birds were only about 2.5 lbs on average when harvested in 1925, but today they are about 5.47 lbs at harvest. Despite the fact that today's birds are almost three pounds larger than their 1925 counterparts, they reach this larger weight in 64 fewer days. Today's broilers produce proportionally more white meat, which is in larger demand from consumers. Today's broilers are not just more efficient, but possess a more consistent carcass. By using very similar breeds, feed, and production methods, today's cooks know precisely what chicken meat will taste like, and harvesting machinery has been engineered around the design of these uniform carcasses. Eating chicken is not the "adventure" it used to be, which is lamented by some, but consumers have made it clear that they like the new bird. Per capita consumption of broilers increased from 23.6 lbs per person in 1960 to 89.5 lbs today—in part because chicken meat is healthier, tastier, and more uniform, and in part because it is much less expensive. The retail price of broilers was $3.08/lb in 1960 (in inflation adjusted terms) but was only $1.28/lb in 2009. Chicken meat is almost 2.5 times less expensive today than it was in 1960.

The broiler industry has been especially criticized not for the treatment of the birds that are raised for meat, but the birds designated as breeders—the parents of the birds we eat. These chickens are treated as meat-producing machines; their genetics have been specifically chosen for this job. The chickens we eat were harvested at about 6 weeks of age, whereas the breeder parents will live for one or two years. If the breeders were given plentiful access to feed their bodies would out-grow their feet and they would no longer be able to walk. Breeders are consequently rationed in their feed, only receiving 25–35 percent of the feed they would like to consume.[56] Most of us can imagine what kind of discomfort this regular underfeeding must cause.

At first, these feed restrictions may seem unusually cruel—what is crueler than near-starvation? However, experiments have shown that the birds' bodies adapt somewhat to feed restrictions. The inability to eat all the food they desire causes initial stress to the bird (as measured by the levels of stress hormones found in the blood). However, over time the bird adapts and the stress tapers off. Thus, while deeming these feed restrictions as cruel is understandable, they may not be as cruel as they initially appear. Also, the drawbacks of feeding breeders a restricted diet must be compared to the alternative. When given food *ad libidum*, the animals become lethargic, develop immune disorders, and at some point cannot, as we mentioned earlier, walk.[57] The same problems seen in human obesity is observed in these birds, and the feed restrictions prohibit this from occurring.

Like egg-laying hens, older breeder hens can be aggressive and require beak trimming. They also require their combs to be trimmed, which is usually done without anesthetics. They are typically afforded more space than egg-laying hens are in cage-free systems; breeder hens have two square feet per hen,[58] compared to 1.5 square feet per hen in cage-free egg-laying systems.[59] A number of other activities are rumored to occur in broiler breeder facilities, and it is difficult to determine the extent to which these occur. An example is placing a rod through a rooster's beak so that it cannot fit its head through small openings where the hens' food is placed.[60]

Welfare in breeder broilers must be placed into perspective. Not only must we consider the welfare of each breeder broiler, but also the total number of breeders that exist. In 2007, a total of 57 million hens were used for producing offspring for broiler production. For every 100 hens in a broiler breeder facility there are about eight roosters, implying 4.56 million roosters are serving the 57 million hens, giving 61.56 million total birds. The number of broilers produced in 2007 was 8867 million. Thus, of all the birds involved in broiler production, 0.69 percent are raised in breeding facilities. While the welfare of breeder broilers is likely low due to the feed restrictions, these chickens represent a small portion of the breeder industry, and their low welfare should be discounted appropriately.

Eggs are taken from the specialized breeding farms and are incubated at a hatching facility. The chicks are hatched in a crate within the incubator, which looks like a giant oven or refrigerator. After hatching, the chicks are carried to a broiler house, which is a large, long metal building with a temperature-controlled atmosphere with dirt floors covered with litter made of straw, wood shavings, sawdust, or other similar materials. Tens of thousands of birds are contained in a single house, and when the chicks are young they tend to bunch closely together. Broilers are not caged, which is why animal advocacy groups have focused their attention more on the egg industry than the broiler industry.

A broiler's life is short: most live less than two months. Depending on how the meat will be used the broilers are harvested at four to ten weeks of age, weighing anywhere from 3 lbs to 6.5 lbs. Just prior to harvest each chicken has about 100 square inches (0.7 square feet) of space, although at younger ages

Figure 5.11 The modern broiler farm

space is more abundant (see Figure 5.11 for a pictures of a broiler house when broilers are young). Although the barn becomes quite crowded as broilers approach harvest, feather pecking and cannibalism are almost unheard of.

One reason is that aggressive behavior in chickens does not begin until the bird has become a mature adult, and broilers are harvested before that point. A second reason is attributable to the bird breed. Chickens that are more productive egg layers will tend to be more aggressive than chickens which tend to gain weight faster. There can be problems with "crowding," where the chickens become scared and all flee to the same corner of the building. As they pile on top of one another some animals suffocate. It is our impression that crowding is rare. If crowding was prevalent mortality rates would be much higher than they are. We have also seen farms that utilize an electric wire system to discourage this crowding.

Research suggests that other factors, such as barn temperature, humidity, litter moisture, and ammonia levels, are more influential in determining broiler well-being than stocking density.[61] In the US, 95 percent of broilers are produced under the National Chicken Council Guidelines, which explicitly prohibits beak trimming. Although the mortality rate of growing broilers in 1925 was 18 percent, the mortality rate is now typically less than 5 percent.[62] This low incidence of death is one indicator of animal well-being, and consumers are sometimes surprised at the low death rate, suggesting that they may underestimate the level of care provided by modern broiler farms.[63]

One of the drawbacks of modern broiler operations is that the broilers grow so fast and heavy that they develop leg problems. To assess the prevalence of leg problems scientists watch birds walk and assign them a *gait score* ranging from 0 (normal gait) to 5 (the bird cannot walk on its own). A broiler with a score of 3 or higher is usually considered to have a significant leg problem.[64] Research shows that birds with a gait score of 3 or higher are in pain, because when such birds are given pain relievers they walk more normally.[65] Interestingly, the result of one study not only showed that the gait score is associated with pain, but that birds can make rational choices. Birds were given a choice between two types of feed that were identical, except that one option contained pain relievers and was artificially colored. Birds with a gait score of 3 or higher consumed the food with pain relievers at a significantly higher rate than birds without leg problems.[66]

How prevalent are leg problems in the broiler industry? One study recorded the gait scores of 51,000 birds raised in 176 flocks just prior to harvest (where leg problems are most likely to appear). About 28 percent of birds were assigned a gait score of 3 or higher, and despite the fact that the farmer routinely culled birds with severe leg problems, 3.3 percent of the hens in the study were almost unable to walk.[67] Other studies have found similar results.[68] Thus, leg pain is no minor problem in modern broiler facilities—at least in the last few weeks or days leading up slaughter.

It is important to note, however, that leg problems are likely nothing new to broiler production. For example, in the 1920s one scientist described the breeds of chickens used to produce meat as "slow and sluggish . . . because of their large size and awkward movement the meat breeds do not roam far from the roosting quarters in search of food unless compelled to do so."[69] If these were the traits of meat birds 80 years ago, perhaps we should hesitate before blaming the factory farm for the poor leg structure of modern broilers.

Other problems noted by modern broiler industry adversaries are lighting, air quality, and lack of enrichment. Fowl in the wild naturally roost at night, but farmers prefer broilers to continue eating after the sun has set. Farmers keep the barn lights on for many hours to encourage eating and growth. During the 1980s it was common to allow the birds only one hour of darkness per day; today they are typically given four hours of darkness.[70] There is some evidence that birds are healthier when given more darkness.[71] Barn air is filled with ammonia, which can cause lung problems. There is also some evidence that the genetic makeup of modern broilers leads to a heart problem that kills up to 3 percent of broilers and affects the welfare of another 2 percent.[72] Although broilers have ample access to food and water and have litter for scratching on the ground, some argue that the environment is too crowded and dull. Overall, broilers lead short, unexciting lives. There is evidence of leg pain later in life and other factors that might cause reductions in well-being. However, after reviewing all the obstacles to welfare and the nature of the birds, in our assessment, broiler farms do not cause large-scale suffering.

Broiler Alternatives

Few alternatives to traditional broiler production are feasible, at least, not any that can generate comparable levels of output at similar costs. Nevertheless, there are some changes that could be made within the barn system that should improve animal well-being. One clear example is to breed birds with slightly slower growth rates and with better leg structure and who would thus experience less leg pain. Although geneticists clearly have the ability to breed broilers with fewer leg problems, there are currently few incentives to adopt such practices.

We once had the opportunity to interview an animal welfare representative of a major broiler-producing firm. He told us an interesting story regarding a European buyer who was unhappy with the leg problems in broilers. This buyer indicated he would pay a higher price if the firm could address these problems. Within only a few months, geneticists were able to breed a broiler with substantially fewer leg problems. This story exemplifies the simplicity of addressing farm animal welfare issues, in some cases. If buyers are willing to pay more for meat from happier animals, producers will gladly oblige.

Some free-range broiler operations exist, but as we described with regard to egg layers, such operations are largely unregulated in the US and chicken meat sold with a "free-range" label is likely to come from a system like that described above, only given access to a small, dirt outdoor lot. There are some branded products that guarantee larger space requirements and more enriched outdoor access, but such products are not typically sold in mainstream supermarkets. When they are, the price difference is noticeable. Farms that provide enriched outdoor experiences for the birds are expected to have predator problems. If the predator problem is not managed carefully up to one-quarter of birds could easily be killed by hawks, coyotes, and even skunks.

There are also smaller scale producers selling direct to consumers that have opted for a more drastic alternative to the modern broiler system. One such system is shown in Figure 5.12, which can be referred to as *pastured poultry*. In this system, broilers are housed in a tent structure for predator protection, which is located in a pasture. The birds are given the same access to food and water as conventional broiler units, but they are also able to eat grass and any insects they may find. The tents are moved periodically, but predators remain a problem, as does the weather. The operation in Figure 5.12 is unable to raise broilers in the middle of summer or the middle of winter, and despite the tent structure, a sheep dog lives permanently with the chickens to protect the birds. Producing broilers in such a manner is much more costly. Pastured poultry producers tend to sell whole chickens for $4–$5 per pound, whereas the same chicken meat can be purchased in the grocery store for $0.90–$1.50/lb.

Turkeys

Turkey production very closely resembles broiler production. The same production methods and welfare issues discussed in regards to broilers generally

Figure 5.12 Pastured broiler production

apply to turkeys. Only a few differences are worthy of noting. First, breeders have been particularly interested in raising turkeys with large breasts, so much so that the birds can no longer breed naturally. Artificial insemination is required. Second, turkeys live longer than broilers, being harvested at about 4 to 5 months of age. Third, turkeys produce more meat per bird. For our purposes, however, we assume that broilers and turkeys possess similar levels of welfare and similar challenges and possibilities for improved welfare. We should also mention that we know far less about turkey production than broiler, egg, pork, and beef production. We have never worked at or visited a turkey farm, nor have we interviewed any representatives of the turkey industry.

Animal Industrialization—Hogs

Historical Hog Production

Historically, hogs have been raised in a variety of settings. Egyptians are thought to have raised pigs in kennels not very different in set-up from factory farms of today.[73] A letter to the editor of *The Times* in Britain in 1909 suggested that outbreaks of hog diseases could be attributed to the disgusting conditions in which the animals were kept. The letter described hogs living, "in an unnatural state, closely confined in sites often foul and unwholesome, either in themselves or upon soil deeply saturated with the drainage of generations."[74]

In other places and at other times hogs were less confined and in some cases were not confined at all. In some areas hogs lived in wooded areas living on roots and nuts. In other areas hogs were driven in large herds like cattle in order to take advantage of seasonal variations in feedstuffs. Often hogs would better resemble household pets than livestock, lying about lazily in or just outside their owner's house. Pigs would roam city streets performing the useful function of eating roadside litter. These street pigs were often more of a nuisance than a servant, however, and were officially banned from some city streets. Wall Street got its name from the walls erected on the street to keep out roaming pigs.[75] Egypt still employs pigs for help in managing its trash. A group of Christian minorities in Egypt receive trash from Cairo; after separating recyclables such as glass and paper, they feed the remainder to their pigs.[76]

Most people, in their condemnation of factory hog farms, romanticize the hog farm of 90 years ago. During this time, hogs were almost always raised in farms producing crops and other livestock. Besides providing meat, the hog can be seen as a method of storing grain. One can store grain by making sure it stays dry and protected from scavengers, or one can feed it to hogs and keep the hogs alive until the time the corn, now in the form or muscle and fat, is desired. For example, a nineteenth-century British journalist traveling in the US wrote, "The hog is regarded as the most compact form in which the Indian corn crop of the States can be transported to market."[77]

Many farmers began hog farming with very little investment. One observer from the 1920s stated, "for successful production of hogs expensive equipment is not necessary. A few buildings to shelter the animals comfortably from extremes of cold and hot weather is sufficient."[78] That the hogs needed few buildings is evidence that pigs primarily lived outdoors. As indicated, swine were mainly fed corn, which met most but not all of the animals' nutritional needs. Neither farmers nor scientists knew how to supplement the corn with the vitamins and minerals it lacked, so hogs were allowed access to pasture and were fed other foods such as alfalfa hay, skim milk, and table scraps.[79]

Although hogs spent a great deal of time outdoors they were often kept in large pens or pastures—often just a muddy lot. Breeding animals were often encouraged to exercise so that they would be in better shape for breeding. Feeders would be placed away from the shelters to force the hogs to walk for exercise. Some farmers would allow boars to roam freely with sows and mate whenever they pleased, but that practice was eventually replaced by artificial insemination. When it came time for sows to give birth, they were often moved to an individual farrowing hut similar to that shown in Figure 5.13, with about 50 square feet of space and straw for bedding.

Some sows are not very good mothers and they make up for their motherly ineptitude by having many babies. Sows often had to be watched closely after birth (and piglets removed periodically) as sows easily crush piglets by sitting down and rolling over, sows sometimes refuse to nurse their piglets, and sometimes sows have been known to eat their own offspring. The farrowing huts were designed with side-bars to provide an area where the piglets could go but the mother could not. Even today, farms that use farrowing huts have significant crushing problems—with piglet mortality rates as high as 25 percent.[80]

Piglets were weaned at 2–3 months of age, at which point they entered the finishing stage of hog production. The term "finishing" refers to the process of

Figure 5.13 Farrowing using individual huts in pasture

feeding pigs all the grain they can eat in an attempt to bring them to slaughter weight quickly. The ration was mostly corn, but as with sows, the feed had to be supplemented, perhaps with skim milk. The barrows (castrated male hogs) and gilts (young female hogs) were often provided with at least 32 square feet of shelter space per animal with access to an outdoor lot. The size of the lots and pastures varied from farm to farm, but there was usually some outdoor access and rooting area in addition to the shelter. Once the animal reached 200–250 lbs, it would be harvested. Pictures of a hog farms from 1934 are shown in Figure 5.14, as is a modern-day non-confinement farm that we visited in 2007. Take a close look at Figure 5.14. People often have a naive notion of what "pasture raised" and "outdoor access" mean. With hogs, pasture is not always a pleasant green meadow, but rather a smelly mud-hole. However, hogs like mud and have a proclivity for turning green meadows into barren mud-holes anyway.

Hog Farms Now

Even as early as 1913 hog farmers began experimenting with raising hogs on concrete floors in small, enclosed lots, reporting "the plan of close pens to be the most economical in the end."[81] The main advantage of concrete floors was that they "can be kept sanitary most easily."[82] Maintaining healthy hogs is perhaps the greatest challenge of hog production, and the improvement in mortality and growth rates conveyed by concrete floors more than made up for the costs. Many of the parasites that infect livestock will only thrive in outdoor lots. Taking animals off the dirt and placing them on concrete or litter reduces infection and disease significantly, assuming the concrete or litter is cleaned.[83]

Figure 5.14 A 1934 hog farm (left) and contemporary non-confinement hog facility (right)

Source: Left picture is at wikimedia commons in the public domain ⟨http://commons. wikimedia.org/wiki/File:PigPenacp.jpg⟩.

As early as 1913 hog farmers began raising hogs indoors, and by the 1950s some hogs were raised indoors on concrete.[84] It took several decades for the indoor production method to become the standard, but today, the vast majority of hogs produced in the US are raised on indoor concrete pads for their entire lives. These indoor facilities have substantially improved hog health. Lung-worms in the 1940s affected more than half of all hogs but only 11 percent today. Similarly, the vast majority of pigs were infected with kidney worms in the 1940s, but affect very few today. Trichenella, which caused most households to fear eating pork that was not fully cooked, is almost non-existent today.[85] All of these improvements in hog health, which enhance welfare, can be attributed to the sanitation provided by today's factory farms.

As producers gained more experience of raising hogs inside they developed other practices that improved efficiency. Barns were designed to control temperature and manage manure via slatted floors. Such barns are expensive. A modern hog farm built to house 1200 sows may cost $5 million to build and will last 10–20 years.[86] The average cost of hogs produced often falls as the number of hogs within the building rises. However, sows become bad-tempered if cramming into a tight space with other sows. To keep sows from injuring one another they must either be kept in small groups or individual crates. The result is the modern factory hog farm shown in the top two rows of Figure 5.15.

Let us take a tour of a modern factory hog farm, which we refer to as the *confinement-crate* system. Breeding sows are kept in the gestation crates where each sow remains exclusively in an individual crate that is barely larger than the sow herself (14 square feet).[87] The animal cannot turn around and can have trouble lying down. When they do lie down, their feet often extend into the neighboring crate, which makes it difficult for all sows to lie down simultaneously

	Farrowing stage	Gestation stage	Finishing stage
Confinement-crate system (*factory farm*)			
Confinement-pen system (*factory farm*)			
Confinement-enhanced system			
Shelter-pasture system			

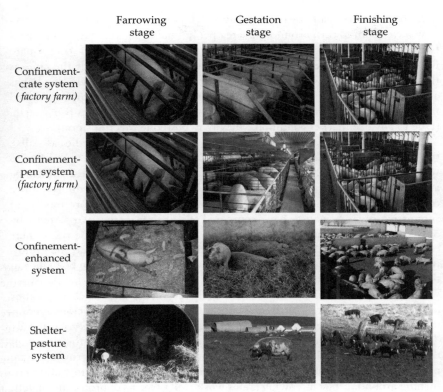

Figure 5.15 Illustration of four hog farm systems
Bottom row © istock

in a comfortable manner. The sows stand on top of slatted floors through which excrement falls into a pit. The building is generally kept at a comfortable temperature and sows are individually fed and provided health care. Breeding is done artificially in the crate. Sows are exceptionally difficult to move and handle and crates greatly ease the management task. Gestation crates reduce labor costs by reducing the number of hours a person must work to care for the same number of sows. The crates also simplify sow management to the extent that the farmer can employ less-skilled—and lower paid—workers.

Not surprisingly, gestation crates have come under intense scrutiny by animal advocacy organizations. Florida, Arizona, Oregon, Colorado, and California have all banned the use of crates in favor of a *confinement-pen* system, which is shown in the middle picture of the second row from the top in Figure 5.15.[88] This system replaces the individual gestation crates with a group-pen, where two to six sows are assigned to one pen where they have room to walk and turn around. There is no bedding or dirt for rooting. The floor is slatted so that hog manure can fall beneath the floor and away from the animal.[89] Though the sows now have room to turn around, the area per animal is not very large. The pens typically allot only 16–30 square feet of room per sow (compared to 14 square feet in a gestation crate).[90] Everything else about the farm is exactly the same as the confinement-crate system. The largest pork producer, Smithfield Foods, has also indicated it will voluntarily phase out gestation crates on company-owned farms by 2017, and they will most likely transition to the group-pen system. (Due to adverse market conditions, they may postpone this transition).

Compared to the crate system (the crates are sometimes referred to as stalls), the group-pen has benefits and drawbacks. The benefit is that the sow is free to walk and turn around. The drawback is that the tight quarters cause sows to be aggressive and inflict injuries on each other. Also, group feedings induce competition for food and result in dominant sows receiving more food than submissive sows. Scientists have compared well-being in the two systems by observing the sow behavior and measuring stress hormones. Some researchers conclude that sow welfare is equivalent in the individual crate and the group-pen system.[91] Others contend that the research suggests welfare to be higher in the group-pen.[92]

A sow will give birth three months, three weeks, and three days after breeding. Shortly before birth sows are moved to a farrowing crate such as that shown in the top left-hand corner of Figure 5.15. The crates are specifically designed to force the mother to lie down carefully and to reduce the chance of the piglets being crushed. Similar to the gestation crate, the farrowing crate provides space of about 14 square feet, prohibiting the mother from walking or turning around.[93] The farrowing crates work well in terms of protecting piglets, reducing the pre-wean mortality rate from 25 percent to 10 percent or lower. Of the 10 percent of piglets that die in farrowing crate systems, half of the deaths can be attributed to crushing by the mother.[94]

Farrowing crates provide other benefits aside from reduced morality. Heat lamps are often affixed to the crates to keep the babies warm. The crate forces mothers who would otherwise neglect their young to nurse the piglets. Because of the beneficial features of farrowing crates, animal advocacy groups typically oppose gestation crates but not necessarily farrowing crates. Farrowing crates present a dilemma: they improve the welfare of piglets but detract from the sows' quality of life. In our experience, there is no technology that promotes the welfare of both the mother and her offspring simultaneously during birth and nursing, and so a trade-off decision must be made.

Once piglets are born, the males are castrated and some farmers clip the piglets' "needle teeth" to protect the sow's udder and to reduce injury from sibling fights. In addition, the pigs' tails are often shortened or "docked" to prevent injuries from tail biting. Castration, teeth clipping, and tail docking are all performed without anesthetics. The piglet's ears are notched in a way that assigns it an identification number. These notches will not grow out, allowing for the piglet to be identified by this number throughout its life. Three weeks after birth the piglets are weaned and sent to a "nursery" where they are kept in small groups and are monitored closely.[95] Eventually, the hogs are placed in larger groups and are moved to a finishing house. From birth to death the pigs live exclusively indoors and in tight quarters. Market hogs have continuous access to food and water. Just prior to harvest hogs weigh about 250 lbs and their bodies occupy an area of about 5 square feet.[96] At this weight and size, each hog has about 8 square feet of space in the finishing barn (see the last column of Figure 5.15).[97] Younger hogs are smaller and have proportionally smaller space.

Are hog well-being levels higher in modern factory hog farms as compared to hog production methods used 50 to 100 years ago? In the past, sows and growing pigs typically had at least 49 and 32 square feet of barn space, respectively. Today, they are afforded only 14–24 and 8 square feet, respectively. Hogs in the past had straw for bedding and outdoor access. Hogs today live on a concrete floors, never venturing outside. That hogs have less space and fewer enrichment activities does not necessarily imply that the average hog is less happy today than its great-grandparent. Much has changed in the way of hog nutrition and health care. The diseases, worms, and nutritional deficiencies that plagued many hog farms in the past are virtually non-existent today. Whether improvements in nutrition and health care outweigh the cramped and barren conditions is unknown, but perhaps this is not the comparison we should be making. Today we have the possibility to use nutritional and veterinary knowledge *in conjunction with* production facilities with more space and enrichment.

Scientific studies support the contention that the modern factory hog farm results in low hog welfare. Recall that the SOWEL model is a mathematical model that calculates a well-being score for a sow production system based on descriptions of a hog farm. One a scale of 0 (worst) to 10 (best), the modern confinement-crate system was assigned a score of 0.66. The only system that performed worse than the gestation crate system was a system where sows were literally tethered in place (score of 0). Many other systems, some of which are discussed shortly, received a higher score. Although the SOWEL model cannot tell us whether the confinement-crate system passes a minimally acceptable level of animal well-being, what it does suggest is that there are other systems in which sows needs are better met.

Pork Production Alternatives

There many different ways to raise hogs, so to help us understand these many systems we group farms into one of the four stylized production systems shown in Figure 5.15. The two systems show at the top of Figure 5.15 (*confinement-crate* and *confinement-pen* systems) might be considered two types of factory farms. The confinement-crate system is the most widely used system in the US, but due to ballot initiatives and public pressure many producers are slowly converting to the confinement-pen system, which uses group-pens for sows instead of gestation crates. The confinement-crate and confinement-pen systems are thus identical except for the manner in which pregnant sows are kept.

Figure 5.15 also shows two alternative systems. First, consider the system at the very bottom of the figure, which we referred to as a *shelter-pasture* system. The shelter-pasture system represents an attempt to maximize hog well-being: virtually all the animals' needs are addressed and a number of preventive actions are taken to reduce pain and suffering. The hogs are provided with the outdoor access to pasture and are given materials (e.g., straw) to carry out their natural activities (e.g., nest building). Hogs in shelter-pasture systems live naturally but without the worries of wild animals, as they have continuous access to shelter, food, and water.

Health problems are likely more frequent and intense in the shelter-pasture system than confinement systems, though we know of no study proving this. The fact that animals wallow in the mud, their own feces, and the feces of other animals can only encourage the spread of worms, viruses, harmful bacteria, and the like. One of the major reasons farmers switched to the confinement systems was to avoid these health problems. With today's advanced veterinary technologies these problems are not as great as they were 90 years ago. Medicine is now available to treat almost every type of worm, and so long as the farm is not an organic farm, animals can be given antibiotics when they suffer a bacterial infection.[98] It is our contention that the health problems associated with shelter-pasture systems are not as prevalent as they were 90 years ago. Hogs can be outdoors and healthy.

The *confinement-enhanced* system is a compromise between the shelter-pasture and the factory farm systems. The confinement-enhanced system seeks to exploit the advantages of raising hogs indoors while providing certain amenities to improve hog well-being. No two confinement-enhanced systems are exactly alike, but consider one such system shown in Figure 5.15. On this farm, sows occupy a spacious farrowing room where they care for their offspring. The sow is free to occasionally leave, and when large enough, so are the piglets. The nursing room contains side-bars to minimize crushing by the sow. In the gestation stage the sow lives in an indoor pen with other sows. All hogs are provided with straw for warmth and rooting, and have access to the outside, which on this farm is a concrete lot.

Our assessment based on a reading of the literature and visiting (and working on) hog farms is that the confinement-enhanced and the shelter-pasture system provide much higher levels of welfare than the factory farm system. The SOWEL model supports this argument, assigning the confinement-crate system a score of only 0.66. Moving from the crate to the group-pen system increases the well-being score to only 2.96. However, the SOWEL model gives systems like the confinement-enhanced and the shelter-pasture systems scores approaching

10, indicating that the systems meet almost all of the hogs' needs, and there is little that can be improved.

The Animal Welfare Institute (AWI) is a non-profit institution devoted to "alleviating suffering inflicted on animals by humans."[99] One manner in which they pursue this goal is to oppose factory farming methods and encourage the use of the shelter-pasture system. The AWI has constructed a set of production standards for all livestock animals, including hogs, and pork produced under the AWI guidelines is eligible to be sold with an Animal Welfare Approved (AWA) label. This book is replete with references to the AWI standards for hogs but not the respective standards for other animals. The reason is that we believe the shelter-pasture system to closely resemble the AWI standards for pork, and one of our graduate students has calculated the cost of hog production in the shelter-pasture system.[100] The respective costs for producing cattle, broilers, and eggs under AWI standards is unknown to us. Also, later we will discuss our research describing consumer demand for pork that could probably be considered Animal Welfare Approved pork, and because we have both cost and value estimates for shelter-pasture pork we study the economics of this particular pork product in detail.

Veal

Animal advocacy groups have been so successful in portraying the historical horrors of veal production that it is generally considered an unethical food, even by some beef cattle producers.[101] Veal consumption in the US has fallen dramatically—down from 5.2 lbs per person per year in 1960 to just 0.4 lbs per person per year in 2008.[102] For every 100 lbs of beef that Americans eat, only 0.7 to 2.37 lbs of veal is consumed.[103] In the context of human consumption of animal products (meat, milk, and eggs), veal consumption is negligible, and thus, veal receives little attention in this book compared to eggs and pork.

Almost all veal calves are born on dairy operations. A cow will only produce milk if she has given birth, and as such, breeding and calving is necessary for milk production. Because the milk that nature has intended for the calf is to be redirected to humans, the milk-drinking dairy calves are removed from their mothers almost immediately after birth. Most of the female calves will be retained by the farmer to subsequently incorporate into the milking herd, while the male calves will be sold as soon as possible. Most of the male calves will be raised like beef cattle and harvested for beef, while about 30 percent of these males will be raised and harvested at an early age for veal. The female calves will also be harvested for beef, but not until they have been milked for several years.[104]

A few decades ago, almost all veal was raised using an individual stall system shown at the top of Figure 5.16 which involved either tethering or stalls similar to the gestation crates used in sows production. The stalls prevented the calves from moving freely. The American Veal Association (AVA) states that the tethered stalls are in little use today, and have been replaced by larger individual stalls which afford more movement.[105] Even in these larger individual stalls, the calves typically do not have room to turn around.

While it is true that maintaining calves in individual pens reduces the spread of sickness among calves, the obvious disadvantage is that they prevent the animal from moving or even turning around. In response to criticism some veal producers have adopted group-pens, as those shown in the bottom of Figure 5.16. Group-pens consist of placing calves in groups where they have freedom to walk and turn around, and enough space to lie down comfortably. As long as the increase in disease incidence from group-pens is not too great, group-pens must provide a substantial increase in quality of life for the calf. However, even with group-pens there is very little movement, and the cages provide a barren environment.

In addition to the small spaces in which veal calves are confined, a number of other welfare problems arise from veal production. In addition to inducing stress in the calf and mother at the time of separation, the absence of the family bond prohibits the calf from maturing properly. The calf will experience greater stress in new environments, will be less able to solve problems, and will show a reduced motivation to solve problems. The calf never sucks from its mother's teat. The first few days after birth the mother's milk contains a substance called colostrum, which is necessary for the calf to receive if it is to develop a proper

Figure 5.16 Veal calves in stalls (left) and group housing (right)

Source: The left picture is made available by Compassion in World Farming at wiki-media commons ⟨http://commons.wikimedia.org/wiki/File:Veal_Cows.jpg⟩ and the right picture is made possible by the American Veal Association.

immune system. The farmer must milk the mother and deliver the colostrum to the calf. Remember that the farmer only has a financial interest in the female calves. They will sell the male calves for a very low price. It is rumored that farmers frequently deliver a disproportionate share of the colostrum to the female calves, thus permanently impairing the male calves' immune system. Moreover, calves have a natural desire to suckle a teat, or a teat-like structure. Veal calves are fed out of a bucket and are denied this natural desire. It is not just the milk the calf wants, but the feel of the teat as it is drinking.

The feed provided to the calf is a center of debate. Veal calves are fed a strictly liquid diet. The animals grow well on the feed, so it provides much of their nutritional requirement, but the feed may not be ideal. Calves have evolved to feed from their mother's milk initially, and then to begin eating grass and other plants alongside the milk. Animal advocacy groups criticize the veal industry for not adding fiber to the animals' diet. While the liquid feed provides most of the nutrients found in the grass, the lack of fiber poses some health and welfare problems. First, the calf's digestive system does not develop normally due to the absence of fiber. This may not pose a problem if the calf is slaughtered at an early age. Second, being ruminants, cattle desire to chew their cud but with a liquid diet there is no cud to chew. They consequently perform odd behaviors with their mouth in what might be an attempt to satisfy this desire. The absence of fiber and chewing cud could (we speculate) produce a feeling similar to malnourishment.

The question of whether veal calves should be given fiber in addition to their liquid diet depends partly on when calves naturally desire to graze. When does a calf begin to eat grass or hay? This question was more difficult to answer than we initially expected. Our experience told us that calves will begin eating grass at around 3 to 4 weeks of age. Interviews with several beef producers confirmed our estimate, though some thought a better answer was 8 weeks. Whether calves start eating grass at 3 or 8 weeks of age, veal calves are not slaughtered until they are 16 weeks old. This suggests calves do indeed desire, and should receive, some fiber in their diet during their lifetime. Indeed, a scientific veterinary committee in Europe suggests calves should be given fiber as early as 2 weeks of age.

The liquid diet given to veal calves is specifically designed to be low in iron, which gives the calves the pale flesh that consumers desire. As a result, one in four veal calves are slightly anemic and one in ten is clinically anemic. This anemia is also caused by a lack of movement due to the size of the stalls in which they are housed. Problems with anemia and iron deficiency are probably not as bad as they were in the past, however. The anemia statistics cited above are from a 1995 paper, and the American Veal Association states that, "veal farmers have advanced tremendously in the last decade in their understanding and management of veal calf iron, so as to avoid anemia."[106]

Veal Possibilities

The most interesting feature of veal production is the ease and simplicity of improving it. That, at least, is our opinion. The fact that the veal industry has not pursued these changes suggests we may underestimate the challenges. It would seem that providing veal calves with room to move and access to the outdoors and pasture is inexpensive and would not correspond to health problems or injuries. Cattle are animals that are easily cared for in pasture lands. They are not vulnerable to predators like chickens, and they do not damage fields like hogs. Cattle graze during cold and hot weather, although some breeds are better adapted for extreme temperatures. Establishing and maintaining a group hierarchy is of far less importance to cows than to hogs and chickens. They seldom fight, their fights are relatively docile, and the membership of a group can be interchanged with little consequence for the social order.

Disease is more prevalent among groups of cattle, but veal producers could follow many dairy producers' lead by keeping calves in individual pens for the early portion of their life and group-pens after the first few weeks of age. Adding iron to their diets might result in a meat color inconsistent with consumers' preferences, but slapping on a label indicating the animal was raised humanely is likely to more than offset this drawback. Finally, adding fiber to the calves' diet with grass, straw, or grain is inexpensive and would not add considerably to labor costs.

All of this seems a low price to pay given the corresponding gain in animal welfare. The calf would be able to walk more than a few feet and would begin to feel nourished. Asking that an animal be able to walk more than a few feet and to feel nourished is not asking much. To the veal industry's credit, they are pursuing some of these changes. Approximately 35 percent of veal calves are currently raised in group housing, and by 2017 most all veal operations will be doing the same.[107] Twenty years ago, figures suggest that virtually all veal calves were raised in individual stalls, so the veal industry is applauded for the improvements they have implemented. Further improvements would be relatively easy, but the industry has yet not made a public commitment to pasture access and a more enriched environment.

Regardless of whether we eat veal or not, as long as we drink milk or eat cheese, we help subsidize veal production. It is therefore important to know what goes on at a veal farm. It is important, however, to not overstate the link between veal and milk. Two 8 oz servings of milk is only 1 lb of milk, and one cow produces about 19,125 lbs of milk each year. Even if you drank two 8 oz servings of milk each day, you are only consuming 1.9 percent of a cow's annual output of milk. Thus, your milk consumption refers to only a fraction of a cow's output, and the veal produced from 1 lb of milk is tiny, implying your milk consumption is only responsible for a small amount of veal production. More-over, veal would be produced even without milk production. Milk purchases do subsidize veal production, but only a by tiny amount. If veal production were outlawed the number of dairy cows would barely change, and if milk produc-tion were banned veal calves would easily be obtained from beef breeds. All things considered, a vegetarian can consume milk and cheese even if they are opposed to veal production, knowing their action subsidizes veal production only by a miniscule amount.

Dairy

Dairy is ubiquitous in our modern diet— in the form of milk, cheese, yogurt, ice cream, and butter. Over nine million cows are milked in the US each day. As has been the case in other agricultural sectors, dairy farms have been increasing in size. In the 1970s, a large dairy farm had 100 milk cows, which were fed grain but also provided with pasture or hay. In later years, larger and more cost-effective dairies emerged which housed cows in large barns or dry feedlots. In 1992 there were only 15 dairies that had over 1000 cows. By 2002 there were 176 dairies of this size. Today more than 35 percent of all dairy cows in the US are housed on farms with more than 1000 head of cattle, and it is not uncommon to see farms with more than 5000 dairy cows in western America.[108] Such changes in farm size and in cattle living conditions have been accompanied by changes in dairy cow well-being.

Cows

Cows must be bred and give birth to produce milk. Because dairy cows are raised to provide milk for human consumption the mother's offspring is taken away shortly after birth, as discussed above. This practice produces stress and frustration for the mother and calf. It is true that the calf would probably not drink all the milk the mother could provide, but separating calves from their mothers each time the cow is milked (normally twice a day) would be impracti-cable in commercial operations of any size. Without working on a dairy farm it might be difficult to understand just how time-consuming separating mother and calf would be. Having worked on two dairies, we can testify to the inconvenience for the farm workers and the considerable stress on the mother and calf this procedure would cause. It is probably better, from a welfare point of view, to separate mother and calf once, at birth, rather than separating them two times a day over many months.

Once a cow has adjusted to the absence of its offspring, it returns calmly to its life given to commercial milk production. Dairy cows receive the most nutritious and expensive diet of any livestock animal. The result is an enormous volume of milk being produced. The most productive cows generate about 8 gallons of milk each day during the milking period, and produce over 2300 gallons of milk a year—enough milk for more than 50 people to drink a pint a day.

Dairy farms use a diverse variety of housing systems. To provide some context consider the pictures in Figure 5.17 showing a small dairy farm managed by Oklahoma State University. The cows are provided with a dry lot with shelter, they have access to indoor stalls with dry sand for bedding, and they are fed on a concrete pad under shelter. The cows generally have ample room and convenient access to both indoor shelter and the outdoors. The only amenities missing from this system is access to grass pasture. To what extent would the cows benefit from pasture access? The answer is that they would value the pasture but not as much as we might expect. Cows might enjoy pasture, but they already obtain all their required nutrients in the feed ration the farmer provides. In fact, cows prefer the ration of hay, silage, and grain to pasture grass. Pasture land would provide a more varied environment and a soft place to lie down outside, however, and cows do value those amenities.

Some farms limit the cows' opportunity for outdoor access and freedom of movement. These dairy farms use a free-stall system, where cows have access to a stall of their choice inside a barn, or they tether cows to a particular stall, which is referred to as a tie-stall system. Both the free-stall and tie-stall systems can be used in combination with occasional outdoor or pasture access, but in some cases the cows live exclusively indoors. The picture on the right of Figure 5.17 shows a free-stall design where cows have a choice between individual stalls with some sort of bedding or a concrete floor. Although concrete is the most common type of flooring, many dairies use rubber mats over concrete or cover the concrete with straw, sand, or sawdust. The free-stall barn is crowded, but the individual stalls provide an area where the cow can relax and not be bothered by others. The stalls do not confine the cows as tightly as gestation crates confine sows. The reason is that crowding cows can cause them to step on other cows' udders, resulting in a significant loss of revenue for the farmer.

In the US about 50 percent of dairies predominantly use a tie-stall design where the animal is tethered to an individual stall for significant periods of time. The tie-stall design restricts animal movement and prohibits the cow from grooming itself, which has adverse consequences in terms of well-being. Some producers untie the animals for part of the day or for certain periods of time. Smaller operations are much more likely to use tie-stall production methods than larger operations, and as farms have become larger, the number of cows in tie-stall operations has declined. For example, although 44 percent of cows in the US were milked in a tie-stall operation in 1996, only 22 percent of cows are milked in this manner today.[109]

Figure 5.17 Dairy farm

Sources: Right picture at wikimedia commons under GNU free documentation license <http://commons.wikimedia.org/wiki/File:Herdim.JPG>, Left picture author's original.

Across all types of production systems in the US, 19 percent of lactating cows have no outdoor access, 22 percent are provided pasture, and the remainder has access to a concrete pen (17%) or dry lot (40%) in the summer. Outdoor access is more limited in the winter. Almost one-third of all lactating cows do not have outdoor access in winter, 4 percent have access to pasture, and the remainder has access to a concrete pen or dry lot.[110] Outdoor access provides the cows with more space, enrichment, sunshine, etc. However, as we previously noted, access to pasture and outdoors may not be as essential for improved animal welfare as one might initially think.

Even when cows are densely housed, they are generally amicable and rarely fight or harm one another. About 78 percent of cows are milked in a parlor-type system where they are taken into a milking barn. This process does not seem stressful for the animal. In fact, cows know exactly what to expect and often voluntarily return to the same stall each milking. Moreover, the process of milking itself is probably pleasant because cows experience pain if they are not milked and their udders swell too much. Additionally, forcing the animal to walk from the paddock to the barn provides beneficial exercise.

We have already mentioned a practice used in some sectors of the dairy industry that potentially lowers cow well-being: the use of tie-stalls. We now turn to more general problems that affect cow welfare. Cows often experience bone and skeletal problems as a result of their genetic makeup due to breeding that places a higher value on milk production than skeletal health. Some Holstein dairy cows look like a huge udder attached to skeleton. It is often difficult for a dairy cow to maintain her own body weight despite highly nutritious feed. Their bodies are programmed to focus nutrients on milk production and reproduction, the cow's own biological condition being a third priority. As a result, milk cows are culled at only 4 or 5 years of age (after about three milking cycles). Estimates suggest that between 14 percent and 25 percent of all dairy cows suffer from some form of lameness, which is arguably one of the biggest welfare problems in dairy cows.[111] Lameness is influenced by breeding but also by living conditions; constant living on concrete floors, while increasing sanitation and reducing muddiness, can adversely affect cow hoofs and legs.

Like bone structure, udder structure is also affected by genetics. Dairy cows have disproportionately large udders. In addition, the cows are milked intensively—two to three times a day, every day, for most of the year. The result is that cows can develop mastitis—an inflamed udder. Cows with mastitis have red and swollen udders caused by the infection, which can lead to clots or pus in the milk. Mastitis is painful and potentially deadly for dairy cows, and is estimated to affect 16.5 percent of all dairy cows (up from 13.4% in 1996).[112] When cows are milked at the Oklahoma State University dairy farm, the first thing the milk-hand does is to hand milk the udder to visually inspect for mastitis.

The prevalence of lameness and mastitis is influenced by the use of the growth hormone rBST, administered to about 30 percent of US dairy cattle. Administering rBST boosts milk production but has adverse consequences for cow health. Like genetics that emphasize milk production over skeletal health, rBST will boost milk production by redirecting nutrients away from skeletal health. The use of rBST, then, is no more humane or inhumane than the act of breeding cows to produce maximum quantities of milk. We disagree with the view that rBST leads to unsafe milk, as rBST and regular milk are chemically identical. The use of rBST has caused consumers to question the safety of milk,[113] at a time when many people are known to suffer from Vitamin D deficiency.[114]

Approximately one-third of US dairy cows have two-thirds of their tails removed—a practice referred to as tail docking.[115] Docking, according to those who practice it, is said to contribute to the cleanliness of udders and to prevent the spread of disease as milk-hands move from one cow to another. Docking causes pain at the time of docking and can lead to chronic pain; the absence of a full tail also means that the cow is unable to free itself of flies. Tail docking is often performed not when cows are younger (when pain is minimized), but when cows are two years or older and almost always without anesthetics. However, some studies have shown that docking provides no benefits in terms of cleanliness or prevention of disease transfer.[116] It is for these reasons that the EU and the state of California have banned tail docking, and why the American Veterinary Medical Association opposes the practice. Why, then, would farmers dock tails? Probably because it makes milking more convenient. Some automatic milk-hands approach a cow from behind (traditional automatic milk-hands approached the cow from the side), and tail docking simply makes it easier to see the udder and attach the device. It might seem shocking that farmers should chop cow tails simply because they are a nuisance, but we would suggest this practice is no worse that pet owners docking their dog's tail just for looks.

Lactating dairy cows are handled by humans every day, thus the animals' welfare is influenced by human handling. Studies have shown that milk production is affected by the behavior of human handlers, and that cattle given a name and treated as individuals yield more milk.[117] A lactating cow can live a peaceful or a stressful life, and the difference is significantly affected by the attitude of the people handling her.

Dry Cows

Soon after a dairy cow has given birth it will be re-impregnated using artificial insemination and will be pregnant while providing milk. Two months before her next calf is due she will be given a "dry period" where she is no longer milked. Cows at this stage are more likely to be given more space, outdoor access, and pasture. In summer, 39 percent of dry cows are able to graze in pastures; that percentage in winter falls to 12 percent. Of all the dry cows in winter, 22 percent are not given any outdoor access, this number falling to 11 percent in summer. Comparing the housing of dry to lactating cows, dry cows receive more outdoor access and pasture, though the difference is not marked. During this time the cows are mostly left alone, and provided high quality feed and health care.

The COWEL Model

As with the egg and hog industry, the aspects of farming that affect the welfare of dairy cattle are numerous and complex. Fortunately, as with the egg and hog industry, researchers have developed a mathematical model that accepts farm descriptions as information and calculates a number corresponding to the well-being of the cows. Referred to as the COWEL model, its structure is similar to the FOWEL and SOWEL models discussed previously. As one might suspect, farms which do not tether animals to a stall for extended periods of time, use non-slippery floors, dry resting areas, large space allocations, access to pasture, and the like are rated higher. Cows that are deemed to have a lower level of well-being are those confined to tight spaces, have difficulty finding dry sleeping areas, have little shelter, must wait in long lines on slippery surfaces before milking, and are handled harshly by handlers. Generally, an intuitive judgment on which farm settings are most conducive to better animal welfare coincides closely with the COWEL model.[118]

Dairy Calves

Calves are immediately removed from their mother so that all the cow's milk can be used for human consumption. The calf will receive formula-milk instead. The farm retains almost all of the female calves for subsequent breeding and milking, so they often receive excellent care. Each female calf is housed in an enclosure that looks like a large plastic dog house, with a small outside area (see the left picture of Figure 5.17 for an example). The pen is nothing like the veal stalls described previously. Each female calf has room to walk and turn around, comfortable bedding in the shelter, outdoor access, and is given nutritious food. The individual pens are used because calves are especially susceptible to sickness at this age, and separating the calves from each other prevents the spread of disease and allows the farmer to monitor the calf's eating levels. After a few weeks the calf will be transferred to a group-pen, and some weeks later it will be moved to a pasture or dry lot.

Male dairy calves are treated very differently due to their inability to produce milk and the prevalence of artificial insemination, meaning that they are not needed on the farm for breeding. In the US most male dairy calves enter the beef supply chain. Some farmers specialize in purchasing male dairy calves and feeding them until they can be placed in feedlots with beef cattle. Male dairy calves are less likely to spend time in pasture; they live to about 1.5 years of age and are harvested when they weigh about 1100 lbs. In recent years, dairies have increasingly used sexed semen when artificially inseminating cows. The sexed semen alters the female/male birth ratio from 50/50 to about 80/20. Although sexed semen is more expensive than traditional semen, dairy farmers are increasingly adopting the new technology as it allows them to reduce the prevalence of the much lower valued males. Whether this makes dairy production more humane depends on our own points of view.

Beef

For beef cattle to have survived in America's youth, they had to be tough. This is especially true for the cattle raised in the Western US. Breeds like the Longhorn were popular in areas where predators were prevalent, and where cattle would have to be driven for many miles to reach the consumer. Ranchers would turn their cattle out onto large tracts of land with no animal care. The animals lived like wild creatures, except for when they were rounded up periodically for sale. The animal breeds were also domesticated, which made them easier to handle than wild animals. The welfare of these animals was dependent upon the weather and the number of animals that the ranchers released per acre. When the weather was favorable, there would be few cattle deaths from the cold and the cattle would be well-fed with the plentiful grass. During harsh winters or drought many cattle died and many of those that survived suffered.

It was commonly thought that many animals did indeed suffer because the ranchers gambled on good weather and released large numbers. To protect the ranchers, laws regarding animal cruelty did not include intentional starvation of these cattle.[119] Animal advocates who argue against the intense confinement of animals are often criticized by the livestock industry as wanting to return to the days when farm animals were left to fend for themselves against predators and the weather. However, this is unfair, and from interviews with various animal advocates, we have found that while they want the animals to have access to pasture, they also argue for them receiving shelter.

Whatever hardships beef cattle faced in the past, few cattle are now left uncared for on large tracts of land, and ranch managers make better decisions about how many animals to stock per acre today. One of the main criticisms of animal rights groups about modern livestock production is that animals are not allowed to live natural lives. However, in the case of beef cattle, these cattle do live much of their lives very much like they might live in the wild—excepting that they are provided with predator protection, food security, and good health care. In fact, a mother cow will often live her entire life in this manner. The beef industry in the US is *huge*. For example, on July 1, 2009, it was estimated that there were 32.2 million beef cows (recall that there were only about nine million dairy cows), 4.5 million heifers for replacement beef cows, 35.6 million calves, and about 11.6 million cattle on feed in feedlots.[120]

Beef production methods vary across farms, especially regarding the timing of certain events, but consider a stylized example of a cattle operation where calves are born in the early spring. Mother cows are maintained in a herd, typically on pasture. Some beef cows are bred using artificial insemination and some are bred naturally. Cows are typically bred so that they will calve in February or March, and some farms use artificial hormones to ensure that all cows will conceive (and thus will birth) at approximately the same time. Calves are typically born naturally, but birthing is usually closely monitored. In college, one of the authors had a job that required him to stay the night at a barn during calving season; every two hours he checked the herd to determine if any mothers needed assistance birthing. Once a calf is born the herdsman will perform a few activities on the calf (e.g., castration, if male, and installing an ear tag), after which the mother and calf will be left alone to graze for up to seven months. During this time they will only be handled in order to move them to different pastures, weigh them, or perform some sort of health care, like vaccination.

There is only one reason that beef cows are left alone to graze naturally while hogs and chickens are kept exclusively indoors: *economics*. The most economically efficient way to produce cattle is to raise the mother and calf on grass. Cattle can be raised on relatively unproductive land—land that would not be viable for growing corn or soybeans, for example. Cattle have the ability to take

Figure 5.18 Cow–calf pair, stocker calves on wheat, and feedlot

Sources: All pictures are available in public domain at wikimedia commons <http://commons.wikimedia.org/wiki/File:DBAZBenedicts_AZ_feedlot_blackbirds3.jpg> and <http://commons.wikimedia.org/wiki/File:Cow_with_calf.jpg>

grass, inedible to humans, and convert the energy to meat. Most calves are scheduled to be born between February and May in most regions of the US, just before the summer, when grass is most plentiful. Calves begin eating grass between 3 and 8 weeks of age, and as they grow they will continue to nurse and will increase their consumption of grass. The animals are usually housed in pastures large enough that space, crowding, and fighting are not a problem. Periodically, cow welfare is compromised if bad weather strikes; however, cows are robust animals easily able to withstand what would be harsh weather to humans. In uniquely bad weather, such as a blizzard, deaths occur that could have been avoided if the cattle had access to shelter, but these events are rare.

Assuming a typical farm where mother cows give birth around February or March, when October arrives, calves are weaned from their mother, and farmers begin to refer to them as *stocker calves* (Figure 5.18). In Oklahoma, stocker calves are moved to wheat pastures where they will graze on the growing wheat. In late spring the stocker calves will be removed from the wheat so that the plant may form its seed and be harvested. At this point stocker calves are at least 1 year in age and are sold to a feedlot for "finishing." Some calves never become stocker cattle, and are delivered to a feedlot immediately after weaning.

The feedlot marks the end of what is largely a natural life, but the cattles' life in the feedlot does not necessarily mean they suffer a lower level of well-being. Feedlot cattle are placed into dirt pens with other cattle, with a concrete pad on one end where food is delivered. These so-called finishing or fed-cattle are given a highly nutritious mixture of silage, forage, grain, and other supplements until they are about 16 months of age, after which they are harvested. Generally, 50 to 250 cattle are held in one pen, the cattle have ample room to move, and aggression is not typically a problem. The space afforded each steer/heifer in a feedlot is around 250 square feet when at their largest size, which is about ten times the amount of space provided to a finishing hog on a per-pound basis.[121] The most frequent injury occurring in feedlot cattle is due to *riding*, which occurs when one cow mounts another as if to breed. Both male (castrated or otherwise) and female cows will ride. These are large animals, and if a cow falls due to riding, a substantial injury can occur. Although riding is perhaps the biggest cause of injury in feedlot cattle, such injuries are rare. Riding is easy to spot, and cattle that are prone to riding others can be removed from the pen and harvested earlier, or placed in an individual pen. The feed given to heifers is sometimes supplemented with a hormone referred to as MGA, which prohibits heifers from coming into heat and this helps to reduce the frequency of riding within heifer herds.

The practice of finishing cattle in feedlots emerged for a variety of reasons. Perhaps the most prominent is that the energy-rich feed causes cattle to grow quickly, and US consumers prefer the taste of corn-fed beef. Cattle could be left to pasture until ready for harvesting, but it is generally more cost-effective to leave this grazing land for mother cows. The efficiency gains from the feedlot sector not only allow farmers to produce more beef from a given amount of land but it also reduces the resources required to produce a given amount of beef. There are a huge number of cow–calf producers spread all across the country, but there are only a few processors who are located in the Midwest. Feedlots serve the function of assimilating calves from many regions, grouping the calves into similar groups that will be ready for harvest at similar times, and funneling the calves to the beef processors.

Most of the literature that we have looked at about factory farming written by animal rights activists contains very little about beef production. That is because there are very few animal welfare problems that exist with beef production. If such literature mentions beef cattle at all, it is usually in regard to the feedlot. Some writers argue that cattle in feedlots stand and sleep in their own manure. This is partially true, but not as bad as it sounds. Productivity can fall as much as 25 percent if the feedlot is muddy for a long period of time.[122] Consequently, feedlot operators have an economic incentive to reduce muddiness, and for this reason most feedlots are located in dry areas of the country, like Western Kansas, Oklahoma, and Texas. Dry weather reduces animal sickness, and cattle grow better with lower mortality. Moreover, despite the argument of some well-known critics of factory farming, humans are *not* getting sick from the manure in feedlots,[123] as it is extremely difficult for manure from the feedlot to come into contact with meat.[124] Though some data suggest that *E. coli O157:H7*—a bacteria found in the digestive tract and manure of cattle which causes food poisoning—is more prevalent in cattle that are finished on grain rations as opposed to foraging (hay and grass), those data were apparently based on observations of only three cows. Moreover, numerous studies attempting to corroborate this relationship failed to verify the link, and many other studies have found that the amount of *E. coli* in cattle manure is the same, regardless of whether the cattle are in the feedlot or in pasture.[125] Consequently, we would argue that the food safety charges leveled at feedlots are not substantiated by research.

It is sometimes argued that cattle well-being in feedlots is low because cattle are cramped with no pasture. However, as we have indicated, compared to hogs and chickens, feedlot cattle are *not* cramped (most of the time cattle stay close together, leaving most of the pen empty), and the benefits of pasture access for the cattle is debatable. Cattle strongly prefer the feed provided to them in feedlots to grass. If cattle were given access to both pasture and the feedlot feed ration, our interviews with farm managers and our own experience lead us to believe that the cattle would elect to consume most of their nutrients from the feed ration, and utilize the pasture only to combat boredom and to find a more comfortable place to rest. Anyone who has raised cattle can attest to their love for corn and soybeans. If you feed cattle grain at the same place and at the same time- of- day, the cattle will promptly congregate at that location when they see you coming. We have even seen a mother cow who was giving birth, on seeing the feed truck approaching, rise to her feet and walked to the trough to eat—her calf's head and front feet protruded, waiting for the mother to push the rest of its body through!

Some people are critical of the diet of feedlot cattle, claiming the diets routinely make the cattle sick. Cattle are designed to slowly digest large amounts of foraged food (e.g., grass and hay), which are low in nutrient density. The cereal grains fed to feedlot cattle (90% grain, 10% hay), however, are nutrient-dense and are digested quickly. As a result, some people claim feedlot diets are "unnatural," an argument that is a bit misleading. Ruminants in the wild actively seek grains to eat—the same type of grains fed in feedlot. However, in the wild grains are sparse and cattle's ancestors had to rely on digesting large amounts of grass to live. The "unnatural" feedlot diet can cause digestion problem, such as bloating. Opponents of feedlots argue that cows are constantly sick because of their "unnatural" diet and must be administered antibiotics on a routine basis just to live. This is misleading and exaggerated. We will discuss antibiotics shortly, but in brief we would argue that while cattle fare better with antibiotics they would still perform well without the antibiotics. Moreover, no one is more aware of the potential problems from the feedlot diet than the feedlot managers. Consequently, they go to great lengths to prevent such sickness and are mainly successful in their efforts.

A survey of over 28 million feedlot cattle in the early 1990s revealed that the mortality due to the cereal diet was only 0.061 percent.[126] Our conversations with a number of experts suggest that only about 2 percent of cattle experience any form of bloating, and most of the bloating is mild. Loss by death in feedlots as a whole is small, around 1.5 percent, indicating that there are many feedlot features that are conducive to animal health. When bloating is observed, the feed ration is altered and the condition quickly subsides. Another complaint about feedlots is that the cattle are not provided shade in the summer. This is true. Feedlots rarely provide shade, even during west Texan summers. How detrimental is this to welfare? Our opinion is that it would improve animal welfare to provide shade, but that it is not so big a problem that overall welfare is compromised. We have worked alongside the cattle in these feedlots, and though at times we might have liked some shelter ourselves, the absence of shade provided us with no problems, and humans are a lot less hardy than cattle.

There is also the problem of flies and dust in feedlots, the Midwest being so dry. But there is little that can be done to ameliorate this situation. Watering down the pens to reduce dust could arguably negate the many advantages to the cattle of the dry weather, reducing animal welfare overall. However, it does have to be said that the large number of flies in a feedlot, presumably attracted to the manure, does serve as a constant nuisance to the animals.

We have a generally favorable view of animal well-being in feedlots, but some disagree. Detractors of feedlots might put it that the animals stand in their own manure, in the hot sun, eating an unnatural diet that makes them sick, a diet they can only survive on with the regular use of antibiotics. Conversely, someone with a more positive view of the feedlot might put it that feedlot cattle live in pens with adequate room and are generally kept dry. Although the diets are somewhat unnatural cattle love to eat the ration, and sickness is quite rare. Antibiotics are used, but removing antibiotics would only mean feedlots with sicker and slower growing cattle. Despite our favorable view of feedlots, we do contend that the cattle would be more content if both dust and flies were managed better, and if cattle were provided shade in the hot summer.

Grass-fed Beef

Those who dislike feedlot production methods might prefer to buy meat from grass-fed cattle, beef that can be increasingly found in grocery stores and farmers' markets. This beef is derived from cattle fed a diet consisting almost entirely of forage—grass and hay. Because *all* cattle eat grass at some point in their life, there has been some controversy as to what constitutes grass-fed beef. After years of deliberation, the US Department of Agricultural (USDA) finally ruled that to be labeled as grass-fed, "Grass and forage shall be the feed source consumed for the lifetime of the ruminant animal, with the exception of milk consumed prior to weaning."[127] "Grass-fed" beef cannot come from cattle that have been fed grain.[128]

Note, however, that the USDA labeling standards for grass-fed beef do *not* mandate that cattle be raised in a pasture. Grass-fed beef can just as well come from cattle raised in a dry dirt lot exactly as feedlot cattle are raised, except that they receive hay instead of grain to eat. Moreover, grass-fed labeling standards do not prohibit producers from feeding cattle antibiotics or growth hormones, which is something consumers of grass-fed beef often object to.[129] Thus, if you purchase "grass-fed" beef, it may imply that it came from an animal raised in pasture for its entire life or it may mean that it spent the last 100–200 days of its life in a feedlot. The only way to know the actual production methods used is to investigate the company or farmer selling the beef.

This brings us back to a point we made earlier, that most consumers strongly prefer the taste of grain-fed to grass-fed beef. One study showed that in a blind taste test, only about 20 percent of US consumers prefer the taste of grass-fed to grain-fed beef. And, on average, across all consumers, people are willing to pay about 30 percent more for grain-feed beef than grass-fed beef.[130]

Controversies with Modern Livestock Production Practices

A discussion of the production methods used on modern livestock farms would be incomplete without touching on some of the controversial topics raised by polemics of factory farms: use of antibiotics and hormones, natural foods, organic foods, and the environment. Each issue is briefly addressed in what follows.

Antibiotics and Hormones

Critics of modern livestock agriculture often point to the use of antibiotics and hormones on the farm. First, it should be noted that poultry (laying hens or broilers) and hogs are *not* administered growth hormones in the US. Claims to the contrary are false. Feedlot cattle and growing pigs are routinely administered low doses of growth hormones, either in their feed or via implants. *We are aware of no scientific evidence showing that administering subthereapeutic hormones to livestock has adverse consequences for humans.* Indeed, although the EU has banned domestic use of growth hormones in cattle production since 1985, and has banned US beef imports since 1989, the World Trade Organization found that there was no compelling safety reason for the trade barrier, and they ordered that the EU must make compensation payments to beef-exporting countries, unless the ban is lifted. There is some evidence that some growth hormones lead to less tender beef products, but no evidence or no real reason that it should lead to unsafe beef products.

It is common to hear horror stories. For example, some people often argue that elevated levels of hormones in beef cause human girls to mature early. Consider a few facts. Beef from cattle administered growth hormones contain only 0.4 ng more estrogen per 3 oz serving of beef (the primary hormone compound used to promote growth in beef cattle) than cattle who have never received the hormones. Is 0.4 ng per 3 oz serving a really a significant amount, if we consider that 4 oz of raw cabbage contains 2700 ng of estrogen, and that 3 oz of soybean oil contains 168,000 ng of estrogen? The birth control pill that women voluntarily take has many hundreds of times more estrogen than is in a serving of beef.[131] We would argue, then, that hormone use in beef production is not making girls mature earlier and, furthermore, for similar reasons it cannot be causing human illness.[132] Hormone levels can be higher in beef derived from a bull that was not given growth hormones than in a steer that was given hormones. Given these facts, the debate about growth hormones pales into insignificance.

Beef cattle, pigs, and broilers are routinely administered antibiotics in their feed. The antibiotics are not given in response to sickness, they are given to all animals, healthy and non-healthy alike. Animals grow faster when given subtherapeutic antibiotics; they are healthier, suffering few illnesses and experiencing lower mortality rates. Antibiotics use is heavily regulated and farmers are required to cease administering antibiotics a certain number of days before the animal is harvested to prevent any antibiotic residue from appearing in the food. Antibiotics are *not* used in egg production because chickens lay an egg almost every day. *There is absolutely no health risk involved in eating meat from animals given antibiotics, and because of mandatory withdrawal times, there are no antibiotic residues present in the meat.*

The one potential danger with the use of subtherapeutic antibiotics on the farm is antibiotic resistance. There is a risk that certain bacteria may become resistant to the antibiotics, and that these bacteria may mutate into a form that infects humans. This is a real risk, but it must be put into perspective. Voluntarily, and in response to public pressure, livestock industries have dramatically reduced the use of certain antibiotics that are also used in humans. Most experts believe that resistance resulting from doctors over-prescribing antibiotics to humans presents a much greater risk to human health than the use of antibiotics in livestock production. Moreover, though banning subtherapeutic antibiotic use on the factory farm can reduce total antibiotic usage,[133] it could actually increase total antibiotic usage on the farm because without subtherapeutic use, animals are sicker and consequently require greater use of full therapeutic antibiotics.[134]

Given the state of scientific knowledge, what explains the common belief that antibiotics and growth hormones infect our food and threaten our health? There are probably a variety of reasons that relate to the way humans interpret and deal with low probability risks in agriculture, but unfortunately, one reason is that some groups so fervently oppose modern agriculture that they will sacrifice honesty to scare consumers.

Organic Food and the Animal Welfare Institute

The "organic" label in food generally refers to food produced under a production system designed to promote biodiversity and soil biological activity. Organic food is typically grown without synthetic pesticides, fertilizers, or genetically modified ingredients. However, we would argue that on the whole this label is largely the result of a marketing gimmick. We would argue that organic foods are no more nutritious or higher in quality than most non-organically produced food.[135] Furthermore, it is our opinion that organic farming is not more sustainable or environmentally friendly than non-organic farming. There are reasons to suppose it may be better *or* worse for the environment, but given how politically charged the issue is, one should not expect a definitive truth to emerge in the near future.[136] Even if the production of organic food does help small farmers, it is not clear to us how the people running small farms can be considered any more important than the workers and shareholders comprising corporate farms.

Animals produced under organic systems probably experience higher levels of well-being than animals in non-organic systems, but the difference might not be as marked as many believe. A farmer must undertake a number of activities to achieve organic certification, many of which are unrelated to animal welfare. For example, to produce organic pork all the corn and hay fed to hogs must be raised organically (meaning no pesticides, commercial fertilizer, and the like). Farm animals are unlikely to be affected in terms of well-being based on whether they are eating organic or non-organic grain and hay. Organic production prohibits the use of antibiotics (both therapeutic and subtherapeutic), which almost certainly lowers well-being as more animals become sick without access to antibiotic treatment. Animals that become sick on organic farms are either allowed to remain sick and potentially die or are segregated, given antibiotics, and are sold at a lower price in the non-organic market. Too many sick animals, we believe, do not receive proper treatment because the farmers fear the loss in income they will experience in having to sell the animal on the non-organic market. Animals need certain amino acids for prime health, and these amino acids are usually provided through synthetic amino acids added to the animal feed. These synthetic amino acids are not allowed under organic standards. Consequently, organic producers have a difficult time meeting their animals' dietary needs, and the animals suffer. A number of animal scientists in the US feel organic production is cruel for this reason.

Organic certification for livestock does include provisions for ensuring minimum standards of care. Animals in organic systems must have some access to the outdoors and to natural sunlight, but as previously mentioned the outdoor access is relatively unregulated. For many organic egg-laying operations, producers will simply have an open door and a little outdoor area surrounded by chicken wire. The area will be small and barren. Ruminants (e.g., cattle and sheep) raised for organic production must have access to pasture but the pasture allocation can be small. The outdoor area must have shade and animals must be provided clean, dry bedding. The US organic certifications state that the animals must be maintained in a manner that, "provide for exercise, freedom of movement, and reduction of stress appropriate to the species. Additionally, all physical alterations performed on animals ... must be conducted to promote the animals' welfare and in a manner that minimizes stress and pain."[137] Obviously, organic production prohibits the use of battery cages in egg production and gestation stalls in hog production.

Farrowing stalls may be used for organic pork production as certifications allow confinement of the animals because of the animals' stage of production. (As we discussed above, farrowing stalls are bad for the mother but good for her offspring.) Laying hens may have beaks trimmed and hogs may have their tails docked in organic systems. Organic chicken operations can use fast-growing broilers and thus are prone to the same lameness and leg problems associated with non-organic systems. In short, laying hens and hogs, and to a lesser extent broilers, raised in organic systems do experience a higher level of welfare as compared to non-organic systems. There is probably not much difference in the well-being of cattle in organic and non-organic production systems.

Those who are interested in purchasing food with the highest guarantee of animal care should seek products sold under the Animal Welfare Approved (AWA) label. Such products can be difficult to find but represent the best in terms of animal care. The standards attempt to ensure animals a pleasant life by identifying every animal need and setting standards to ensure these needs are met. The standards are specific but reasonable. For example, in the case of space constraints for pregnant sows the AWA standards dictate the amount of space that should be available within the shelter, in the resting area, and in the outdoor space. The AWA standards are more stringent than organic standards. For example, organic standards allow tail docking and beak trimming, while the AWA standards do not. The AWA requires broiler birds to possess the genetics that reduce the degree of leg problems, while organic broilers do not. Beef sold under the AWA label must always have access to pasture, even if they are fed a high-grain diet.

The AWA standards attempt to ensure high levels of well-being while remaining cognizant of costs of production. For example, when the AWA standards address pig castration they state that pig castration can be performed by a competent person. They could have required that a veterinarian perform the castration, but that would be prohibitively expensive.

After exhaustive research on the issue our conclusion is that animals raised under the Animal Welfare Approved standard receive superb care—more so than under virtually any other labeling scheme. Some AWA standards are a bit unnecessary, an example being the requirement that only "independent family farms" can produce food under this label. Why the AWA would want to prohibit a corporation from treating animals with compassion is unclear. It is also unclear why a person who works for a corporation matters less than a person who works on a family farm. Aside from these small issues, the AWA standards clearly contribute to improved animal well-being. Whether the improvements in animal well-being are worth the cost is an entirely different matter.

A final comment about the organic, AWA, and other such labels is warranted relating to consumer skepticism over the possibility of producers "cuting corners" by using practices that lower costs while still looking to sell at a premium under a brand. This skepticism is not entirely unfair. We have visited hog farms selling under a brand claiming to sell antibiotic-free pork, but the farmers told us plainly that antibiotics are used. We have visited hog farms claiming to produce under standards dictated by one of the humane labels but it was transparent the operation did not meet the standards: there was no dry bedding for sows and no dry place for sows to lie. However, these are just a few exceptions to the many farms we have visited that are in compliance with the standards. Both the AWA and organic standards require routine farm inspections, which does help to minimize non-compliance.

The Environment

Paul McCartney of The Beatles encourages others to join his cause and save the environment. He has argued, "If anyone wants to save the planet, all they have to do is just stop eating meat. That's the single most important thing you could do. It's staggering when you think about it. Vegetarianism takes care of so many things in one shot: ecology, famine, cruelty."[138] He has even called for a "meat-free day" to battle global warming.[139] This is not a rare view among vegans. Many who care for the environment also contend that meat production is a significant contributor to environmental pollution and global warming.

It is true that excess animal waste can lead to adverse environmental outcomes if not properly managed—but waste disposal on large confined feeding operations is *heavily* regulated. Even with these regulations we do not doubt that there remain some adverse environmental outcomes, although our assessment is that the costs of such problems are small relative to the benefits of livestock production.

Here, we want to make a simple point that seems to be forgotten by McCartney and others. The production of almost *any* good results in some amount of pollution. This is as true of agriculture as it is of auto manufacturing. Let us start by accepting McCarney's premise: an omnivorous diet creates more environmental pollution than a vegetarian diet. Then, let us couple that with a fact: a vegetarian diet is probably less expensive than an omnivorous diet.[140] All other things being equal, this means that a vegetarian will have more disposable income than a meat eater. What do vegetarians purchase with the money they would have spent were they to eat meat?

Suppose a newly converted vegetarian family buys an HD television as result of their improved purchasing power. The fact that the family will now probably watch more TV (and will thus stay inside longer, running the air conditioner), might mean that they use more coal, and this is one of the largest contributors to global warming. All things considered, a family's decision to become vegetarians might lead to more global warming. Of course, a family might decide to become vegetarian and *not* buy an HD TV, but the effective saving is real and the extra money will go somewhere. Even if the extra money is stashed in the bank, the bank will loan out the money to entrepreneurial individuals trying to create a business manufacturing something that results in pollution.

The point is that the environment is affected by *all* the goods we purchase. If a vegetarian diet were equally expensive as an omnivorous diet, this would suggest that substantial processing of the vegetables was entailed between the farm and the fork. The processing of vegetables adds value to the vegetables, and just like converting corn into meat adds value, they both create pollution.

Even if meat-eating results in more greenhouse gases compared to a vegetarian diet, it does not imply that meat-eating leads to global warming. For example, many households tend to grill meat outdoors, which places less stress on air conditioners, which rely on coal energy, which is a greenhouse gas contributor. The point of this is not to be naive about the relationship between food and the environment, but to bear in mind that environmental pollution stems from all the things we enjoy. It may be that an omnivorous diet is more polluting than a vegetarian or vegan diet, we do not know. What we wish to contribute to the conversation is that all those disciples of the vegan diet do not know whether it is less polluting either.

Thinking carefully about the relationship between food and the environment can muddy what we naively thought were simple concepts, and produce results counter to our intuition. It is often assumed that grass-fed beef has a smaller carbon footprint than conventional beef, because the latter entails feeding cattle for months on a high-energy corn diet. Corn production requires intensive use of farm machinery and fossil fuels, much less than pasture—so we believe. However, pastures are often improved by using synthetic fertilizers and hay production also entails intensive use of farm machinery and their associated fossil fuel use. It takes longer for grass-fed cattle to reach their harvest weight, resulting in larger emissions from burping and flatulence. For these reasons, grass-fed beef might arguably have a larger carbon footprint than conventional beef from corn-fed cattle.[141]

Groups concerned about greenhouse gas emissions should be keenly interested in the emissions that take place at the farm. Often, global warming alarmists obsess with the concept of *food miles* instead. They seek food produced locally under the assumption that food which travels a smaller distance from the farm to the fork will incur lower greenhouse gas emissions. However, emissions resulting from the transportation of food are small compared to the emissions that take place on the farm. Studies have shown that consumers interested in reducing emissions from food production should—instead of purchasing local food—replace a portion of their meat and dairy production with chicken, fish, or eggs, as the latter incur fewer emissions at the farm.[142] Within these categories of animal foods, livestock raised under factory farm conditions is usually associated with a lower carbon footprint than their organic counterparts.[143]

The amount of pollution that stems from the production of a good can be viewed in one of two ways. One view considers the total pollution emanating from the total production of a good. Vast amounts of meat and dairy products are produced every day and it is impossible to produce a vast amount of anything without some levels of pollution. The question is whether the value created in the process outweighs the costs of pollution.

A second view considers the pollution emitted per unit of production. Instead of asking what is the total pollution emitted from all meat production, the second view asks how much pollution is created per pound of meat produced. This definition has many favorable qualities in modern, factory-farm-style food production. Using this second view, livestock agriculture has become much more productive over time, and increases in productivity, by definition, imply that fewer inputs are required to produce the same level of output. Lower inputs typically translate into less pollution. Dairy production today can result in the same amount of output as a dairy in 1944 using only 23 percent of animal feed and 10 percent of the land, and producing only 24 percent of the manure and roughly half of the greenhouse gas emissions of a 1944 farm.[144] Greenhouse gas emissions are lower for feedlot beef compared to grass-fed beef,[145] factory-farmed chicken compared to organic chicken, and factory-farmed eggs compared to organic eggs.[146]

A lower per-unit value of pollution does not imply less overall pollution. If the price of farm output falls as farms become more efficient, consumers may be purchasing more of the product and—while pollution per unit falls—total pollution may rise. In order to ensure total pollution falls as per-unit pollution falls, the price must be prevented from falling below its former level. This could be achieved by taxing production. And again, such taxes should only be administered if the value of reducing greenhouse gas emissions is larger than the value society would receive from the lower food prices.

The word *sustainability* is often used when discussing the relationship between the farm and the environment. A sustainable farm is often deemed to be a farm that could continue using the same management practices for thousands of years. That definition, however intuitive, is not wise. If farms in 1650 were required to continue using the same "sustainable" practices continually, we would not have our contemporary and magnificent food supply.

Farming is a business, and the only way a business can survive for decades is to continually alter its management practices in response to changing conditions. For sustainable agriculture we do not need to prescribe practices we believe can be repeated indefinitely. Instead, we would argue that the only prerequisite for sustainable agriculture should be a government that allows farms to change with the times. No one understands better than farmers how their practices need alteration as changes dictate, and we would argue that the most important consideration is that we need to make sure we are not standing in their way.

Natural Foods

It is fashionable to seek *natural* foods, and there are some good reasons for doing so. American diets could use a hefty dosage of unprocessed fruits, vegetables, and grains. The trend towards whole wheat bread with a higher fiber content will likely have positive health benefits. The term *natural*, however, requires some perspective as there is little natural about most of our food in the first place.

An agricultural economist has written, "It is not an exaggeration to say that all of agriculture is intrinsically a struggle against nature."[147] It is true: everything about agriculture is an attempt to improve upon the natural world. The genetics of the natural world do not serve us well, so since 10,000 BC we have directly and indirectly modified the genes of plant and animal genes—through domestication. Natural fields, valleys, and meadows do not serve us well, so we fell the trees and plow the land. Wild foods are difficult to extract nutrients from and are not always very tasty. Thus, we harvest grains and grind them to a powder and bake that. We grind meat into ground meat as well; otherwise much of animals' meat is somewhat tough to eat.

Much has changed in the last 60 years of agriculture. Technology has progressed in mighty leaps, changing the food landscape and the human–food interface. And arguably it would be very strange, considering how much our lives have changed in almost every other area, if the only feature of society that did *not* change in the last 60 years was our food! Would it not seem odd that replacing typewriters with computers was not associated with any improvement in food processing? Would it not seem suspect if our improved ability to treat injury and disease was not also associated with safer food? Why lament the fact that modern meat, dairy, and egg processing has made food more tasty to consume and less likely to make us sick? To expect our food not to become more *un*natural as humans progress is absurd.

Food processing can produce some harmful side-effects, and not all sorts of processing are good. Though we can preserve fruits and vegetables to last decades, fresh fruits and vegetables are still healthier. The point is that to yearn for *natural* food, when put in a proper perspective, is illogical. The question is not whether we desire natural or unnatural food, it is which type of unnatural foods we should produce. Though fresh fruits and vegetables are the most important part of a healthy diet, those fruits and vegetables *are anything but natural*, and we would argue that *we are the better for it!*

However, when it comes to the farm animal welfare debate, the idea of *natural* food has a special meaning. This is because the animals used to produce our food do possess natural instincts. They want to perform many of the activities they would perform in the wild. To deny them this opportunity is to create frustration for the animal. Desiring natural animal foods, if that means allowing animals to behave naturally, is logical, if one cares about the welfare of animals.

But let us be cautious of animal food products labeled as natural. When the USDA certifies meat as natural this certification implies only that the meat possesses no artificial coloring, no synthetic ingredients, and is only minimally processed. An example would be the pork chop. By carving out a pork chop and packaging it up, doing nothing else, the pork chop is simply a cut of meat and can be labeled as natural, even if it was raised on a factory hog farm. If a consumer wishes to purchase food products from a farm that allows animals their natural habits, they must conduct independent research on the farm. Only by identifying the farm from which the animal was raised and the living conditions on the farm can we determine if the food meets your definition of natural.

Food Safety and Food Health

Raising farm animals under different environments could impact on the safety of food. We have already discussed how a conversation with an egg farmer revealed that neither he nor his employees would eat eggs from a cage-free system because they felt those eggs were more likely to come from a sick bird than cage eggs. There is some evidence that cage-free eggs are less likely to be contaminated with salmonella, and some evidence to the reverse is also true.[148] Evidence on how alternative production systems affect food safety is conflicting and sparse. Our general impression is that all foods sold are very safe to eat, and will continue to be so even if farm animal welfare is improved.

Animal rights activists are known for their lectures on the health benefits of a vegetarian and vegan diet. Meat-eaters are quick to respond with assertions of nutritional deficiencies of vegan diets. Which diet is truly healthier: vegan or non-vegan? There is no absolute answer to this question, and, we would argue that anyone who answers the question without hesitation is more of a salesperson than a health expert. A vegan who does not take special care in choosing foods with a high-protein element, vitamins B12 and D, iron, and calcium is likely to develop serious health problems.[149] Non-vegans who neglect fruits and vegetables are likely to also develop health problems. Equally, both vegan and non-vegan meals can constitute healthy diets.

Consuming animal products helps to ensure individuals receive all the nutrients they require. When reasonable amounts of meat, dairy, and eggs are consumed in conjunction with fruits, vegetables, and grains, the person is almost certain to be receiving adequate nutrition. Individuals seeking to remove meat, dairy, and eggs from their diet can do so and still be healthy, but they must pay greater attention to the nutrient content of their food.

Summary

In this chapter we have looked at the pig, the layer, the broiler, the dairy cow, and the beef cow. The production of each animal has been discussed, with particular attention to the features that are favorable and unfavorable for good farm animal welfare. There is a wide literature available on the issue. What makes our book different is that we do not have an agenda to promote or change current livestock agriculture. Our ultimate objective is to inform the consumer.

In this chapter, we tended to focus primarily on the everyday life of farm animals. Animal advocacy groups will often mention a myriad other issues such as the transportation and slaughter of livestock. These issues, while important, are temporary experiences for the animal. We sought to describe the everyday life of farm animals, not the single worst days. Animal advocacy groups, such as the HSUS, have produced a series of white papers documenting every potential problem with raising animals for food. These white papers do not contain myths or lies; their arguments are carefully supported by scientific studies. What is potentially misleading, and what we attempt to rectify, is that the negative facets of livestock production are not described alongside the positive facets. There are some advantages to every livestock production system and there are times when livestock do not suffer and are very likely to be happy in each of the systems. What is missing from the arguments by animal advocacy groups is the context. Activists deliberately disregard the context in which animals experience pain or are caged. Without context, discovering truth is impossible.

To illustrate why this context is important, we could easily document all sorts of abuses that occur from the existence of humans. Humans commit environmental pollution, they abuse kids physically and sexually, they murder, lie, cheat, and the list goes on. Despite all this, most of us feel humans should continue existing. Similarly, the mere fact that farmed animals suffer sometimes is not sufficient reason to conclude that animals should no longer be farmed.

Knowing how to raise farm animals is no easy task and sometimes our intuition is wrong. Farmers often grow frustrated by attempts to regulate their activities by people who know very little about farm animals. Animal advocacy groups can sometimes be misguided. Mike Rowe hosts the popular TV show *Dirty Jobs*. In one episode he visited a sheep farm and the day's activities included castration. The farmer castrated the sheep by cutting the testicle sheath with a knife and ripping the testicles out by hand. Before the visit, Rowe had contacted PETA asking which castration method was most humane. PETA recommended the band method, in which a tight rubber band is placed around the scrotal sack. After a few days, testicles eventually fall off the body. Rowe asked the farmer if he could use the more "humane" method. He did, and the results were obvious: the sheep experienced more temporary as well as permanent pain using PETA's method.[150]

Farmers are sometimes misguided too. We have had farmers confidently and honestly assert that hogs and hens could not be happier than they are in gestation crates and battery cages. Some have become so desensitized to the animal suffering and the daily death that occurs on farms that they forget animals are sentient creatures. We have little doubt that many farm animals, especially pigs and layers, can be provided with a better life.

Farmers are, in a sense, caught between a rock and a hard place because they cannot provide animals a substantively better life without losing money. Rather than pretending or deluding themselves that the animals cannot be made better off, farmers should acknowledge the simple truth that many farm animals can be made better off, but that consumers must be prepared to pay higher prices in return. In a sense, it is not the farmer's choice but the consumer's. Animal rights groups should also acknowledge that there is a limit to the level of well-being that can be provided. Animals are farmed for only one reason—and that is because humans want to consume their meat, milk, or eggs. If food products are not affordable there will be many fewer animals, and although some animal rights activist would no doubt rejoice at this outcome, the animals themselves might prefer to continue living. Thus, in many ways, it is also in the animals' interests for food to be affordable.

CHAPTER 6

Talking with Philosophers

How Philosophers Discuss Farm Animal Welfare

"One could not stand and watch very long without becoming philosophical... Was it permitted to believe that there was nowhere upon the earth, or above the earth, a heaven for hogs where they were requited for all this suffering? Each one of these hogs was a separate creature. Some were white hogs, some were black; some were brown, some were spotted; some were old, some were young; some were long and lean, some were monstrous. And each of them had an individuality of his own, a will of his own, a hope and a heart's desire; each was full of self-confidence, of self-importance, and a sense of dignity... a black shadow hung over him an a horrid fate waited in his pathway. Now suddenly it had swooped upon him, and seized him by the leg. Relentless, remorseless, it was; all his protests, his screams, were nothing to it—it did its cruel will with him, as if his wishes, his feelings, had simply no existence at all; it cut his throat and watched him gasp out his life. And now was one to believe that there was nowhere a god of hogs, to whom this hog-personality was precious, to whom these hog-squeals and agonies had a meaning? Who would take this hog into his arms and comfort him, reward him for his work well done and show him the meaning of his sacrifice?"

Upton Sinclair in *The Jungle* (1906)

The intellectual foundations of the animal advocacy movement can be traced to the writings of philosophers and ethicists (hereafter, moral philosophers). The most prominent and influential writings on animal welfare and animal rights have come from moral philosophers, and as such, any serious analysis of the topic of animal welfare requires a look at how they think about the issue.

Moral Instincts

To illustrate some of the challenges involved in ethical reasoning, consider the following scenario.

> Julie and Mark are sister and brother, traveling through Oregon on Spring Break. One night during the trip, Julie and Mark decide to have sex. Being responsible, they used birth control pills and a condom. The sex was enjoyable, but they decided it would be the last time they made love. They kept their night of passion a secret for the rest of their lives, and it never complicated their relationship. If anything, they became closer.[1]

In all likelihood, you believe Julie and Mark's sexual encounter constitutes an immoral act. You are not alone in your beliefs: sibling fornication is frowned upon by virtually every human culture. But, why was Julie and Mark's act immoral? Intimate relations between siblings are typically discouraged by societies to prevent inbreeding, but in this case Julie and Mark used ample birth control. Their actions did not hurt anyone, and both could be said to have benefited from the pleasure of the experience.

What is the *logical* reason you believe their actions to be immoral? Most of us cannot say, logically, why the action is immoral, but we continue to maintain that it is indeed immoral. The inability to justify a judgment, however, need not imply that you are unsophisticated, and many times there are good reasons for gut feelings. Indeed, research shows that most moral judgments are made in this way—with our intuition.[2] Morality is a system of beliefs about how people should behave; it is a system of norms that guide and aid human relations in an often harsh world.

Morality and norms are transmitted from one generation to another, and are often anything but *ad hoc*. Offspring from two siblings will tend to suffer from the ills of inbreeding, which implies both that people attracted to their siblings will have offspring less likely to survive and pass along their genetic predispositions, and that people who witness the results of sibling mating will learn of its consequences and will want to pass along this knowledge to others. As a result, nature and culture have instilled in us a strong instinct to breed only with those outside our immediate family, and conversely we have an instinctive disgust for sibling intimacy. These instincts emerged and withstood the test of time for good reasons.

In our story, Julie and Mark used contraception, which seems to nullify the justification for the instinctive disgust. Social norms are general rules that often help us make prudent choices, but sometimes those general rules do not apply to a specific situation. In the Bible, Jesus may have advised us to turn the other cheek, but it is doubtful he wanted us to allow someone to continually pound our face for no reason.

Our moral intuitions extend beyond sex and include intricate perceptions about proper relationships among fellow humans. When we are cheated, we desire to exact revenge even when it is costly. This desire is predictable and keeps people from cheating one another in the first place. Likewise, we often desire to reciprocate acts of kindness because we feel it is the right thing to do, and these motivations make us act more kindly that we might otherwise.

Intuitions about the proper treatment of animals often run counter to the arguments made by philosophers. For example, when attempting to address the philosophical arguments of treating human and animal suffering equally, Richard Posner, a prominent lawyer and economist who serves as a judge on the US Court of Appeals, has argued that a human infant is superior to a dog, "is a moral intuition deeper than any reason that could be given for it and impervious to any reason that anyone could give against it."[3] He goes on to say,

> I do not claim that our preferring human beings to other animals is "justified"
> in some rational sense—only that it is a fact deeply rooted in our current
> thinking and feeling, a fact based on beliefs that can change but not a fact that
> can be shaken by philosophy. I particularly do not claim that we are rationally
> justified in giving preference to the suffering of humans just because it is
> humans who are suffering. It is because we are humans that we put humans
> first. If we were cats, we would put cats first, regardless of what philosophers
> might tell us. Reason doesn't enter.[4]

Moral intuitions can often make us impervious to the logic of philosophical argumentation, and in regard to animal well-being we often have conflicting emotions. Humans have developed a sophisticated system of empathy that prompts us to feel pity for other humans in pain—empathy which extends to animals as well. Other emotions prompt us to naturally view animal as prey and food. Indeed, for humans living 100,000 years ago all animals were prey. It was the eating of meat, especially cooked meat, that helped provided our ancestors with the rich nutrients they needed to form large and sophisticated brains.[5] Ironically, our highly developed brain that allows us to imagine and empathize with animal suffering is ultimately due to the fact that our ancestors hunted down animals and consumed their highly nutritious meat.[6]

Of course, the fact that our moral intuitions favor certain actions does not alone certify that those actions are moral. The simple fact that humans evolved as omnivores does not imply that it is ethically correct to remain omnivorous in the modern world. Moral intuitions are not important because they dictate what is moral and amoral, they are important because they demonstrate that ethics is a complex issue. Morality involves a mixture of logic and emotion, and this mixture can result in an infinite combination of the two, which is why moral debates are rarely settled.

Although we have argued that intuitions drive much of what we believe to be morally and ethically correct, and are often good guides in doing so, it is useful to take a step back and re-evaluate the beliefs that canopy our moral intuitions. To arrive at an informed opinion of how we should treat farm animals, it is important to consider the subject carefully and review evidence and ideas with an open mind; however, we cannot expect clear-cut answers. Logic is a tool, not the answer. It is hard to imagine that most vegetarians eschewed meat consumption based solely on the logic of a given philosophy. Instead, many vegetarians intuitively felt meat eating was wrong—intuitions which were later justified through moral philosophy. Of course, the same can be said for those who argue against animal rights. As Richard Posner put it, "Philosophy follows moral change; it does not cause it."[7] Similarly, Oliver Wendell Holmes, a US Supreme Court Judge from 1902 to 1932, argued that the logic dictated by the judges are formed *after* their decision, not before it. Judges, he argued, form their decisions based on experience, impression, culture, and in some cases self-interest.[8] The true reason for a decision cannot really be explained because it is based on innate feelings that have no language for expression, so when handing their verdict judges devise explicit, logical reasons for the decision. The justification given for a verdict is used to justify, not form, the decision. Modern research has confirmed Holmes' conjecture by showing that judges' decisions do not really emanate from the justifications that they give.[9] Such is the case for many life decisions.

In these matters relating to farm animal welfare and whether or not we should continue to raise and eat them or what they produce we must draw our own conclusions. We can do our best to use logical argumentation to arrive at our decisions and we can look at the evidence. However, as the Nobel Prize winning economist, Friedrich Hayek, once stated, "if we stopped doing everything for which we do not know the reason, or for which we cannot provide a justification . . . we would probably soon be dead." Nevertheless, ethical philosophy can be helpful in forming and clarifying one's opinion, especially in this case, where instinctive morals provide conflicting guidance. If we are to assert that our decisions are informed, we must listen and consider the reasons put forth by animals rights activists and their detractors. Aristotle once said, "It is the mark of an educated mind to be able to entertain a thought without accepting it." Similarly, to form an educated opinion about how farm animals should be treated, we must be willing to entertain the ideas of those on all sides of the issue, regardless of our initial presuppositions.

The Ethics of Farm Animal Welfare

Having argued in Chapter 4 that animals have the ability to feel pain and pleasure, we now enter the arena of moral philosophy.

You = 11,500 Sheep

Reason magazine once published an article entitled, "You = 11,500 Sheep." The title was derived from the results of a nationwide telephone survey we conducted with over 1000 US households who were asked questions about farm animal welfare. In the survey, we asked people to respond to the following statement.

> *If a technology were created that could eliminate the suffering of 1 human or X farm animals, it should be used to eliminate the suffering of the 1 human.*

We varied the number, X, that we chose to insert into the question, across respondents. The numbers we randomly selected from were: 1, 10, 50, 100, 500, 1000, 5000, or 10,000. Respondents were asked whether they agreed or disagreed with the statement.[10] Had you been a participant in the survey, for example, you might have been asked the question at a value of X = 5000 animals. Would you agree with the statement, believing the suffering of one human is more important than the suffering of 5000 farm animals, or would you disagree?

It turns out that when X was only 1 farm animal, almost 90 percent of respondents agreed with the statement and chose to eliminate the suffering of one human instead of the one farm animal. However, as X increased the number of people agreeing with the statement fell. In fact, when X was equal to 10,000 farm animals, a little more than 50 percent of respondents agreed with the statement and chose to eliminate the suffering of one human instead of the farm animals. That roughly 50 percent agreed and 50 percent disagreed when X was 10,000 implies that any person selected at random is likely to be roughly indifferent to agreeing or disagreeing to the statement. This led us to conclude that the "average person" in our sample equated the suffering of one human with suffering of roughly 10,000 farm animals. Using this line of logic and a bit more math yields the more precise result that people, on average, believe the suffering of 11,500 farm animals equals the suffering of one human.

The article summarizing our findings was published online and several readers posted comments. While many comments were no doubt intended to be humorous, people's reactions to the story provide insight into how people think about the ethics of animal treatment. Consider just three of the comments posted online.[11]

> "I'd be curious to see how the numbers would skew if the question were rephrased substituting one's family members instead of just some random stranger."
> "I would personally strangle every chimp on earth, with my bare hands, to save one street junkie with AIDS."

"How many persons of low IQ should suffer to prevent a person of high IQ from suffering?"

The first comment pondered how people balance the suffering of family members versus strangers. It is instinctive and natural for humans to value some humans above other humans, and then humans in general above animals. Although we like to assert the equality of all humans, we are far more likely to donate a kidney to a family member than to a stranger. All humans may be equal in an abstract sense, but in our practical day-to-day lives we all value family and friends more than strangers. The online comment makes the point that our ability to sympathize with someone or something depends on our similarity and proximity to that person. Historically, humans lived in small clans where each individual member's well-being depended crucially upon the cooperation of their fellow clansmen, while at the same time being harmed by competition with other clans. And at that time, humans hunted and ate animals, so it is no surprise if we instinctively value the life of a human much more than that of an animal.

Most people value family members over strangers and strangers over animals. Furthermore, all animals are not considered to be equal. For example, for most people, dogs are preferred to cattle and cattle are preferred to snakes. Dogs, it seems, are a particularly valuable animal to people in many Western cultures. In fact, 75 percent of dog owners consider their pet as a child or family member.[12]

Nonetheless, many of the same people who bestow family status on their dogs will still happily eat from the meat of hogs, which are just as smart and capable of suffering as dogs. Is such behavior inconsistent? Philosophically speaking, if it is not ethical to eat dogs, should it be unethical to eat hogs? Most of the time people are not actively philosophizing, but rather they are relying on social customs, norms, and instincts to guide their decisions. Moreover, when one appeals to the notion of self-interest, the reasoning becomes clear. Dogs are often given human status because dog owners get something in return: a dog's love and affection. Hogs often acquire a lower status because what humans get from them is something quite different: food. When predicting human behavior, assuming that people act in their own best interests is a better bet than assuming people abide by a consistent moral philosophy. People, in general, love their pets and they love to eat meat; the fact that humans place a high value on their dog's life and a low value on a hog's life makes perfect sense, in so far as satisfying atavistic desires.

Animal rights activists spend considerable time and effort to expose the inconsistency in how people treat the suffering of humans and the suffering of different animals. As demonstrated by our survey results, most people would seek to avoid the suffering of one human over that of thousands of farm animals. However, for some people, the suffering of one human is of equal importance as the suffering of one animal. Jerry Vlasak, from the North

American Liberation Press Office, for example, has publicly stated, "I don't think you'd have to kill—assassinate—too many...I think for 5 lives, 10 lives, 15 human lives, we could save a million, 2 million, 10 million non-human lives."[13] This leads us on to consider the next issue, that of speciesism.

Speciesism: A Man is a Woman is a Chicken

Animal rights activists have pointed out that our moral intuitions to treat animal suffering as less significant than human suffering bear an uncanny similarity to that of, for example, some men who treat women as being of lesser value than men, and some white people who treat black people as being lesser than white people. The term *speciesism* (or specism) was coined in the 1960s in discussions about acts of preferring one species to another.

Peter Singer's book *Animal Liberation* is thought to have laid the philosophical foundation for modern animal rights activism. Singer's arguments as we understand them, are as follows. Men and women are different, and that is of no doubt. Consider any objective measurement, whether it be reading skills, math skills, ability to see color, or talent for sports, and you will find, he says, that males and females will not be equal. Yet our society, for the most part, considers men and women as equals. Specifically, men and women receive equal consideration in voting, in court, and in their inalienable right to life, liberty, and the pursuit of happiness.

Consider the third comment to the *Reason* article stated in the previous statement, in which it was asked, "How many persons of low IQ should suffer to prevent a person of high IQ from suffering?" We can only assume that the question was asked in jest as few would believe that people of different intelligence levels should be given different consideration. As Thomas Jefferson stated, "Because Sir Isaac Newton was superior to others in understanding, he was not therefore lord of the property or persons of others."[14] Using a similar line of reasoning, Peter Singer asked, "If possessing a higher degree of intelligence does not entitle one human to use another for his or her own ends, how can it entitle humans to exploit non-humans for the same purpose?" After all, a cow is likely more intelligent than a newborn human.

Why is it, then, that we believe it proper to give all humans equal rights and equal consideration? If men and women are really different in so many ways, what is it that they share in common that compels us to treat both sexes equally? Singer argues that it is largely the equal capacity to feel and suffer. Moreover, Singer asserts that suffering is what should matter in a moral philosophy. If one wants to behave morally, he argues, one should behave in a manner that minimizes the amount of suffering in the world and maximizes the amount of happiness. Because 81 percent of Americans believe farm animals have the same ability to feel pain and discomfort as humans,[15] should we not then grant animals equal consideration as humans? As the philosopher Bentham once stated about animals, "The question is not, Can they reason? nor Can they talk? but, Can they suffer?"[16]

Singer's argument is that because both humans and farm animals can suffer, and suffering is bad, the suffering of farm animals should receive equal consideration as the suffering of humans. Suffering is suffering regardless of who or what is experiencing it. Singer is not saying humans and animals should be treated equally—only that their *suffering* demands equal consideration. Furthermore, Singer's philosophy suggests that if you are going to make efforts to make people happy by giving them what brings them happiness, you should also give hogs, chickens, and cattle whatever makes them happy. So while humans value the possibility to go to college, for example, farm animals would value pastures with plenty of room, shelter, and so on. The commonality between sexes is the commonality between species—suffering and pleasure, misery and merriment. If suffering can be measured and described in units, a unit of suffering by a rat is the same as a unit of suffering by a human—so Singer argues. In light of this argument, someone who thinks one person's suffering is equal in importance to the suffering of 11,500 farm animals would, according to Singer, be a speciesist. Singer posits that the well-being of one person should count only a little more than the well-being of one farm animal.

Utilitarianism

The next time you approach the meat counter in the grocery store to purchase a whole chicken to prepare for dinner, you might pause and ask yourself a question: *should* I eat it? By using the word *should* there is an implication that this is a moral decision—that there is a right and wrong choice. You know you like the taste of chicken and that the price is easily affordable; the only thing that could possibly stand in your way is morality. What harm is done, what ethical code is broken by eating the chicken?

It might be tempting to say that no harm can be done by your purchase because the chicken is already dead. However, one person deciding not to buy just one chicken automatically means that that chicken has become harder to sell, even if by a little. When chickens becomes more difficult to sell, the price grocery stores are willing to pay to the chicken processors falls and ultimately the price the farmer receives falls, and as a result, fewer chickens will be produced. This outcome is not an instantaneous reaction to the decision not to purchase a chicken, but there is an effect—even if very small—that eventually trickles back to the farm. Over a long time horizon, the quantity of meat produced must equal the quantity of meat consumed, and the mechanism that ensures that these two are equal is the price. So our individual food purchases do matter.

Back to the question: is it *moral* to eat a chicken? One way to answer this question is to employ the philosophy of utilitarianism, which asserts that the ultimate goal of a moral person is to maximize the total amount of happiness in the world, which equals the sum (or perhaps the average depending on the type of utilitarianism advocated) of the happiness experienced by each body capable of experiencing happiness.[17] An action is deemed morally justified if it increases aggregate happiness. Note that although we use the word happiness, what is really implied is well-being. Utilitarianism is often summarized by the phrase, "the greatest good for the greatest number." In most cases utilitarianism is only extended to consideration of people, so that the moral objective is to maximize the happiness of all humans. Yet the first philosopher to make utilitarianism a school of thought, Jeremy Bentham, was quick to include animals as well. The idea of including animals in the utilitarianism equation has been popularized by Peter Singer, causing adherents to the utilitarianism school of thought to argue, "In order to justify eating animals, we would have to show that the pleasure gained from consuming them, minus the pleasure gained from eating a vegetarian meal, is greater than the pain caused by eating animals."[18]

To summarize, utilitarianism is the idea that the morality of an action is judged by its consequences, and in particular, that the sum of "bad" experiences by all people, as a result of the action, should be subtracted by the sum of the "good" experienced by all people. If the mathematical result is positive, the action is deemed ethical. If the result is negative, the action is deemed unethical. At this point, we should note that utilitarianism is *one* moral philosophy. There are many other ways of thinking about the morality of a situation. Moreover, although there are logical reasons for doing so, one need not include animal feelings in utilitarian calculations. You can implement whatever form of utilitarianism you want—and indeed philosophers have done just that. There are average and total utilitarians, negative and positive utilitarians, act and rule utilitarians. We will not delve into all these differences, as it would be akin to trying to describe the differences between Anglicans, Methodists, Lutherans, and Baptists—and the debates are just as heated! The point here is that although utilitarianism is a widely adhered to philosophy, there is diversity in how it is applied in any given situation.

Let us accept the utilitarian premise that the morality of an action is judged by its effect on the total happiness of all people *and* animals. Given the assumption that animals can suffer, including animal happiness in a utilitarian calculation has some appeal, but note that just because we take a chicken's happiness into account does not necessarily imply that a chicken counts as much as a human. The utilitarian goal is to maximize a mathematical equation with many "+" and "−" signs, where the "+" and "−" accounts for the happiness and suffering realized by the population of humans and animals. If drinking milk makes people happy, then the equation contains a "+" sign for each unit of happiness each person receives drinking milk. If dairy cows live happy lives then the equation contains even more "+" signs, but if dairy cows suffer and would be better off had they not been born, the equation contains "−" signs for

Frank eats a chicken raised on a factory farm.
Consequently, the change in total happiness (in the
long-run) in the world is ...

Change in Frank's happiness	Change in chicken's happiness
Case 1: +20	*Case 1*: −15
Case 2: +10	*Case 2*: −15
Case 3: +10	*Case 3*: −5
Case 4: +10	*Case 4*: +5

Figure 6.1 Demonstrating utilitarianism

each cow that suffers. However, while utilitarianism seems quite simple in theory, we will see it is not so simple in practice.

Going back to the chicken example, suppose someone called Frank is considering whether he should eat a chicken. For argument's sake, let us suppose that for every extra chicken Frank eats, an extra chicken is born, raised, and harvested.[19] Thus, in a sense, Frank's decision of whether to eat the chicken is a decision of whether the chicken is even born at all. In Figure 6.1, we depict four different scenarios to illustrate the utilitarian philosophy. In the first three cases, we see scenarios where Frank receives pleasure from eating the chicken and the chicken leads a miserable life. From the point of view of the chicken, its suffering outweighs any pleasure it experiences being alive: the chicken would be better off not born. The last case considered describes a situation where the life of the chicken is overall pleasant, and its life is worth living even if it is raised solely for the purpose of being harvested for food.

Case 1

Frank enjoys eating chicken. By eating chicken Frank's life improves +20 happiness units—let's call these units *utils*. A *util* is an abstract concept used to measure the extent to which an action causes more or less pleasure or more or less suffering. For example, most would agree that placing your hand in a fire causes greater suffering than placing your hand on an ice cube. This is like saying a hand in the fire produces −100 *utils* and a hand on an ice cube produces −5 *utils*.

Frank benefits by 20 *utils* from eating the chicken. The chicken, however, is assumed to live a horrible life. Whatever pleasures the chicken experienced on the farm are outweighed by the suffering it experienced. If the chicken could contemplate and commit suicide, it would do so. When summing all the pleasures and suffering, the chicken's life generates −15 *utils*. This does not mean the chicken never enjoyed anything. The chicken might have received, say, +30 *utils* from having readily available food and water and protection from rain and predators. However, the chicken had its beak trimmed, which caused it great pain. It was constantly chased and pecked by other bully birds. It never saw the sunshine except on the drive to the slaughtering plant. In the later weeks of life it was crammed in tightly with other chickens. Experiences in the slaughtering plant were certainly not pleasant. The sum of these bad experiences equals, say, −45 *utils*. Adding the up the good (+30) and bad (−45) in life yields a calculation of −15 utils.

Frank benefits from eating the bird, but the chicken suffers. Utilitarianism requires us to sum the *utils* over all people and animals, and when we do so, the total happiness in the world equals Frank's happiness of 20 *utils* plus the chicken's happiness of −15 *utils*, resulting in a sum-total happiness of 5 *utils* resulting from Frank's action of eating the chicken. All animals and humans considered, according to utilitarianism, the world is made a better place by Frank consuming the chicken. Consuming chicken, then, is the moral thing to do—even if the chicken suffers.

Case 2

Consider the second scenario depicted in Figure 6.1. In this case, the chicken suffers just as much as it did in Case 1, but now Frank's happiness from consuming the bird has fallen. Frank may enjoy eating chicken less for a host of reasons. For example, the chicken may be less healthy or nutritious, or perhaps better substitutes have been developed. This last point is important as one should keep in mind that Frank's *change* in happiness from eating the chicken is calculated relative to Frank's next-best eating alternative (i.e., how his happiness would have changed had he eaten something else). The introduction of meat substitutes, which are increasingly tasty, serves to decrease the benefits from meat consumption, not because meat is less tasty, but because its substitutes are tastier. Another possibility is that Frank learned that the chicken lives a bad life, and because Frank is empathetic, he becomes less happy to eat a chicken that has been unhappy.

Although Frank's happiness from chicken consumption declines in Case 2 relative to Case 1, the suffering experienced by the chicken is unchanged. Comparing Frank's benefit to the chicken's cost, we conclude that total happiness in the world is reduced by 5 *utils* (Frank's benefit of +10 plus the chicken's suffering of −15 = −5). In this situation, a utilitarian must conclude that it is immoral for Frank to consume the chicken.

As Case 2 illustrates, the introduction of meat and dairy substitutes can change meat consumption from being seen to be a moral to an immoral act. Under the utilitarianism philosophy, what constitutes an immoral act changes according to technology. Likewise, if animal activists are effective in convincing people that farm animals are unhappy, *people* are also likely to become less happy with animal production. Although we are not accustomed to believing that the morality of an action depends on scientific technologies or on changes in social norms and people's preferences, the contrast between Case 1 and Case 2 shows it can—at least from the utilitarian perspective.

The past two decades have ushered in a number of interesting meat and dairy substitutes. Grocery stores regularly carry milk-like beverages made from soybeans and rice alongside real milk. US sales of vegetarian substitutes for milk, meat, egg, cheese, and other animal products increased 45 percent between 2000 and 2005. Sales of meat and dairy substitute products reached sales of $1.2 billion in 2005 and were projected to hit $1.7 billion by 2010.[20] Although the increased consumption of such products is attributable to a number of factors, let there be no mistake that one of the reasons is that food manufacturers have figured out how to make such products much tastier. As vegetarian alternatives become closer substitutes for meat and dairy products, the benefits we receive from consuming animal products, instead of their next best alternatives, falls. As this benefit falls the consequent calculus demanded of utilitarianism makes consuming animal products less ethical—assuming the animal is better off dead.

Even more fascinating is the research into in-vitro meat; this is meat grown in the laboratory from cells.[21] Simply take a few cells from an animal, place it in a nutrient-rich medium, and watch it grow like mold! In-vitro meat is not connected to a brain, and thus there is no pain or pleasure. If no pain or pleasure is generated then it doesn't enter into the utilitarian calculation. The possibility of tasty and healthy in-vitro meat would revolutionize all the arguments against the morality of eating meat. To encourage in-vitro meat development, PETA has offered a $1 million prize to whomever can grow the first in-vitro chicken meat available for public consumption by June 30, 2012.[22]

Of course we should not forget that meat producers are continually working to make meat tastier, more consistent in quality, and cheaper. Indeed, the drastic rise in chicken consumption over the past 40 years is largely attributable to improvements in quality and consistency as well as improved understanding of the relative health benefits of eating chicken instead of beef. Such changes would increase the number of *utils* derived from eating chicken.

Case 3

In the previous scenario, Frank received some pleasure from consuming the chicken, but the pleasure was not sufficient to cancel out the suffering the chicken experienced, and the utilitarian view concluded that eating the chicken was unethical. Many people may subscribe to this particular view. They salivate at the thought of a pork chop, but hate to think of how the hog was raised, and resolve to eat a salad instead. There are likely many others that contend Case 2 is not realistic because it is a mistake to expect that any pleasure humans receive from meat consumption to outweigh the horrible suffering of some farm animals.

The Humane Society of the United States (HSUS) and other groups have sought to reduce what they perceive to be suffering on farms. One issue aggressively pursued, as we discussed earlier, is the banning of gestation crates for sows in states like Florida, Oregon, Arizona, Colorado, and California through public referenda, legislative action, and activism. The activities of the HSUS are not just carried out by vegetarians. In California, for example, over 60 percent of voters voted to ban gestation crates in pork production and cages in egg production, and in that state fewer than 3 percent of voters are vegetarians. There is clear evidence that many people who want to eat meat and eggs still do not want the animals to suffer.

Given this context, consider the outcomes in Figure 6.1 as we move from Case 2 to Case 3. That is, Case 3 might mimic a policy that requires a small improvement in space per chicken but does not mandate any other changes. In such a case, chicken well-being improves from −15 to now only −5, *but* the bird continues to lead an undesirable life. Performing the utilitarianism calculus, the benefits of eating chicken again outweigh the suffering experienced, and it is once again considered ethical to eat meat.

The movement from Case 2 to Case 3 is what concerns Rutgers law professor Gary Francione, who has become known for his efforts in developing a project called The Abolitionist Approach. Although he is an animal advocate, Francione is *against* the efforts of groups like HSUS and PETA who seek to improve farm animal welfare by, for example, banning gestation crates. The bans improve the lives of farm animals, but Francione insists that the animals' lives are so poor it would be better off if they never existed. That the elimination of a gestation crate means a sow is being treated somewhat more humanely, alters the utilitarian calculations (as demonstrated by the comparison of Case 2 to Case 3), and thus induces more consumers to consume pork. As a result, Francione argues that efforts to marginally improve the lives of farm animals might increase the number of people who believe that consuming the animal food product is ethical. The question now becomes: *does the reduction in animal suffering outweigh the new suffering introduced by an increased demand for the product?* No one really knows the answer to this question, but Francione's guess is that the end result is more suffering. As a result, when HSUS advocated Proposition 2 in California, which would force farms to adopt (presumably) more humane methods, Francione did not support the measure. An extreme take on Francione's view is that if an animal is to be raised for food, it should be raised in the most horrid, inhumane environment possible. The hope is that the tremendous suffering caused by the miserable animals will shock some consumers, inducing them to become vegan; the vegan-conversion then prevents the existence of a number of animals who would also have lived miserable lives. There is little difference between opposing measures that would improve animals' lives and supporting measures that cause animals to suffer. Francione's bet is that the suffering of animals today will pay dividends in terms of saved suffering in the long run.

Case 4

Many farm animals live in good conditions and lead lives that—at least comparable to wild animals—are undeniably pleasant. That, at least, is our opinion about, for example, pigs raised under the welfare guidelines compiled by the Animal Welfare Institute. Hogs living in such systems are never caged for long periods of time and always have access to outdoors, including pasture. They are given plentiful space and measures are taken to prevent fighting among pigs. It is hard to imagine an unmet need or missing amenity that hogs in such systems do not have. This is the case depicted by Case 4 in Figure 6.1.

Going back to Frank, in the Case 4 scenario, he enjoys eating the meal and the chicken had a good life. Everyone wins! No doubt, some may claim that if Frank really cared about the chicken he would not eat it. But that is not the issue addressed by utilitarianism. In our example—and in real life—the chicken would never have existed if Frank did not want to eat it. And according to utilitarianism applied to Case 4, Frank is morally justified in eating the chicken.

The Limits of Utilitarianism

Utilitarianism is an appealing philosophy and on some level almost everyone practices some limited form of utilitarianism. Indeed, when considering our own individual actions we almost always perform some sort of utilitarian calculation and decide whether to undertake an action based on whether it will, on net consideration and in the long run, produce for us a good or bad outcome. Another beneficial feature of utilitarianism is that it forces attention on the consequence of actions. Although we may choose not to undertake an action even when utilitarianism suggests we should, we would do well to remember that the consequences of any act are real. The troubles with utilitarianism start when we begin extending the concept beyond the boundaries of our individual actions; we can arrive at some counter-intuitive and unappealing conclusions about whether an action is moral.

To illustrate, recall our previous discussion of in-vitro meat. Should Frank eat in-vitro or natural meat? Assuming the price and eating experience of both types of meats are identical, a utilitarian calculation suggests Frank should eat the in-vitro meat in Cases 1, 2, and 3 because Frank receives the same *utils* he did before and because now there is no animal suffering. So far, so good. But now consider Case 4. In this case, utilitarianism would suggest that it is now immoral to eat in-vitro meat because it would imply the non-existence of a farm animal that lived an overall good life. Frank should eat regular meat just to give the animal a meaningful life—if Frank is a moral utilitarian. In fact, even if in-vitro meat is a little cheaper, the positive *utils* experienced by the farm animal may outweigh the costs imposed on Frank by paying more for meat, and Frank may be morally obliged to eat the natural meat anyway. Even if the animal was slaughtered, even if it was castrated without anesthetic, and even if it experienced a hot and cramped ride to the slaughter house, so long as the positive emotions the farm animal experiences during its life outweigh the negative, eating natural instead of the in-vitro meat might be the ethical choice for a utilitarian.

If this conclusion comes across as a bit awkward, you are beginning to understand the drawbacks of utilitarianism. Utilitarianism, carried to its logical extremes, can arrive at conclusions that just do not "feel" right. Suppose that hogs are indeed truly happy animals when raised for meat. This implies that utilitarian vegetarians should buy pork, even just to throw it away! By vegetarians refusing to use their money to buy this meat, the hogs are never brought into existence to live an overall pleasant life. Utilitarianism might require vegetarians to buy meat because the costs imposed on them would be less than the benefits to the animals.

One may respond to this criticism of utilitarianism by acknowledging that it is an imperfect moral philosophy, or by asserting that we are not using utilitarianism in conjunction with realistic assumptions. Vegetarians may assert that it is very unlikely that farm animals raised for food will be treated in a way that their life is, overall, worth living. How can one raise an animal for profit and simultaneously treat it well? It is our understanding that this is essentially Singer's position. Singer also adds that once people become accustomed to the dietary change they can be just as happy with vegan meals. Singer argues, in regard to the difficulties associated with not eating meat, that, "...these are minor human interests that we should not allow to outweigh the more major interests of non-human animals."[23] The problem with this argument is that it is sheer assertion with little grounding in empirical evidence. Our calculations suggest quite the contrary: eating meat is no minor issue for most Americans. Analyzing data on people's actual food choices reveals that the value people place on meat and dairy consumption far outweighs that for any other food group, suggesting that meat has no real substitute as a food.[24] A consistent and forthright utilitarian cannot simply discount the *utils* gained from activities one does not like.[25]

Indeed, one of the biggest problems with utilitarianism applied to animal welfare issues is that it causes us to make a judgment about whether farm animals would be better off dead. The logical conclusion of the argument that farm animals have lives that are not worth living is that we should immediately kill the more than nine billion farm animals alive in the US today (or turn them all loose to fend for themselves, assuming they would be better off wild than on the farm). Few people would be comfortable with this course of action. Also, consider the fact that it is not unreasonable to suspect that many (and perhaps most) wild animals lead uncomfortable lives—more uncomfortable than those of farm animals. Competition for food and shelter leaves many animals hungry and at the mercy of the weather. Wild animals are also vulnerable to worms, fleas, ticks, and skin problems that are largely absent on the farm. If humans should not raise livestock for food, are we also obliged to kill as many wild animals as possible to end their suffering too?

This discussion demonstrates a number of difficulties with utilitarianism, and there are even more. For example, if hogs are generally happier creatures than humans, and if the Earths' resources can support more hogs than humans, a utilitarian would have to conclude that a hog-filled earth devoid of humans may be preferable to the current situation of an the Earth containing more humans than hogs. Utilitarianism also makes no allowance for individual liberty or property rights. If I get more pleasure from robbing you than you feel sadness from being robbed, I am a morally justified utilitarian thief. We could go on with further examples, but the point has been made: the problem with utilitarianism is that we sometimes do not like its conclusions. For example, in protest against a Singer speech at the University of Minnesota, a mother of a retarded child wrote, "Peter Singer openly promotes the ideology that my son has no more value or relevance to society than a turkey or a dog."[26] If we disagree with the resulting utilitarian logic, we either need to rethink the assumptions behind our calculations, or seek a different moral philosophy.

Tom Regan, a moral philosopher from North Carolina State University, ardently disagrees with Singer's utilitarianism logic. Regan argues that under utilitarianism it *could* be ethical to consume animal products. Not able to accept the consumption of hamburgers as a moral activity, Regan dismisses utilitarianism as a moral theory which is self-evidently dysfunctional. Because consuming animal flesh is a *possible* moral activity under utilitarianism, Regan concludes utilitarianism is defunct. Rejecting utilitarianism, he must look to a moral theory which makes the raising of animals for food immoral. The easiest way to do this is to adopt a rights-based theory of morality. Regan states, "Such a result clashes with our reflective intuitions about the wrongness of killing. Killing a moral agent is so grievous a moral wrong, we think, that it can only be justified under very special circumstances... The hedonistic utilitarian's position makes killing too easy to justify."[27]

A Rights-based Theory of Morality

The rights-based theory of morality is simply a process whereby each individual must decide for themselves what a person's, or animal's, rights are. For example, many animal rights activist would include the right of animals to be free from being owned or consumed by anyone.

We are continuously exposed to rights-based philosophies. In the ongoing health care debate in the US, some people assert, for example, that Americans should receive equal health care regardless of the cost, and claim that "Americans have a right to health care." Proclaiming something as a right is at times reasonable; however, if not careful, a philosophy of "rights" can come dangerously close to arguing that something or some people have the right to something simply *"because I say so."* To be sure, philosophers intensely debate the origin of various perceived rights and attempt to justify them or disprove the justification for them. However, the idea of rights is so absolute—and so easy to use—that it can discourage intellectual engagement.

Despite these arguments, Tom Regan's rights-based approach to animal rights, which is elucidated in the book, *The Case for Animal Rights*, is anything but intellectually lazy. Regan makes the case that animals should have a special moral status by presenting a great deal of empirical evidence and through the use of logic. Regan does not just say animals should be given rights *"because I say so,"* but rather based on facts about the nature of animals and reasoning about when we choose to extend rights to other people.

In fact, rights-based morality has a firm hold on our thinking. When we earlier argued that utilitarianism could potentially justify, for example, theft, we chose to reject that conclusion because we believe that we have a "right" to our possessions. The founding fathers of the US certainly believed in the idea of rights. The US Constitution does not assert that the benefits of life, liberty, and the pursuit of happiness should be weighed against the cost: it says people have a *right* to life, liberty, and the pursuit of happiness.

The ubiquitous use of rights-based morality leads to statements being made that are seen to be beyond debate—*"we hold these truths to be self-evident."* Whether certain truths really are self-evident, however, is often the subject of debate. While the US founding fathers thought it was self-evident to extend certain rights to humans, apparently it was not so evident to them that these rights be extended to people who were not white and men. It took many years of argument, violence, and debate for it to become "self-evident" that such rights should be extended to all people, regardless of their gender or color.

Whether we give animals a given right is largely a matter of opinion, and while we can measure opinions, it is difficult to attest to their correctness—especially when competing people have competing opinions. As economists, using some form of utilitarianism comes naturally, as it provides a framework for measuring and weighing benefits and costs. That said, we are not advocating utilitarianism per se—but we are arguing that one should at least consider the costs and benefits of an action, as these are undeniable realities. Most economists adopt a particular form of utilitarianism by referring to a compensating principle that says: an action is desirable if the benefits exceed the costs, *and* if the benefits of the "winners" can (conceptually) be redistributed to the "losers" such that no individual is made worse off.

We posit that most of the arguments about farm animal welfare are debates about benefits and costs, or at least can be constructively framed as such. Saying animals have a "right" to a life free of suffering is similar to saying we strongly desire that animals live without suffering. We can measure your desire to prohibit animals from suffering (and your willingness to trade this desire against other competing desires), but the rights argument puts up a stumbling block that prevents analysis.

Again, we emphasize that just because one uses utilitarian calculations as a tool does not mean one fully accepts utilitarianism and the conclusions it suggests. For example, if we were to find that racial profiling was an activity in which the benefits exceeded costs, modern societies still oppose racial profiling because it reflects a sense of racism most modern societies reject. We *feel* people have a *right* not to be treated differently because of their skin. In truth, there are some things we just know are moral or immoral even if we cannot demonstrate exactly why. Just as people became accustomed to the idea that black people have the same rights as white people, we may one day believe that animals have some of the same rights as humans.

A Case for Speciesism?

A vast majority of the philosophers who have entered into the animal rights debate have concluded that eating animals is immoral. Animal advocates have been big proponents of the use of philosophy to justify vegetarianism and animal activism, but does this mean that all philosophical reasoning will lead to a conclusion that it is wrong to eat animals? So far meat and dairy consumers have had little need to justify their dietary preferences because they have had the weight of habit, culture, and inertia on their side. However, given the rise in animal activism, there appears a need for a more intellectual defense of meat-eating.

We begin this section by asking whether there is any good reason for being a speciest, and then, without necessarily advocating any of the positions, we will sketch out several positions which suggest that it can be morally justifiable to eat meat. After all, it is unlikely that writers espousing veganism arrived at their choice of diet through rigorous logic alone. Rather, they *felt* that eating animal products was wrong. While most ethical decisions are probably initially arrived at intuitively, trying to explain why we make certain decisions will help us digest the question further, and can sometimes lead us to change our minds. Ethical philosophies are better described as the final stories justifying our decisions, after the question has been attended to by the intuitive portions of our mind. The next section will be of interest to those people who choose to eat animal products and are looking to justify their intuitive belief that this is an acceptable practice.

Why We Are Speciesist

The typical approach taken to justify a change in the way humans treat animals is to point out that animals and humans are very much alike, so much alike in fact, that humans are animals. If humans are nothing more than animals, then only concerning ourselves with the suffering of humans makes little sense. It would be as unethical as being racist or sexist.

The underlying assumption of such a view is that there is nothing truly special about being human. The following quote from a well-known animal rights advocate is representative of the presumption: "To avoid speciesism, we must identify some objective, rational, legitimate, and nonarbitrary quality possessed by every Homo Sapiens that is possessed by no nonhuman... But none exists."[28] The mere observation of the taxonomy of animals suggests that there are obvious, discernable differences between a *Homo sapien*, Sus scrofa (pig), *Gallus gallus* (chicken), and a *Bos Taurus* (cow). Geneticists have obviously identified ways of classifying beings into different species such that all beings in any one species posses a totality of characteristics that are possessed by no other species. If this were not true, then the word *species* would have no meaningful use or definition.

The criteria used to demarcate humans from farm animals are not simple and involve a complex combination of physical, intellectual, and sentient traits. What proponents of speciesism point to is *one* area in which animals and humans are the same—the ability to feel pleasure and pain—and we are asked to ignore all the ways humans and farm animals are different. In many ways, this is a noble and logical idea. It is difficult to disagree with the assertion that animals feel pain and suffering in similar ways to humans, and that reducing this suffering is as desirable as reducing the same suffering in humans. People may disagree on whether animals can perceive pain with the same intensity as humans, but to whatever extent that they do experience pain, most would agree that pain is pain.

Given that the goal of reducing suffering and promoting happiness is desirable in both animals and humans, animal rights activists would ask why we would tolerate the caging of sentient chickens and hogs in small, barren cages for their entire lives while we would never do so with humans. They would also ask why one would raise and kill an animal for meat when the thought of cannibalism is immediately revolting. It almost seems their arguments hit a home-run when they challenge our treatment of pigs and laying hens compared to our treatment of mentally disabled people and infants, when the former, the animals, are more intelligent and perhaps more sentient than the latter.

There is an answer to these questions, and the answer is found by considering the actual decisions that humans must make, their impact on the animal community, and the outcomes for humans and animals of those decisions.

What would be the outcome if everyone adopted a vegan diet? Except for the few people who keep livestock as pets, the millions of hogs, chickens, and cattle would no longer exist. There can be no real *animal liberation*, as liberation is tantamount to virtual extinction. Would millions of hogs, chickens, and cattle prefer to not exist than to live in a world where they are raised for human consumption and profit? It is not unreasonable to suspect some of these animals would chose extinction, or, given that animals cannot understand such a question, the negative emotions are greater in number and more intense than the positive emotions they experience, such that as a whole they lead miserable lives and would be better off dead. However, it is not unreasonable to think that some of the animals would choose to live, and do experience more positive than negative emotions.

Animal rights supporters are quick to point out that we grant equal rights to minority groups and women because they are considered "equal" to the rest of the human population, and if we came to that conclusion by looking to the capacity of these minority groups and women to suffer then logic would dictate that animals should also become one of these "equals." However, granting additional rights to animals has consequences distinctly different than granting equal rights to minority group or female humans. Once human slaves were freed, for example, they were, in the main, able to adequately care for themselves—to thrive, in fact. If it became illegal to raise animals for food, the consequence would be that, except for a few animals being kept as pets, and the few animals that find a place to live in a feral state, such a law would essentially imply extinction of many animal breeds. This is quite a different outcome than occurred from freeing slaves. Whether or not it is acknowledged, the destiny of animals is under humans' control. We decide whether to preserve an animal habitat for buffalo to live in a wild state, and we decide whether to place sows in gestation crates.

With the absence of livestock labeled as property, one might argue that the habitats the livestock once inhabited, including the vast croplands reserved for growing animal foods, would now be available to wild animals. Though livestock might cease to exist, wild animals would take their place. Yet, this assumes that wild animals have pleasant lives, and this assumption is far from proven. Anyone who watches wildlife documentaries can attest to the trials of the natural world. Until an extensive study is conducted on the well-being of animals in the wild versus that of livestock, the question of whether wild creatures taking the place of livestock is desirable remains a matter of speculation.

Finally, there are a number of practical matters that cause our treatment of infants and the mentally disabled to differ. Laws that would allow the mentally disabled to possess fewer rights than those who are not disabled is a scary thought. Who gets to choose who is disabled and who is not? Would you not be scared that your unborn child might be mentally disabled, and denied some of the basic rights given to the rest of humankind?

Humans value other humans not just for what they are but what they could be. It may be true that a pig is smarter than an infant, but the infant has more potential. A human infant will grow up to perhaps take pleasure in reading Shakespeare, while the pig will continue to aspire to no greater pleasure than, say, rolling in mud. A human with transient global amnesia may have chicken-like consciousness, but amnesia is sometimes healed. The amnesia victims are valued not just for what they are but what they may someday be. A mentally disabled person may indeed be no more sentient or intelligent than a cow, and the mental disabilities almost never ameliorate. Yet, mentally ill people are *humans* and still share a common culture and set of social norms with other humans. Other humans feel a connection with the disabled because they are human. Moreover, the laws extended to the mentally disabled must be the same laws extended to the rest of society, out of fear that attempting to draw a line between the humans who do and do not have rights will eventually lead to tyranny.

Humans are not humans just because of their mind, but every other feature that arises from their unique DNA. We care more about other humans because *we* are humans. We care more about family members because they are *our* relatives. Identification with a group matters, and because our capacity to sympathize has limits, group identification serves to help ration this scarce resource. To require that we must care about every human or other sentient being would likely mean that we would meaningfully care about no one.

The human ability to create and sustain complex social societies is a result of our ability to relate and deal with fellow humans. One of the key mechanisms by which human society operates is based on a rule of reciprocity.[29] We are kind to others because we expect them to be kind to us. We avoid causing harm to others because we do not want them to cause us harm. We may not want to provide animals with similar rights to our own because we can never expect gratitude or similar treatment in return. The same could not be said of extending rights to black people, women, or other disadvantaged humans. Although some animals can display various forms of affection, it is generally beyond the ability of animals to understand the intentions behind human actions and to reciprocate. The failure of *all* farm animals to be capable of being fully reciprocal in respect of our actions is no minor matter, and it is certainly not a matter that humans would ignore in dealing with one another.

Some Philosophical Foundations for Meat-eating

We have introduced several philosophers who oppose owning livestock and pets and who abhor meat-eating. Let us now consider several philosophical lines of reasoning that might condone the practice.

First, as we have already pointed out, *utilitarianism* can justify owning animals and eating meat. The primary condition that must be met for meat-eating to be justified on utilitarian grounds is that farm animals must, when all things are considered, live lives worth living—experience more pleasure than pain. While many animal advocates argue that modern factory farms do not meet this condition, we can certainly imagine production systems that might.

A second justification is something we refer to as an *exchange-based morality*. One of the difficulties many people have with utilitarianism is that it can easily justify things like "steeling from the rich to give to the poor." Many libertarian-minded individuals espouse ethical and moral rules along the following lines: (i) Each person is an end unto him- or herself; (ii) each human determines his or her own happiness; and (iii) each human is entitled to his or her own life and to the results of his or her own labor, and no other human may infringe upon those without consent. Under this line of thinking, an action between people is moral if it results from an un-coerced, mutually agreed upon exchange. Alas, the cognitive capacities of the hog prohibit a definitive answer as to whether they are willing to engage in a voluntary exchange, but as we must do with much of philosophy, we must try and decide—on the animal's behalf—whether it would willingly enter an exchange.

Pet owners engage in an exchange with their dog. The dog is provided with comfortable housing, ample food, daily walks, and ample medical care. In return the dog gives companionship and entertainment. A pet owner and his or her dog engage in mutually beneficial and voluntary exchanges that enhance both lives. What about pigs? The pig is provided with shelter, food, water, comfortable temperature, and protection from predators. In return the pig gives the human its life—its meat. But this is hardly a voluntary exchange—did the hog engage in a trade that was of its own free will and by its own consent? It is hard to say. The hog owed its existence to the fact that people want to eat pork. Is the hog willing to accept having a short and uneventful existence for the sake of simply having life at all? Would the hog trade ample food and shelter and a certain but short life in the factory farm for the random and capricious conditions of the wild? We can never know for sure, but under certain conditions we might presume that the hog is indeed willing—that if they could say so, they would choose life in a factory farm over having no life at all, even if this means that after this meager existence they will then have to give themselves up for humans to eat their meat.

No doubt an animal rights proponent would argue that the presumption is invalid, but the activist is simply exchanging one presumption for another: that the hog would rather willingly never exist than live on a factory farm. Both positions are based on presumptions that cannot be validated empirically. The truth is this: farm animals can never be placed in a situation where their lives are solely determined by their own actions—their lives are invariably affected by the decisions of humans. Dealing with farm animals will *always* entail some degree of care from humans and human presumption about what is in their interest.

This leads us to a third philosophical justification which we refer to as the *stewardship argument*. As we have noted, some animal rights groups are even opposed to owning animals as pets. They only want animals to exist in a wild state, but nature imposes its own distinct forms of cruelty. Wild animals, though adapted to their environment, face many obstacles to receiving adequate nutrition, and most face constant pursuit as prey. In our assessment wild animals do *not* generally have a high level of well-being. In one of a series of intriguing essays an author writing under the pseudonym Alan Dawrst gives a meticulous account of all the manners in which wild animals suffer, and an assessment is made of the extent to which different wild animals suffer. Dawrst's overall assessment is that, "wild animals, taken as a whole, experience more suffering than happiness."[30]

Let us now combine this view—that wild animals suffer—with that of a thinker like Gary Fancione, who argues that it is impossible for humans to hold animals as property and for those animals not to suffer. The implication of these two views is that it is undesirable for animals to exist either as wild animals (if you hold our view that wild animals lead mostly miserable lives) or as property (if you agree with Fancione). The only way to alleviate suffering, therefore, is to destroy the institutions of livestock and pet ownership, and then engage in an all-out assault on wildlife. Only when humans remain as the last living creatures will suffering be minimized. Whether it be pigs destined for slaughter or dogs destined for doting, nature's hand apparently has no choice but to be cruel. Consider another possibility—one that is quite opposite from the view that humans are the cause of animal suffering and/or must eradicate all wildlife. *It is possible that the only way in which animals can enter this world and to experience more happiness than suffering is for them to be under the stewardship of humans.*

We end this discussion with a final line of reasoning often used to justify meat-eating—*Christian theology*. Although some might be reluctant to admit it, theology—the rational and systematic inquiry into religious questions—bears much similarity with philosophy. One thing can be sure—neither philosophy nor theology can be classified under the heading of "science," but this need not imply that neither can be used to increase knowledge.

Many of the animal rights activists go to great lengths to dismiss Christian beliefs on the nature of humans and animals. The distain for the Judeo-Christian view by many animal rights activists is, in many ways, rather transparent utilitarian because a belief in this God would undercut the premise that there is nothing particularly different about humans and animals. Once this premise is undercut, so too are the many of the arguments for animal rights.

It is commonplace for animal rights activist to focus criticism on the Judeo-Christian belief that God gave humans dominion over animals. What is often forgotten is that the Judeo-Christian doctrine asserts that not only were humans given dominion over animals, but they were also given responsibility for the care of animals. Any careful reading of the Bible makes clear the idea that humans have a moral obligation to care for those creatures with which they have been entrusted (see Proverbs 12:10 or Exodus 20:10.)

Proponents of animal rights make much of the analogy to slavery in talking about the animal–human relationship. It is instructive, then, to consider the arguments of people like Rodney Stark, who persuasively argues that slavery was ended not by humanist or enlightenment thinking, but by Christians who began to recognize that slavery was inconsistent with the underlying message of Jesus.[31] It is true that some Christian societies might have been callous in their treatment of animals, but as with slavery, Christianity provides those societies with a moral foundation for arguing for improvements in the way they treat animals.

Judeo-Christian beliefs about the relationship between humans and animals can adequately rationalize the position that it is proper to raise and use animals for human purpose, but that we should be good stewards in so doing. In a rebuttal of Posner's "human-centric" arguments for our use of animals, Singer rightfully asks, "Why then should humans incur any costs in order to reduce the suffering of farm animals?"[32] A Judeo-Christian answer might be that we should take care of our animals because we have a moral obligation to do so, not because humans are the same as animals, but because we are much more.

Who Wins?

Hopefully our brief tour through ethical philosophy has helped inform and clarify some of the thinking on animal welfare. However, we are not naive enough to believe there is a single system of thought with which we can all reach agreement regarding the proper relationship between man and animal. As the judge Richard Posner put it, "The philosophical discourse on animal rights is inherently inconclusive because there is no metric that enables utilitarianism, Romanticism, normative Darwinisim, and other possible philosophical groundings of animal rights to be commensurate and conflicts among them resolved."[33]

While we find some compelling arguments in many of the philosophies discussed, we will not promulgate a grand moral philosophy which promises to make food choices easier. For us to conclude that the views of animal rights activists should be dismissed out of hand is to deny the truth in some of their arguments. Conversely, to conclude that animal rights activists are correct while the preferences of the majority of consumers are wrong is equally unattractive. We would suggest that it is for each individual to weigh up the evidence and carefully consider the arguments provided by the philosophers as well as other experts in animal welfare and husbandry before eventually reaching his or her own conclusion.

In addition, we must always bear in mind our tendency to use intuitive reasoning in deciding ethical and moral issues. Neuroscience research such as that described in Jonah Lehrer's book, *How We Decide*, shows that when the brain makes decisions about what is ethical or unethical, the part of the brain associated with rationality is ignored. Instead, morality judgments are made the same way aesthetic judgments are made. Such findings have implications for understanding the motivations behind the arguments of those engaged in the animal welfare debate. People who oppose or support animal rights are probably not doing so based solely on intensive research and dispassionate logic. Furthermore, it is common for people to make snaps judgments about these matters relating to animal care and use, and it can take a lot for many people to ever subsequently change their mind.

We do not mean to suggest that it is impossible for people to change their minds or for them to alter their stance in response to new information, but we would argue that the formulation of individuals' morality is based on many complex activities of the mind. We will now move on to look at how morals form and evolve.

The Ethics of the Average American

We have so far mainly focused our attention on the views and ethics of experts and philosophers, but in a pluralistic and democratic society it is necessary to take into account the ethics of the general population.

Consider the following responses derived from a nationwide telephone survey we conducted with a random sample of over 1000 Americans.[34] Respondents were asked several repeated questions in which they were presented with two randomly selected issues and were asked which issue was of more concern. For example, a respondent may have been asked, "Which issue are you more concerned with, the well-being of farm animals or the financial well-being of US farmers?" The responses were used, along with some statistical modeling, to assign a "relative importance score" to seven different societal issues, including the well-being of farm animals. The results are in Table 6.1.

The importance scores can be interpreted as follows: of the seven issues in Table 6.1 respondents considered the following issues were most important: 23.95 percent said human poverty, 23.03 percent said US health care, 21.75 percent said food safety, and so on. The important point for our discussion is that Table 6.1 shows that farm animal welfare came up as the least important of the seven societal issues studied. Issues directly related to human welfare (poverty, health care, and food safety) were each more than five times more important to the respondents than the well-being of farm animals. These results re-enforce what we illustrated previously: people care far more about their own well-being than that of others, and then they care more for that of other people than they do about the well-being of farm animals. This does not mean that people do not care about farm animals or that they would not be willing to pay something to improve animals' lot in life—only that human well-being is seen as more important.

To delve into this issue more deeply and to provide additional evidence on what the public believes regarding some of the ethical issues in the farm animal welfare debate, we conducted in-person surveys with about 300 people in three US locations (Wilmington, NC, Chicago, IL, and Dallas, TX), after providing some extensive and objective information about modern agricultural production practices. The sample is not an exact or perfect representation of the US population, but it is unbiased, attentive (we paid people about $75 to ensure they were attentive), and reasonably well-informed (because of the information we gave them).[35]

Given that virtually all Americans believe it is ethical to eat meat, since about 97 percent of all Americans eat meat, the interesting question is not whether people think animals should be raised for food but rather how those animals should be treated. Participants were presented with the statements shown in Table 6.2 and were asked whether they agreed with the statement. One surprising result was that almost a third, 31 percent, believed that animals have a soul—but they justified eating meat based on the belief that God created and intended livestock to be a human food.

Only 15 percent of the respondents believed that, "God gave humans animals to use as we see fit." Only 6 percent of the sample was atheist or agnostic, and of

Table 6.1 Importance of Farm Animal Welfare Compared to Other Social Issues

Social issue	Importance score
Human poverty	23.95%
US health care system	23.03%
Food safety	21.75%
The environment	13.91%
Financial well-being of US farmers	8.16%
Food prices	5.06%
Well-being of farm animals	4.15%

Note: The numbers associated with each issue indicates its relative importance compared to other issues, where all scores are normalized to sum to 100%. The scores were taken from a telephone survey of over 1000 Americans.

Table 6.2 Percentage of 263 Randomly Selected Americans Who Agree With Each Statement

Statement	Percentage who agree with statement
Animals have a soul	31%
God gave humans animals to use as we see fit	15%
God wants humans to be good stewards of animals	64%
Humans and animals are just different products of evolution	28%
Humans are part of the biological world, not the masters of it	45%

the majority who were religious, 64 percent believed that "God wants humans to be good stewards of animals." This last finding is probably not news to the HSUS, as they have created an Animals and Religion program with resources and activities for those who want to combine a zeal for promoting animal care with serving their God. When a majority of citizens indicate that caging farm animals for their entire lives is inhumane, as was the case in the ballot initiatives in Florida, Arizona, and California, it might be said that current methods of producing eggs and pork are seen as unethical to a majority of Americans for religious reasons.

Close to one-third of the sample believed that "humans and animals are just different products of evolution," a belief that surely runs counter to the belief that humans are given dominion over animals. We conjecture that individuals who believe that animals and humans share a common origin should be more likely to reject speciesism. Finally, contrary to the predominant Christian view that the biological world exists to serve humankind, about half of Americans feel they are *part* of the biological world, not masters of it.

Table 6.3 presents the answers to a different set of questions in which respondents were presented with three statements and were asked which *one* statement best described their attitudes towards the treatment of farm animals. The answers leave little ambiguity as to the ethical beliefs of the average American, based on this study. Humans believe they have no obligation to ensure animals have a happy and contented life. While most respondents believe animals should not suffer, they state that there is a limit to their altruism towards farm animals. Chickens, hogs, and cattle do not need to be happy; they just need to be free of suffering. The most telling result is that 28 percent of the sample believed that animal feelings are not important. Compare this to the 15 percent of respondents who feel "God gave humans animals to use as we see fit." Should these percentages have been more similar? Not necessarily, as one statement is framed in a religious context and the other is not.

Table 6.3 Attitude of 263 Americans Towards How Farm Animals Should Be Treated

Statement	Percentage who say statement best describes their attitudes (they must choose one)
The feelings of animals are not important	28%
Farm animals should not suffer, but society has no obligation to make sure they are happy and content	69%
Farm animals should be guaranteed a happy and content life	1%
No response	2%

Taken together, the results suggest animal welfare is of little concern to the average American consumer. Close to one-third of our sample subscribed to the view that the feelings of animals are of *no* importance—none! This is a finding that will reappear periodically throughout this book. For instance, in a subsequent chapter, we will find that about one-third of Americans would not pay a penny more for food produced by methods that ensured higher animal well-being. Only a small minority of individuals believe animals should be provided with a happy and contented life. Roughly two-thirds believe animals should not suffer, but that humans have no obligations to animals beyond that.

In summary, Americans either do not care about the lives of farm animals or only want to seek to ensure that farm animals do not suffer. This places the farm animal welfare debate into a different light than typically seen. The issue is not whether hens and hogs are "happier" in one system or another. The major issue is whether the animal suffers in one system or another. The issue of how to determine when an animal suffers might come down to the utilitarian argument of whether a farm animal would be better off dead than alive. Although a great deal of complexity remains, these survey results are important. If farm animal welfare policy is to be set according to the wishes of the citizenry, its primary goal should be to ensure that animals do not suffer. There will be some who will vehemently argue that animals should be treated more kindly and given a happy life, not just one free of suffering. But it seems that they make up a small portion of the population, and it does not seem prudent that government policy should be dictated by a small minority—not in a Republic.

CHAPTER 7

Talking with Economists

How Economists Discuss Farm Animal Welfare

"What economics is good for, besides preventing really bad decisions and dumb policies, is providing a language and framework for thinking about complex matters in an organized and rigorous way..."

Russell Roberts in *EconTalk* (2009)

Some of the most prominent writing on animal welfare has been by philosophers, ethics specialists, and lawyers, and their stated goal is to tell readers what they *should* do and eat. As economists, our intention is different. We want to equip the reader with the information and tools necessary to *make up their own mind* about the treatment of farm animals—both as it relates to individual decisions and to the decisions we make as a society through public policy.

Economists can contribute to the animal welfare debate by articulating the economic consequences of individual and government actions. Often, policies cause outcomes contrary to the policy-maker's intentions, such as minimum wage laws that particularly affect the poor. Economists can identify situations where outcomes and intentions clash. We do not imagine the world as it should be, but rather take it as it is and ask what might happen if different courses of action are taken. This realistic take on life often causes economics to be labeled the *dismal science*.

This juxtaposition of our view-point and that of philosophers is vividly seen by turning again to Peter Singer's writings. His popular book, Writings on an Ethical Life, begins with a suggestion as to why you should read the book, or any writing by an ethical philosopher, for that matter. The thrust of the motivation is that, just as a medical doctor is better qualified to perform heart surgery than a dentist, an ethical philosopher is better qualified to make ethical decisions than a non-philosopher. Ethicists spend years scouring and contemplating the writings of ancient and contemporary ethical philosophers. Surely, an ethical philosopher who has spent years studying the farm animal welfare debate is more qualified to make decisions about what constitutes ethical food than the ordinary consumer?

"[I]t would be surprising if moral philosophers were not, in general, better suited to arrive at the right, or soundly based, moral conclusions than non-philosophers. Indeed, if this were not the case, one might wonder whether moral philosophy was worthwhile" (Singer, 2000, p. 6). Singer's assertion seems reasonable, but as we pointed out in the last chapter, even philosophers cannot always agree on the best moral outcome. We live in a diverse society and have little hope of reaching universal agreement on the ethical and moral way of treating farm animals. Philosophers play a healthy role in helping people form their preferences, but this is not economists' role. In our role as economists, we have two objectives. First, we want to understand the world *as it is* by looking at people's actual choices and their associated consequences. As scientists, we seek to understand how the world behaves using the same general methodology as chemists and physicists. Second, we use this scientific knowledge to help predict the outcome of individual and collective choices. Thus, we also view our role as economists as being something of a *consultant*. If an individual or policy-maker has a particular goal, economists can help predict whether a particular course of action will help achieve the goal and the costs entailed in doing so.

Consequently, when we decided to begin studying the farm animal welfare debate there was a tacit understanding by the authors that one of our aims would be to help people and policy-makers decide how *they* want to raise farm animals. There are ample writings seeking to dictate to readers on how farm animals *should* be raised; we want to know how *you* want to raise farm animals and illuminate the consequences of your choices. If our investigations reveal that consumers care little about the welfare of farm animals, or that the high costs of improving well-being cannot justify the improvements in animal care, this is what we will report. The point here is that the authors' preferences (as economists) about how farm animals should be treated are largely irrelevant in a discussion of the economics of farm animal welfare.

Humble, Practical, and Large

The study of economics is much the same today as it was described in 1932 by Lionel Robbins, as "the science which studies human behavior as a relationship between ends and scarce means which have alternative uses."[1] Robbins, who was once head of the economics department at the London School of Economics, packed a great deal of insight into such a brief phrase. To illustrate, consider an example.

One of the most contentious issues today is global warming, and economics is at the center of the debate. People in our society have *ends*, which is just another way of saying we have desires that include basic needs like food and shelter, wants like sports cars, and abstract goods such as religion and community. We want everything; we want to meet all our basic needs, drive a sports car, watch an opera, and still mitigate global warming. As Robbins pointed out, however, we have *scarce means* with which to achieve these ends. We must either learn to live with slightly higher temperatures and sea levels or reduce greenhouse gas emissions by undertaking actions that result in paying more for food, electricity, cars, and televisions. If we spend more money on these things, it means we have less money to spend on other things—like charities aimed at fighting global hunger. There is a trade-off between global warming and our consumption of other goods.

A central premise of economics is that people make these trade-offs for themselves. As Robbins put it, "It follows that Economics is entirely neutral between ends; that, insofar as the achievement of any end is dependent on scarce means, it is germane to the preoccupations of the economist. Economics is not concerned with ends as such."[2] This is why at times we are critical of animal welfare advocates and at other times we are critical of the livestock industry; the economic approach is a commitment to a way of thinking, not to a particular preference, in this case, relating to animal welfare. This is one of the main advantages of economics when applied to animal welfare: it provides a coherent framework to analyze situations when different people care about different ends. Claiming a "right" to a particular end (e.g., animals have a "right" not to be legal property) often halts discussion, but thinking about how people who care about different ends interact in free society is an intellectual exercise we believe to be worth pursuing.

This line of reasoning has important consequences for how we address the farm animal welfare issue, because the question, "What should we do about farm animals?" is partially up to the individual. Later in the book, we will discuss research we have conducted with hundreds of people who made real-life decisions about the treatment of farm animals. The intent is not to advance our own particular preferences for animal treatment, but rather to lend a voice to the average citizen.

It is in this sense that we consider economics to be *humble*. This humility follows from two lines of reasoning. The first is that as economists, we generally operate under the premise that each individual (not the economist) is most knowledgeable and best suited to know the best course of action for him- or herself. Even if we lived in a society with a benevolent dictator completely immune to political pressure from the right or the left, it would be virtually impossible for this dictator to obtain the information necessary to adequately plan for all consumers' desires. Second, our humility follows from a *practical* understanding that pursuing utopian dreams often produces bad results. We can envision a utopian civilization, but the world in which we actually live tends to operate by particular principles. To deny such principles is to follow the failed path of central planning which supposed that certain ends or ideals could be obtained in spite of the underlying reality of individual behavior.

Despite years of training and the natural hubris that comes with obtaining a PhD, we are skeptical of our own ability to make decisions for others, and equally skeptical of the ability of other experts to do so (especially of people who think they are qualified to make such decisions). Consequently, the economic approach leads us to be skeptical about the potential for improving people's lives by telling farmers how they should raise animals and telling consumers what they should purchase. This is especially important when it comes to food. Dictating changes in livestock agriculture can have consequences for food safety and human health. For example, a committee of "experts" might decide that all eggs should be produced in an expensive free-range system, but this decision might cause eggs prices to increase to such a level that many people will cease eating eggs. Will many people suffer nutritional deficiencies as a result? We do not know. It is this uncertainty that motivates us to include consumers' and producers' preferences when calculating effects of animal welfare policies. It is the recognition of uncertainty that makes us humble.

Economics is a science that is *large* in the sense that it must account for many considerations at once. People's decisions are not made in isolation and their decisions affect many others. People are often interested in the ethics of their personal food-purchasing decisions, asking themselves how farm animals are affected if they abstain from pork, chicken, or beef. The answer to this question typically focuses on how the individual hog, chicken, or cow is raised. The impact of a single individual's choice, however, is complex. The effect of abstaining from eating poultry on the number of animals raised for chicken meat depends on the size of the population of farm animals *and* on the supply and demand for chicken. Changes in an individual's diet have an effect on the price of, for example, chicken in the grocery store and at the farm, and the number of cows and pigs that are raised, the price of corn, the profitability of farming, and so on. Economics is *large* in that it must account for all of these complex factors. Indeed, many of the discussions in this book are meant to trace out and calculate the effects of policies and consumption choices.

But even when we do our best to calculate the consequences of an action we must always keep in mind that because economics is *large* and outcomes are *uncertain*, people often fail to see the full consequences of their decisions—most especially the decisions of government. That is, policy initiatives often have unintended consequences that are unforeseen at the time a bill is passed. Frédéric Bastiat wrote in 1848 that, "There is only one difference between a bad economist and a good one: the bad economist confines himself to the visible effect; the good economist takes into account both the effect that can be seen and those effects that must be foreseen."[3] Stated differently, the mark of a good economist, when presented with a policy designed to have some effect, is to ask, "And *then what happens?*" It should stand to reason that if we are to consider ourselves good economists, we must look for "what is not seen" when *any* policy initiative is proposed.

This is exactly what one very good economist—Dan Sumner at the University of California—did when analyzing the effects of the aforementioned Proposition 2 in California. The stated purpose of the proposition (the effect that is seen) is to prohibit the use of animal production practices that tightly confine farm animals. However, Sumner pointed out that California is a substantial net importer of eggs produced from other states; they import eggs from Iowa, Minnesota, Utah, Missouri, and Michigan among other states. Passage of Proposition 2 will serve to increase the cost of production to California egg producers, putting them at a relative cost disadvantage to out-of-state producers who do not have to comply with such regulations. Thus, the (unseen) result of the passage of Proposition 2 is that it will likely cause egg-laying hens in California to exit the business and will cause retailers to import all eggs from out of state. The end (unseen) result is that very few chickens live better lives; we have simply shifted where they live—from California to Iowa. The lesson is clear: when a policy initiative is proposed, we must look beyond good intentions and the stated purposes of the policy and ask, "and then what happens?"

Trade-offs Are Necessary

One of the most important lessons from economics is that resources are scarce and as a result, trade-offs are necessary and real. Scarcity implies an imbalance between resources on the one hand and our wants on the other. Scarcity exists because many resources are limited. For example, there are only 24 hours in a day. We cannot choose to spend all day writing a book *and* playing with our children.

Scarcity is not some abstract economic concept; everything we hold dear is scarce. Despite the fact that water is necessary for life, diamonds are more expensive than water. Why? Diamonds are scarce and water is plentiful. Like diamonds, there is only so much iron, coal, oil, and land. Through technological advancement we can often find ways to get more from these scarce resources by increasing productivity, but at any point in time to employ a resource to achieve one end implies that it cannot be used for another.

When considering the options for improving the lives of farm animals, emotions can lead us to assert that livestock care should be improved without carefully considering the trade-offs. This is a problem common to many public debates: global warming is a prime example. It is widely believed that greenhouse gas emissions should be reduced, but it is less widely known that the money spent curbing those emissions is better directed towards delivering micro-nutrient supplements to the developing world. At least, that is the view of some economists, who predict that better micro-nutrient uptake would save millions of human lives immediately.[4] Even if you disagree with the economists' assessment, you should at least acknowledge the trade-off. If money is spent fighting global warming there will be less money to use on other things, and the same can be said for improving farm animal welfare.

Giving hogs access to the outside requires more land, fences, and metal to modify barns. Providing hens with more space requires more barns and energy for each egg produced. Raising broilers who suffer fewer leg problems requires utilizing less efficient birds, which means that more corn must be harvested for each pound of meat consumed, and thus more natural gas to produce more nitrogen fertilizer to apply to increased corn plantings. Most improvements to animal welfare require more grain, more land, more energy, and more labor. If we use more land, more energy, and more labor to improve animal welfare, that means less grain, less land, less energy, and less labor for everything else. Each additional bushel of corn used to feed a hen implies less corn for other things we might value, like feeding starving people in, say, Ethiopia.

Although the presence of trade-offs is probably obvious, we mention them because they are often ignored when forming public policy. Individuals who go to work and pay bills immediately recognize their own income constraints. Making personal choices acknowledging these income constraints encourages wiser decisions. When we imagine what "society" or government should do, the income constraints often receive less acknowledgment. Supporting public projects without recognizing the tax money required to fund those projects encourages foolish decisions. Ignoring the existence of trade-offs does not prevent their existence. It is obvious that using a credit card as if the bill will not have to be paid back is foolish. It is not so obvious that support of government programs as if they are not funded with taxes is also foolish—economists exist to remind us of our resource constraints.

When the presence of trade-offs is ignored it is often because we are offered a "free lunch" by politicians, businesses, and activist groups. However, we emphatically stress that there are no free lunches. During the debate leading up to the vote on Proposition 2 in California, several pundits argued that voters could, in essence, have a "free lunch." It was argued that hen living conditions could be drastically improved with negligible increases in egg prices. We do not doubt these estimates nor do we know, in this particular case, whether the benefits of Proposition 2 exceed the costs.

We are quite certain, however, that there is no free lunch. Even if the passage of Proposition 2 only increased egg prices by 1 cent per egg, Californians consumed 7.5 billion eggs in 2002, meaning they would pay almost $7.5 million more annually on eggs as a result of Proposition 2.[5] It was estimated that in a single egg-producing county (Merced County, California), passage of Proposition 2 would decrease economic output by $70 million and would result in a loss of 50 jobs.[6] Surely some people ignored these trade-offs when they voted in favor of Proposition 2. Our research shows that the average American considers the financial well-being of US farmers to be twice as important as the well-being of farm animals.[7] The fact that a majority of voters approved of Proposition 2 suggests that they may have been unaware of the economic trade-offs being made.

The key question is whether it is worth it to provide animals a better life, given the costs. To answer this question we must first articulate what we mean by "costs." Resources, whether they be grain, time, coal, or money, have *opportunity costs*. When a resource is used to produce one outcome we necessarily forego the option of using the resource differently. When farmers choose to use their land to build a hog farm they necessarily give up the opportunity to use the land to produce corn, soybeans, wheat, or even to build a house. Because land is finite, the cost of a larger hog barn is the foregone income from soybean production. One of the beauties of a market-based economy is that the prices we see for land, corn, and coal tend to reflect all the opportunity costs.

But, where do the prices for these resources come from? They come from individual farmers and people making choices about whether the price they see is worth paying, given the other things that could be done with their finite incomes. Thus, the price premiums for cage-free eggs are reflective of thousands, and even millions, of individual interactions between people each trying to decide the best way to manage the scarce resources under their control. Why are cage-free eggs often sold at 120 percent price premiums over conventional eggs?[8] Shouldn't we mandate that everyone buy cage-free eggs or that farmers sell them cheaper?

To answer such questions, we must admit that there is no way that anyone can account for all the different reasons why cage-free eggs are more expensive than conventional eggs. In fact, it is impossible for any one person to fully understand how any good is made and how its production will change in response to a change in price or a regulation.

These concepts are masterfully illustrated in Leonard Read's essay *I, Pencil*, written in 1958.[9] In the essay, he asserts that no one person knows how to make a pencil. Indeed, the process is almost magical—no one knows exactly the right number of trees to chop down, how much graphite to mine, how much paint to buy, how many machines are required; and yet, walk in a store anywhere and you will find a pencil available for purchase. How are the activities of the lumber jacks, miners, painters, machinists, and retailers coordinated? It is not by a pencil czar but rather by the price system. The price of wood, graphite, paint, and ultimately pencils is taken into consideration by the pencil-maker when deciding how many pencils to make. And, in some small way, the price of pencils influences the person who decides how many trees to plant.

Pencils are made with graphite but graphite is also used to make golf clubs. If golf becomes a more popular sport more graphite must be used to produce golf clubs and less will be available for pencils. Pencil-makers do not need to keep tabs on the popularity of sporting events or negotiate with Tiger Woods to understand that they must utilize less graphite. The pencil-makers simply see the price of graphite rise and adjust their production accordingly. They might make fewer pencils or utilize a production process that reduces the need for graphite, or both. Prices coordinate widely disparate processes and contain the information necessary for producers and consumers to plan. Prices are how consumers and businesses communicate.

No single person knows the "right" number of pencils to make or the "right" price at which to sell them; and likewise no one knows exactly how hogs, chickens, or cattle "should" be raised. All the activities necessary for food production are coordinated by market prices. Thus, no one can foresee all the changes that will take place if the livestock industry improves animal care. What we do see, however, is market prices, and those market prices reflect a wealth of information that can be utilized to help us make decisions.

These observations imply that the trade-offs involved in farm animal welfare can be measured, in part, by market prices. Suppose that improving the lives of 1000 chickens requires giving up 10 bushels of corn, 2 bushels of soybean, 3 hours of unskilled labor, 55 vials of medicine, 10 lbs of steel, and 4 gallons of diesel fuel. The *cost* of this animal welfare improvement can then be determined by the market price of corn multiplied by the 10 bushels, plus the market price of soybeans multiplied by 2 bushels, plus the price of unskilled labor multiplied by 3 hours, and so on. It is important to understand that the costs measured in this manner are not just the cost incurred by the business who must improve animal welfare. The costs measure the trade-off to society at large. One bushel of corn could be used in a number of different ways: feeding a starving child or selling nachos at a ballgame are just two examples. What is the "value" to society of all these possibilities? There is only one way of measuring the foregone bushel of corn to society, and it is the market price of corn.

It is tempting to say that if farm animals suffer their suffering should be reduced regardless of the cost. *But who has the right to claim this?* A person may claim that using 10 bushels of corn used to make 1000 chickens better off is well worth it, but that one person is not alone in consuming corn. Corn by-products are used to make antibiotics, aspirin, toothpaste, and even the gypsum draw-wall board used to construct the inside walls of houses, all items used by a wide variety of people. Then there is the scenario where one person believes everyone's power bill should rise 5 percent so that electricity can instead be diverted and given to the poor in Louisiana. Do any of these people have the right to impose their views on everyone else? As economists we seek to *represent* available options and their consequences. Therefore, to determine whether it is worth giving up certain goods to improve farm animal welfare, we must determine the value of all foregone resources, and the best representation of that value is market prices.

The most frequent counter-argument to this line of reasoning is that market prices do not perfectly represent all costs and values. For example, one could argue that the market price for corn is distorted by government subsidies for ethanol production. True, but what value for corn should we use to measure its relative scarcity instead? If we do not use the market price of corn to reflect the value of additional corn that must be used to improve animal welfare, what should we use? Should we let the president of the PETA decide the value of corn or the President of the National Corn Growers Association? The value assigned to foregone corn should best represent all the varied uses of corn in society, and no figure better represents this scarcity than the market price.

Making a trade-off means giving something up to get something else that you want. Unfortunately, some analyses of trade-offs only focus on one side of the scale—what you give up (the cost). For example, the United Egg Producers hired a consulting company to measure the impact of a nationwide transition from cage to cage-free egg production. The researchers did a creditable job of estimating the *costs* involved in the transition, but they did not acknowledge the other side of the trade-off—the potential *benefits*.[10] This is like saying that no one should purchase a candy bar because the grocery store refuses to give them away for free. The reason you would buy a candy bar is because you enjoy eating it. Paying the grocery store price is the trade-off you are willing to make to get the treat. Likewise, the cost of a nationwide transition to cage-free production is only relevant in comparison to the benefits of the transition. The benefits of a transition to cage-free production are represented in the benefits to the animals and to the humans who now know that the hens experience a better life. Thus, to know the benefit of animal welfare improvements we must know what people want, which is the topic to which we now turn.

Preferences for Farm Animal Welfare

We do need to know why people want the things they do, to know that people make choices and these choices reflect a set of preferences. Indeed, the foundation of economic analysis is based on the assumption that a person's choice between two options reflects their best efforts to make themselves better off, given their beliefs about how consumption of the good will affect their well-being. Unlike the ethicist who tells us what we *should* prefer, economists look at people's choices to see what people *are seen to* prefer, and it is these *revealed preferences* that provide an indication of what people really want.

The idea that people's behavior is driven by their preferences is obvious, but the argument made by many popular advocates of vegetarianism is that these preferences can be changed. How strongly do consumers prefer or desire meat? Vegan advocates have argued, "Those who switch to a vegetarian diet will, over time, enjoy their food at least as much as they did before."[11] This is not an isolated sentiment, "there are many other tasty foods besides those that include meat . . . we are not being asked to choose between eating . . . meat or harming ourselves by depriving ourselves of the opportunities for the pleasures of the palate."[12] Nonetheless, statistics show that people, taken as a whole population, choose to spend more of their household food budget on meat (25.5%) than any other food category (next highest is cereal and bakery products at 15.9%).[13] Americans choose to eat twice as much meat as they do fruits and vegetables. In fact, people's preference for meat is such that the price of meat would have to increase 118 percent to induce people to eat the same amount of meat as fruits and vegetables.[14]

These calculations suggest quite a different picture than that stated by some vegan advocates: eating meat is no minor issue for most Americans—it is by far the most preferred food source. It may be that over time palates would adapt so that people would be just as happy with a vegan meal. But, what is more persuasive: the assertion that vegan meals are equally enjoyable as meals including meat, or the fact that people's actual choices reveal that meat is the most popular category? When faced with the conjecture of an animal rights author or hard data on actual choices, economists will choose the hard data every time. Here is a prime example of the difference between economists and ethical philosophers; if economists want to know how much you enjoy meat, they look at what you choose rather than attempting an inference based their own perceptions.

Ideally, we observe and measure people's preferences by seeing what they choose in the market-place, and such analysis is at the heart of many economic studies. This approach, unfortunately, has limits when studying the issue of farm animal welfare. First, there are many people who care about farm animal well-being who do not eat eggs or meat, and thus, as they already do not buy farm animal food products at all, they cannot use their food purchases to communicate which farm production processes they prefer. As we discuss in detail in Chapter 10, this implies that animal welfare is a public good, and that in effect the choices we observe in a regular market-place are a poor reflection of people's preferences for improved animal well-being.

Second, although the market-place is rapidly changing, it remains difficult for most consumers to find crate-free pork or free-range poultry, and as such, observed choices in the market-place do not reveal people's values for improved animal well-being. Third, many people do not know how farm animals are raised and they might make a different set of choices if more fully informed.

For these reasons we turn to research approaches that rely on directly or indirectly asking people their values for improved animal welfare by creating specially designed markets to measure the preferences of interest. We still measure willingness to pay (WTP) for animal welfare by looking at how people make choices with their own money. The difference is that the purchases are not made in a real market-place, but in one which we have created in an economic laboratory. As economists we recognize the need to ask WTP questions in a way that forces people to recognize the real monetary trade-offs that coincide with their choices. But before we discuss how to accomplish this feat, and other challenges associated with measuring people's values for improved animal well-being, we musts first take a step back and ask what we mean by a consumer's "value" or "preference."

WTP as a Measure of Preference

People often bristle at the idea of assigning monetary value to certain goods, especially goods that intimately affect a person's life. An example is the *value of a statistical life* that government uses to measure the value of a life saved; this value, incidentally, tends to be estimated at between \$4–7 million per statistical life.[15] The idea of placing a value on a life may sound crude, but our everyday actions reveal that we ourselves place a finite value on our lives. Every day you choose to get behind the wheel of a car you increase your chance of dying. The fact that you continue to drive *reveals* that you place a higher benefit on getting from Point A to Point B than on preserving your life. It is not economists who value a life at \$4–7 million; rather, this is the value arrived at through research related to human behavior. When economists measure the costs people incur to avoid risking death, such as accepting a lower wage for a safer job, we find that, on average, people value their lives in the \$4–7 million range.

If the value of a human life is finite, the value of an animal's life is finite too. However much the reader may dislike placing a finite, monetary value on preventing animal suffering, the fact that we would not forego all of our income to protect one chicken from suffering suggests that a value exists and is finite. Denying that a value exists does not preclude it from actually existing. Again, we can *say* whatever we wish, but it is what we *do* that reveals our preferences. Economists measure value by people's WTP. A person's WTP is measured by the maximum amount of money that they would be willing to give up in order to acquire a good; it is the price of a good that makes them exactly indifferent to having and not having the good. WTP is a dollar metric that fully reflects competing desires, market prices, and finite resources. Some people might object to characterizing, say, the value of a life in dollars, but there is nothing special about the metric of dollars; we can measure relative preferences using any unit of conversion—bricks, hours, or cars; dollars simply provide a measure that is easily used to compare against other goods.

What is the value of a hamburger to a homeless, penniless man? Infinite, you might say. Alas, the value is equal to his WTP, which is zero. We *always* want more than we have and it is only the constraint imposed by our income that prohibits us from buying all the goods we desire. Should that homeless man have the right to a good meal? Hamburgers do not fall from trees. To feed the hungry man we would have to take scarce resources from someone else. These resources have a measureable value and redistributing them in such a manner has economic consequences that can be stated in dollar terms. We could give the man a hamburger, but *someone* has to pay the cost. Thus, we say the economic value of a hamburger *to this man* is zero.

You may be willing to buy the man a hamburger, a fact that reveals *societal* or *collective* WTP is likely something greater than zero. Nevertheless, the fact remains that WTP (whether it be the homeless man's WTP or the sum of several individuals' WTP) is the only scientific definition of value because it is the only value which is measurable and which accurately reflects the reality of scarce resources. We can imagine a world where everyone has as much food as they wish and we do not have to give up anything in return for food, but this is not the world in which we live. Manna does not fall from heaven. Food is scarce, and the value of food is defined by the rate at which people are willing to trade other resources for food. Because money represents a medium of exchange among all resources, a person's WTP is a scientific measure of value.

People often express a variety of objections to using WTP as a guide to making decisions. One of the objections is that rich people are able to pay more than poor people, and thus using WTP as a metric is not "fair." Richer people can pay more for everything; including animal welfare, cars, houses, and the environment. Thus, this is not argument against using WTP as a scientific measure of value per se, but is really an argument against capitalism itself or any other such system where incomes are unequal.[16]

It is also important to point out that richer people often must pay more to produce the same change in well-being as a poorer person. Finding an extra $1000 lying on the street will certainly make a person making only $10,000 per year much happier than a person making $100,000 per year. It is far from clear whether a rich man who values something more and pays a higher price is better off than a poor person who values that item less and pays a lower price. Indeed, if WTP is measured correctly, and if a rich and a poor person pays according to their respective WTP for the good, both are made happier by the precise value of zero. As we argued earlier, if a person pays an amount equal to their WTP for the good, they are indifferent between having the good and not having the good.

If one's income is solely the product of luck, then WTP would indeed be a less desirable metric to use. WTP would still provide information on people's preferences for goods, but it would fortuitously provide some individuals with a larger claim to those goods than others. However, while there is no doubt that luck has some role to play in income distribution, a person's effort is a major factor. People with a greater desire for goods that money can purchase make a greater effort to obtain money. In this sense, WTP should disproportionally favor those with high incomes. Those with higher incomes are wealthier because they desired material goods with greater intensity, so of course their measured value for material goods is larger. Others may value leisure more than materials goods, and hence work less.

Using WTP as a measure of value is useful because it is color-blind, gender-blind, and so on. Measuring value by dollars precludes us from arbitrarily—or more frightening, non-arbitrarily—deciding which people should receive greater or lesser weight than others. Measuring the value of a good by how many dollars people will forego to receive it allows us to determine the importance of a good to society without knowing exactly who benefits from the good (making us the exact opposite of politicians).

The last motivation for using WTP as a measure of value is that, though imperfect, other measures are more imperfect. We embrace democracy not because it is perfect but because it is less bad than other systems of government. Likewise, we embrace WTP as a tool because we have no better alternative. Although WTP is a scientific and conceptually straightforward concept, measuring WTP poses a number of challenges.

Preferences Depend on Information and Context

WTP changes with the setting in which people find themselves, and changes as people acquire new information.[17] People's emotions and beliefs are so affected by their surroundings that preferences are often dismissed as irrelevant theoretical artifacts by some social scientists. Research shows that people find the same food tastes better if it is sold under a brand name. If you give someone a randomly determined identification number and then ask how much they value a good, those who receive a higher identification number will tend to value the good by a greater amount.[18] The mere act of asking someone whether they want to purchase a good has been shown to alter preferences for the good.[19] The value people place on a good is not a fixed, stable number. Value changes according to the environment and the context in which questions are phrased—no matter how advanced the science of marketing becomes it will not alter the fact that humans are flexible creatures.

These problems can be exacerbated for a complicated and, to most people, unfamiliar issue like animal welfare. Take, for example, the issue of valuing a change in the well-being of one versus ten versus 1000 farm animals. Joseph Stalin reportedly stated that "the death of one man is a tragedy, the death of millions is a statistic," and this argument appears to be equally true for animals as humans.

Hokget was a dog stranded on an abandoned ship in 2002. The dog was on board an Indonesian tanker that had had an engine-room fire. The fire knocked out communications, forcing the crew to drift for 20 days before a cruise ship rescued them. The captain had raised the dog on the ship and when the crew was rescued they had to leave the dog behind. The Hawaiian Humane Society heard about the abandoned dog, and the Humane Society of the United States raised $50,000 in donations to save the dog. Private rescuers spent five days searching for the dog—to no avail. A Japanese fishing boat later spotted the abandoned ship and tried to save the dog, but Hokget was afraid and would not come out of his hiding place. The tanker drifted into US territorial waters, where the Coast Guard was able to rescue the dog.

The story was a news sensation and tugged at the nations' heart-strings, but did it really make much sense to spend so much money to improve the welfare of a single dog? At the same time that Hokget was adrift at sea there were millions of dogs in animal shelters waiting to be adopted, many of which were later euthanized because of shelter overcrowding. Most shelters are in bad condition; the dogs often do not get walked nor do they have a blanket on which to sleep. How many of these dog's lives could have been improved with the $50,000 spent on Hokget? One might say that our reaction to a single stranded dog was irrational. Emotion interfered with our ultimate objective of helping dogs.

Humans appear to have a built-in desire to help the single over the many— this is a *size-insensitivity* problem. Studies have shown that people will donate more money to help save 4500 lives in a refugee camp consisting of 11,000 people than they will donate to save the same amount of lives in a refugee camp of 100,000 people. These studies have also shown that people will donate more money to save one child dying of cancer than eight children dying of cancer. This is why the Save the Children charity allows you the opportunity to sponsor *one* child, rather than telling you how many lives you could save. They know how your emotions work, and they exploit it—for the benefit of the children.[20]

Our preferences are not only affected in non-intuitive ways to an increase in the number of people or animals helped, but can also be influenced by how we are asked about these preferences. People often express different preferences depending on whether they perceive themselves in the role of *consumer* or in the role of *citizen*.[21] For example, in a grocery store, an individual might never buy cage-free eggs if priced at a $1.50/dozen premium over conventional eggs, but at the same time they might vote for a policy to ban cage eggs with full knowledge that the price of eggs will rise by $1.50/dozen. When measuring the public's preferences for farm animal welfare, should we focus on the consumer or citizen measures?

Economists often wish to measure the value of a good that is not traded in a market. With no market prices to observe, economists resort to simply asking individuals their WTP for a good. The value can vary greatly depending on whether the question is hypothetical or whether respondents must pay real money. The difference in what people say they will pay and what they will actually pay is referred to as *hypothetical bias*. In hypothetical survey settings, people typically *say* they are willing to pay about two to three times more than they will actually pay when a real monetary transaction is required.[22] When answering hypothetical WTP questions, it is as if people forget that that resources are scarce—something they rarely do when shopping with their own money.

Paternalistic Preference Measurement

The sensitivity of value to the setting causes some to infer that people have no stable preferences at all, but instead are simply responding to external cues. As a result, some people call for paternalistic policies—policies and institutions to help save people from the foolishness of their own actions. We do not doubt that there are instances when public well-being might be generally improved by a paternalistic policy. But equally we do not doubt that using results from a poorly designed hypothetical survey would be ill-advised.

Rather than attempt to make decisions for people who at times appear irrational, we seek to develop methods that assist people in rationally stating their preferences and values—something we refer to as *paternalistic preference measurement*. Our intent is not to dictate people's preferences or choices but rather to help ordinary consumers discover their own preferences for farm animal welfare and express those preferences in a way amenable to economic analysis. You will learn more about the specifics of our approach in Chapter 9.

In this chapter, we make the general case for paternalistic preference measurement and briefly offer some counter-arguments to some of the aforementioned anomalous behaviors. The findings of seemingly ill-advised behavior have caused some to mistake economists' views about people's behavior. It is true that people often make choices to their own detriment. This is to say that, in hindsight, people do not always make the "right" decision. However, it is not hard to imagine that people make the choices that they believe best suit them at any given time. People can change their beliefs or their preferences, but at any particular point in time, they do the best they can to make the choices they believe will be most satisfying. To argue to the contrary is to claim that people's choices are random or capricious, which any amount of introspection would reveal as false. Humans are purposeful creatures. It is also worth noting that despite representations to the contrary, there is nothing in the economic approach that suggests people only care about themselves. As the sociologist Rodney Stark put it, "Thus although rational choice theories restrict behavior to that which is consistent with a person's definition of rewards, it has very little to say about the actual content of those rewards. This leaves all the room needed for people to be charitable, brave, unselfish, reverent, and even silly."[23]

For a moment, allow us to take seriously the argument that all choices are made on a whim—that preferences change with the tides. Who then should make people's decisions? Experts? But, who makes the decisions for experts? After all, experts are human, and are prone to the same limitations as the human subjects who participate in experiments. To allow experts to make our decisions, we must presume that there are those who know more about what makes someone else happy than the person does themself! While we may, in fact, think *we* know what would be better for others, no doubt we would not presume *others* know what is better for us. People may be misinformed about how certain choices will affect their well-being, but the ultimate arbiter of the appropriateness of a decision lies at the individual level, because well-being is individual-specific and subjectively defined. Moreover, who is harmed by incorrect beliefs and thus has the greatest incentive to adjust errors? The individual.

We previously mentioned several problems researchers face when trying to measure people's preferences outside a traditional market. One of these was the so-called *hypothetical bias* problem—that people usually do not do what they *say* they will do. The problem arises, in a sense, because although we would like to observe people's actual choices, there exists no functioning market in which to observe them. Our answer to this problem is to create a market for the goods in which we are interested: markets that involve real economic trade-offs, real food, and real changes in animal well-being. Thus, we solve the missing market problem by creating new markets—albeit highly controlled markets—that are referred to as experimental economic markets. We have spent the past decade exploring the intricacies of measuring values in these markets.[24] We bring this knowledge to bear on the issue of animal welfare by creating markets for pork and eggs produced under different levels of care, and by creating markets for sow and hen lives that are unrelated to food consumption.

We also noted the *size-insensitivity* problem in which people sometimes value improving the well-being of fewer rather than more people or animals—although this phenomenon is most likely to occur in situations where the size of that population is unknown. Research shows, however, that when the number of people helped and the population of needy are *simultaneously* known, the size-insensitivity problem disappears.[25] Thus, people can and will act rationally if given a proper decision context in which to do so. In fact, the size-insensitivity problem may be less of an irrationality on the person's part and more of a failure by the researchers to provide adequate information.

Many of the problems we previously noted arise when individuals are either not given enough information or are not given enough time or incentive to process the information. Animal welfare is a multi-dimensional, complicated problem. To confront the average consumer with information and expect a quick and rational evaluation of cage-free and free-range eggs is probably naive. However, we make rapid and quick judgments daily between choices that involve a complicated set of potential outcomes. For example, the seemingly simple task of deciding whether to walk or to drive to work entails an evaluation of potential weather events, the likelihood of accidents with each mode of transportation, and so on. We frequently manage to make decisions about the relative desirability of new and unfamiliar goods such as new iPhone apps, Diet Cherry Vanilla Dr Pepper, and Blue-Ray DVDs.

The difficulties of eliciting views on farm animal welfare are not attributable to a lack in the human's ability to deal with complex tasks. Instead, the difficulties arise because people rarely encounter decisions involving animal welfare, and even when they do, they do not receive feedback on the consequences of their choice. Several lines of research suggest that much of the anomalous behavior observed in preference elicitation studies arises because people are unfamiliar with the good being sold or with the methods used to ask people about their values.[26]

When studying people's preferences for farm animal welfare, very careful attention must be paid to the information provided and the construction of a context which allows people to think about the problem and learn about their values. Providing an objective, dispassionate description of how animals are raised along with expert evaluation of production systems is a first step in helping people to evaluate foods. The next step is to provide an interactive environment where people can express their preferences and then witness the consequence of their decisions.

Think about the first time you bought a car. Confronting a pushy salesperson can be overwhelming, so many of us brought along a more seasoned friend on our first visit to a car dealership. The friend would have attempted to help us decide which car we wanted and how to pay the lowest price possible in the unfamiliar bargaining environment. One can think about our approach to preference measurement as one in which the respondent likewise brings a seasoned friend to help determine their value for farm animal welfare.

When measuring the public's preferences for farm animal welfare, should we focus on the consumer or citizen measures? The answer is that it depends on our purpose for asking—are we trying to predict grocery purchases or voting outcomes? In cases where people's preferences might be used to inform public policy, it seems to make sense to encourage the citizen side of the person. The problem with the answers that people give when they act as citizens is that they often ignore the resource constraints. After all, policies are paid for by *all* taxpayers, including the taxpayer being asked (and if the individual has a low income, they might not even pay income tax). For this reason, the *consumer-citizen problem* is akin to the *hypothetical bias problem*. They may even be the same problem, we are not sure. What we do know is that when people are asked hypothetical questions and are asked to consider a question from the perspective of a citizen, they are reticent over acknowledging resource constraints. Given our objectives to say something about policy outcomes in a way in which people acknowledge resource constraints, we do what few studies we dare: eliciting preferences for public goods while acknowledging resource constraints by using real money. To accomplish both goals and overcome behavioral biases, we had to invent a new research tool.

In our role as consultant economists, we aim to help people avoid irrational choices about animal welfare. In subsequent chapters, we fully describe how we constructed a logical, mathematical framework for thinking about how our actions affect farm animal welfare and for thinking about WTP to improve farm animal welfare. The framework is not intended to take the place of personal judgments but to assist in the formation of those judgments. Much as a shopper might take a calculator to the store, this tool is intended to assist people in making their choices, not alter their preferences. Like the shopper's calculator, our new research method allows individuals to think deeper about farm animal welfare and it provides a profile of preferences that are more rational than they might otherwise be.

Benefits and Costs

One of the key issues stressed in this chapter is that trade-offs are real and unavoidable. When deciding whether to buy cage-free eggs, you face a trade-off between a desire for improved animal welfare, a desire to eat tasty food, and a desire to spend less rather than more money. One tool available to help us to know how to balance trade-offs, both at the individual decision-making level and at the policy-making level, is cost–benefit analysis. A particular individual decision, say to reduce pork consumption, is deemed desirable if the benefits exceed the costs. A particular societal decision, say to ban gestation crates, is deemed desirable if the sum of all the benefits exceeds the sum of all costs. Saying that an outcome is good when the benefits exceed the costs is, of course, a tautology. However, when government conducts a cost–benefit analysis it is forced to articulate exactly why it makes a particular decision. This makes it more difficult for decisions to be made purely for political reasons. In similar fashion, by conducting a cost–benefit analysis for animal welfare improvements, we are forced to explicitly outline the assumptions and logic behind our decisions. If someone wants to criticize a policy decision, a cost–benefit analysis provides a framework in which a rational debate can be held.

Measures of people's preferences provide the data needed to calculate the benefits of an action or policy. Suppose a regulation is proposed to ban cages in egg production. Each individual in a state or country will have some preference for the policy—a maximum amount of money they would be willing to pay for the regulation to pass. Cost–benefit analysis works by adding up all the citizens' WTP to arrive at an aggregate benefit, and then compares this sum to the costs of the policy to society. If the total benefits exceed the total cost, a policy is said to pass the benefit–cost test, otherwise it fails.

The cost–benefit approach has many advantages. It is quantitative and provides an unambiguous answer. Cost–benefit analysis converts a complex, multi-dimensional problem to a single metric—dollars—such that many competing trade-offs can be weighed against each other. Moreover, the desirability of a policy is judged by the citizenry and their preferences, not politicians or special interest groups. Cost–benefit analysis also focuses people's attention on outcomes rather than intentions. Finally, it relieves the researcher from accusations of bias. When a researcher conducts cost–benefit analysis, it is akin to outsourcing the decision to the citizens affected by the decision. Of course, critics may still accuse the researcher of making assumptions within the cost–benefit analysis that are biased, but now the debate is about the validity of an assumption, in which more data can be used to help resolve these differences.

Almost everyone wants farm animals to have a better life—this is not the issue. The question is whether a given action will actually improve animal well-being and whether it is worth all the other things that must be given up to obtain it. We may choose to ignore the results of a cost–benefit analysis and enact a policy for which the costs exceed the benefits, but this does not mean that the costs imposed by a policy are not real; they must be borne by someone, even if someone chooses to ignore the formal cost–benefit analysis.

Cost–benefit analysis is a valuable tool, but it is not a panacea. It is possible, for example, for a policy to pass the cost–benefit test but still adversely affect a majority of citizens. Such an outcome can occur if a few people derive a very large benefit from a regulation. To illustrate, imagine a four-person world consisting of Sam, Sally, Jack, and Jane. Sam, Sally, and Jack place no value on the regulation, but Jane is willing to pay $1000. The total benefits of the regulation are $0 + $0 + $0 + $1000 = $1000. Suppose that the total cost of the regulation is $800. The benefits are indeed greater than the costs. The policy passes the cost–benefit test. Suppose that the costs of the policy are paid by sending a bill to Jane for $800. Jane is better off by $200 and Sam, Sally, and Jack are indifferent; the world is a happier place. However, if the government could isolate Jane and send her a bill, there would be no need for the government in the first place—any aspiring entrepreneur could just produce the policy outcome and sell it to Jane for a price of $900 (Jane would be happy to make the purchase because she values it at $1000 and the entrepreneur is happy to undertake the activity because she profits by $100).

The nature of farm animal welfare is such that isolating the individuals who benefit from a welfare improvement is difficult. Jane may be a vegan, or may benefit when Sam, Sally, and Jack switch to more humane food, so that Jane does not pay more money when the price of meat rises. Suppose the government does what governments usually do: taxes are increased by $200 per person to cover the $800 cost of the policy. The policy still passes the cost–benefit test (total benefits are still $1000 and total costs are still $800), but only one person, Jane, benefits from the regulation while three are harmed. The policy would not pass in a referendum, but it does pass a cost–benefit test.

Even though the benefits exceed the costs, it does so for only one person. Is the regulation good or bad? The question has no definitive answer; it depends on your opinion about the relative worth of equity and efficiency. As the example illustrates, cost–benefit analysis might be too simplistic to judge policies in all cases, as it may mask inequalities that are important to interest groups and politicians.

Because it is a common situation to have "winners" and "losers" from a policy, as was the case in our example, the economists John Hicks and Nicholas Kaldor pointed out in the late 1930s that a cost–benefit test might be combined with something we call a *redistributive test*: are the gains to the winners large enough to offset the losses to all the losers? In our example, we can see that they are. Suppose the government levies a $200 tax on Sam, Sally, Jack, and Jane to pay for the $800 policy. Now, Sam, Sally, Jack are each worse off by $200 but Jane is better off by $1000 − $200 = $800. In principle, the government could then levy *another* tax on Jane for, say $630, leaving her still better off by $800 − $630 = $170 and redistributing this extra tax income evenly in the form of a $630/3 = $210 tax rebate to Sam, Sally, and Jack. Now, Sam, Sally, and Jack are, on the net, slightly better off from the whole transaction: −$200 + $210 = $10. The redistributive test might seem complicated, but it can sometimes be feasibly implemented when there are clearly identified "losers"—for example, egg producers—from a policy. In reality, the redistribution envisioned by Hicks and Kaldor is almost never implemented, but the fact that it *could* be done in a fashion to make everyone better off may be sufficient motivation to pursue the policy.

One way to judge the seriousness of inequities in a cost–benefit analysis is to look at the *median* benefit and cost.[27] Returning to the above example, the median benefit of the regulation is $0 and the median per-person cost is $200. It is evident that median benefits are less than the per-person costs. Determining whether the regulation is "good" or "bad" now requires further thought and judgment. Cost–benefit analysis does not always relieve the practitioner of making personal judgments, but it can provide useful input to the decision process

Non-Speciesist Cost–Benefit Analysis

You may have noticed the similarities in cost–benefit analysis and the utilitarian philosophy discussed in the previous chapter. Utilitarianism involved adding up the change in "happiness" or well-being of each entity involved in an action. Cost–benefit analysis does the same, with the key difference being that units of "happiness" are measured in dollars. One nice feature of this conversion of happiness to currency is that it allows us to consider issues like the *redistributive test* where, in essence, we can think about transferring some of Jane's happiness to Sam.

This chapter has focused on how economists differ from philosophers, but to wind down this chapter, we return to the utilitarianism philosophy of ethics and ask whether philosophy might be fruitfully combined with some economic analysis. Still, we would ask people to keep one fact in mind: whereas cost–benefit analysis is only a *tool* to the economist, to a committed utilitarian it is a *rule*.

Recall that a utilitarian argues that happiness is happiness regardless of whether it comes from a chicken or a person, and to discriminate between the two is to commit an act of speciesism. The utilitarianism argument is that the measure of happiness should be extended to include the well-being of humans *and* animals, which, in our context, would mean that cost–benefit analysis should be extended to include animal *and* human preferences and costs.

If we want to avoid being called a speciesist, the welfare of animals should enter a cost–benefit analysis *directly*, not just indirectly through its effect only on human well-being. Despite this logical argument, there appears to have been little previous attempt to seriously consider how an economist might go about performing such a calculation or whether there might be any complications to carrying out cost–benefit analysis using animal *and* human preferences.[28] We begin by first noting some of the implications and drawbacks to including animal preferences in a cost–benefit analysis, and then we end by discussing how one might actually measure animal preferences in economic terms.

A non-speciesist cost–benefit analysis would proceed by adding up all the benefits to animals and people, and then subtracting all the costs to animals and people. However, we must keep in mind that some of the benefits people receive from the policy (humans' WTP for the policy) are a direct result of the fact that people do care about animals. People are altruistic toward animals. For example, suppose a sow is willing to pay the animal equivalent of $100 to move from a gestation crate to an open barn. Knowing that the sow would be happier in the barn than in the crate, a human consumer might be willing to pay $10 for the policy. If the hog *really* wanted to make the change and instead was willing to pay $1000, you too would increase your value, to say, $100. Your happiness increases in tandem with the sow's happiness. The key point is this: a person's WTP for an animal welfare policy is directly and inextricably linked to the benefits the animal is expected to receive. What this means is that when we conduct cost–benefit analysis and include human and animal benefits, we double-count benefits. We add in the animal's benefit once and then we add them in again when we add the human's benefits (because the human's benefits also include the animals' benefits).

Normally, all that is required in cost–benefit analysis is to compare the benefits to the costs, and if the benefits exceed the costs, the policy passes the cost–benefit test. As we previously noted, implicit in most cost–benefit analyses is the assumption that a policy which passes the cost–benefit test will also pass the *redistributive test*. A very peculiar thing can happen when one conducts a non-speciesist, cost–benefit test: a policy can pass the cost–benefit test and fail the redistributive test. This can happen because, as we just noted, human altruism acts to double-count gains to animals, which can prevent redistribution schemes from benefiting everyone.

To make these ideas more concrete, consider a simple example. Imagine a simpler world with one person (Jane) and one pig (Porky). Jane loves Porky and her happiness depends on Porky's well-being. As a result, Jane wants to support causes and policies that Porky supports. For every $1 Porky is willing to pay for a policy, Jane is willing to pay $0.50. Imagine now a policy proposal to give Porky a larger pen. Porky likes the idea of more room and more freedom, and figures that he is willing to pay $100 for more space. Because Porky wants more space, Jane thinks it is a good idea too, and she is also willing to pay $50 so that Porky can have a larger pen. A helpful non-speciesist economist adds up the benefits and finds total benefits of $100 + $50 = $150. Suppose the actual costs of the policy are $110. The aggregate benefits ($150) exceed the costs ($110) and thus the policy passes the cost–benefit test. Total welfare would apparently increase by $150 − $110 = $40 with the passage of the policy. Should we proceed with the proposal if Jane (being the human) has to pay the $110 cost?

If the policy is implemented, Porky is better off by $100 (he gets the larger pen and pays nothing), but Jane is worse off. She has to pay $110 for something that was only worth $50 to her. What about the redistributive test? Maybe Porky can help defray some of Jane's burden? Can some of Porky's $100 be redistributed or transferred to Jane such that they are both better off? The answer is no. To see why, imagine a redistribution scheme that takes $60 of Porky's benefits ($60 out of the $100 of benefits) away and redistributes it to Jane. The idea behind the scheme is to see whether Jane can personally benefit from the policy by adding her $50 benefit to the $60 transferred from Porky. Jane's benefits now equal $110, and she is fully compensated for paying the policy cost of $110 while Porky benefits by $40—but this is not how the story would truly unfold.

Remember that Jane's benefit from the policy equals one-half of Porky's benefit. When Porky has to give up $60 dollars, this makes Jane unhappy. When Porky is forced to redistribute to Jane $60 out of his $100 of benefits, he only benefits from the policy by $40. Jane's personal benefit is always one-half of Porky's benefit, and one-half of $40 is $20. Jane's $20 personal benefit plus the $60 transfer is only $80—not enough to compensate for the $110 cost Jane had to pay for the policy. In fact, it is impossible for Porky to pay Jane any portion of his $100 benefit in a way that Jane could be fully compensated for the cost. If Porky pays Jane $95, Porky's benefit is $5 and Jane's benefit is $97.5. If Porky pays Jane $5, Porky's benefit is $95 and Jane's benefit is $52.5. No transfer of money from Porky to Jane in the range of $0 to $100 can fully compensate Jane for the $110 cost. The redistributed test fails, and there is no feasible scheme that allows both Jane and Porky to be made better off from the policy. If the policy is enacted, someone (Jane, specifically) is made worse off.

This result does not just hold in our particular example, it is a general rule in all cases where people exhibit altruism toward animals and must pay the cost of a policy. The key lesson is this: *if we want to conduct non-speciesist, cost–benefit analysis, we should logically ignore human altruism*. In fact, if animals pay the cost of the policy (something that can happen by, for example, reducing space or feed or some other input that has monetary value) we can go a step further and say that we do not even need to know how much humans value the policy. The key lesson in the case where animals pay the cost of the policy is: *as long as the quality or safety of food is unaffected, a policy will pass the non-speciesist, cost–benefit test as long as animals' WTP exceeds the cost of the policy—human preferences can be ignored.*

The implications of these findings are rather interesting in that they suggest that we should either follow the traditional *speciesist* cost–benefit analysis approach and look only at human benefits and costs (recognizing that some of the human benefits include animal benefits) or adopt a *non*-speciesist approach and practically ignore the benefits to humans. We make use of the cost–benefit approach in Chapter 10, and we do so using the traditional *speciesist* approach, for three reasons. First, despite the compelling arguments for directly including animal well-being this logical choice is not as iron-clad as some writers suppose. Second, it is doubtful that real policies will respond to any species besides the one which possesses political power. For the foreseeable future, animal welfare policies will be judged by the benefits delivered to humans. Finally, we are qualified and able to measure *human* preferences and WTP, but knowledge of measuring animal preferences is still cursory. As science progresses and as economists and animal scientists work together, we are optimistic that measuring animal WTP may one day represent no more of a challenge than measuring human preferences. To encourage further thinking on this topic, we end this chapter with our thoughts on the issue.

Measuring Animal WTP

Much of this book has focused on what *people* think about changes in animal well-being. However, if one wants to carry out a non-speciesist, cost–benefit analysis more research is needed on the measurement of animal well-being. Some research has been conducted and scientists are continually contributing to the topic. Scientists have developed models to quantitatively describe how animal well-being changes when an animal's environment is altered. Research has not addressed how these welfare changes can be translated into WTP for use in cost–benefit analysis. In this subsection, we sketch out some initial ideas on how such work might proceed.

First, we must assume that animals, like people, have preferences: they like some things more than others. Animals care how much feed they are given, how much space they have, and as we saw in Chapter 4, animals are willing and able to make trade-offs between the factors.[29] For example, an animal knows whether it prefers minimal food and large amounts of space to a farm with ample food and small space. In fact, the methods used by researchers to understand animal preferences are the same methods used to understand human preferences. A small group of economists have used these methods to show that animals have preferences and make rational trade-offs in the face of changing incentives similar to their human owners.[30]

Take for example, the work of Matthews and Ladewig, who studied how hard pigs were willing to work (by pressing a lever on a nose-plate) to obtain food versus social contact with another pig.[31] By varying the number of presses (effort) required to obtain a commodity (food or socialization), and assuming that effort serves as an analog for price, one can determine a pig's relative preference for different goods. Matthews and Ladewig found that pigs were willing to work very hard for food: as the required number of level presses to get a food reward increased 1 percent, pigs only reduced their willingness to work by 0.02 percent, compared to a decrease of 0.49 percent for social contact with another pig. Pigs strongly desire food, and they want food more than social contact.

Under some restrictive assumptions these so-called demand elasticites can be used to determine the trade-off animals are willing to make between food and social contact.[32] Given that animals are willing to make trade-offs, we can imagine approaching a pig and asking: how much feed would you be willing to give up if the size of your pen was increased by one square foot? To answer this question a hog would determine the extra happiness they would get from the extra space and divide it by the extra sadness they would feel from the lost food: a quantity referred to as the marginal rate of substitution. Of course, we could not literally ask a hog this question, but we can get the same answer by looking at how many times they are willing to press a lever.

The marginal rate of substitution is the most amount of food the hog would give up to get one extra square foot of space without becoming any worse off overall. This is just the same as a person's maximum WTP for, say, an extra square foot of living space in their apartment building. The only difference is that the hog has expressed his value in terms of corn and we expressed our values in terms of dollars. To convert the hog's values to dollars, all that we must do is to multiply his WTP by the price of feed—corn being the most widely used animal feed. To illustrate, under a set of assumptions about the hog's utility function and normalizing the units to unity, Matthews and Ladewig's estimates indicate that a pig's WTP for a one unit increase in social contact (in units of payment equal to the quantity of feed) is: $0.49/0.02 = 24.5$ units of feed. If the units are pounds and the price of corn is \$4/ bu (or \$0.0714/ lb), then the animal is willing to pay \$1.75 (24.5×0.0714) for a one unit increase in social contact.

Without knowing more about the units used in the Matthews and Ladewig study we cannot say anything more precise, but our point here is simply to illustrate that animal WTP can be calculated. As our example illustrates, it is not too far-fetched to develop a feasible method of expressing both human and animal happiness in the same unit, and the most convenient unit is dollars, or WTP. Because an animal is willing to forego corn in exchange for space, and because corn has a dollar-value determined by the corn market, a WTP value can be computed for an animal.

Perhaps a critic may argue that the amount of happiness generated from one dollar's worth of corn to a hog is much larger (or, perhaps, much lower) than the happiness a human receives from one dollar. However, this critique is not unique to the human versus animal comparison; a millionaire receives less happiness from one dollar than someone earning a minimum wage, yet economists find sufficient reason for combining WTP from the rich and the poor.

Animal WTP is a nascent concept, and one we hope will provide fodder for discussion, but it is not a concept we are yet ready to fully endorse for use in cost–benefit analysis. Sufficient evidence exists to conclude that animals have identifiable preferences in food, biological condition, and their environment— these preferences exhibit the same stability and measurability as human preferences. The bigger question is whether animal preferences can be translated into WTP, and if they can, whether those estimates should be compared to or combined with human WTP.

Getting Down To Business

Thus far, this book has discussed many of the current ongoing debates about the well-being of farm animals, but no systematic attempt has been made to resolve any of the issues. It is now time to get down to business, and crunch some numbers to help consumers contemplate what constitutes an ethical diet and to help determine the consequences of animal welfare regulations.

CHAPTER 8

Your Eating Ethics

A Guide to Eating Based on Your Beliefs and Preferences

Pretest

From your point of view, which of the following people are unethical?

1. A meat-eater whose desire for food brings into existence an animal that lives in misery.
2. A vegan whose unwillingness to eat meat precludes the existence of an animal that would live an overall pleasant live.
3. Both of the above.

Your Personal Eating Ethics

Suppose you dislike confining hogs in gestation crates and laying hens in wire cages. But what can you do about it? Unless you are a farmer, the answer is: not much. However, you do decide what and how much to eat, and these choices ultimately affect the number of animals raised and their living conditions. As we argued in the last chapter, our goal is not to convince anyone to adopt a particular diet. Taking the approach provided by the economic viewpoint, this chapter aims to provide a framework for understanding the consequences of your diet to animal well-being.

If you want to have animals live a more comfortable life, and if you are willing to change the way you eat, we can show you how to translate your personal beliefs and perceptions about animal treatment into a dietary plan of action. This does not imply that the relative tastiness of beef, cheese, and broccoli should not influence your food choices. Your happiness should be considered along with the happiness of animals. You may decide to concentrate your limited resources to improve the world in ways other than improving the well-being of livestock. Doing so is no more or less virtuous than the vegans who might make all of their food choices based upon the consequent impact to livestock, but who might ignore the well-being of humans in, say, developing countries. There are some people who want to make food-purchasing decisions based, at least in part, on the well-being of farm animals. If you are one of these people, this chapter is for you. Because we are interested in outcomes, and not intentions, the answer to the question of what you should eat may not be as straightforward as you might think; a vegetarian diet may or may not be the most animal-friendly diet.

Your Decision

It is tempting to think that your food choices have no impact on the number of animals raised for food. After all, there are over 6.5 billion people in the world. However, to think that one consumer's food choices have no impact on livestock is illogical. The Great Wall of China is so large that astronauts can even see it from space. From space, it is impossible to see that the Wall is comprised of a seemingly infinite number of small stones. If one stone were removed from the Wall , the Wall would grow smaller even though this would go undetected by the astronaut. The astronaut's inability to see the stone being removed does not negate the fact that the Wall is now smaller. Just as the Great Wall of China is comprised of many individual stones, the market demand for food is comprised of each individual consumer's demand for food. Just as removing one stone reduces the size of the Wall by exactly one stone, when one consumer changes his or her purchasing habits the market adjusts in turn. It may be hard to see the consequences of our decisions, but let there be no doubt, each purchase decision matters. To deny this fact is to contend that every human becoming a vegan would have no impact on the number of livestock raised.

In this chapter we will ask questions such as: "What if I stop eating eggs?", "What if I switched from cage eggs to cage-free?", "What if I ate less pork and more beef?" First, we must realize that most of us eat *a lot* of animal products each year. On average, Americans eat about 85 lbs of chicken, 65 lbs of beef, 50 lbs of pork, 250 eggs, and 600 lbs of dairy products per person per year.[1] Americans are certainly not alone; annual per-capita consumption in the European Union is 39 lbs of beef, 97 lbs of pork, and 35 lbs of poultry.[2] What happens if we choose to increase or decrease these consumption levels? Although the length of the Great Wall of China is partly dictated by the length of each individual stone, we have to keep in mind that stones are inanimate objects with no concern for the number of other stones comprising that Great Wall. Yet, if you decide to eat less pork, your decision has a direct and indirect effect on the total food demanded and supplied. It is understandably difficult to imagine, but your decision to buy or not buy a package of meat causes changes in the meat consumption of other consumers. These interactions are important when projecting the impact of our grocery purchases on animal welfare. For example, if you stop purchasing veal but another consumer simply takes your place, your personal eating decision will have no impact on farm animal welfare. To link our food choices to farm animal welfare, we must develop a link between our purchases, the market for food, and farm animals.

Suppose you decided to eat 5 fewer pounds of chicken next month and every month thereafter. Your choice means that the grocery store now has 5 more lbs of chicken meat left to sell. How will the store convince other people to buy extra chicken? In the near term, they will likely drop the price to ensure that they do not have to throw away stock they have already purchased. A lower price means more people will want to buy chicken, and the grocery store can ultimately sell the entire stock, if it lowers the price far enough. Lower meat prices are less profitable for grocery stores though. Retailers want to devote the scarce and valuable shelf space to relatively more profitable items, and when the store decides how much chicken meat to re-order from meat processors their order will now be smaller than it previously was. Your decision to consume 5 lbs of chicken meat fewer than previously consumed sets into action a chain of events that will, over the long run, cause the grocery store to buy less chicken, and because it is a mathematical fact that the amount of food consumed must equal the amount of food produced, the decision will ultimately lead farmers to produce fewer chickens.

It is possible to imagine scenarios where the purchase of less meat seems to have little effect. For example, suppose the grocery store had a shortage of meat and will sell out of meat regardless of whether you participate in those purchases or not. Does your abstaining from meat matter in this case? Yes. Though the store will run out of meat with or without you, it runs out faster when you purchase meat, which communicates to the store manager a greater demand for the meat.

If you abstain from eating 5 lbs of chicken each month, the overall amount of chicken that continues to be produced and consumed will likely fall by something *less* than 5 lbs each month, because lower chicken prices induce other consumers to buy more chicken. Will the monthly reduction in chicken sales ultimately be 4 lbs, 3.5 lbs, or will there be no reduction at all? The answer ultimately depends on how willing other consumers are to eat more chicken as the retail price falls *and* how prepared farmers are to raise chickens in the face of falling prices—statistics that economists call demand and supply elasticities. (Further details can be found in the Appendix.) It is the magnitude of the demand and supply elasticities—the degree of which consumers and producers are sensitive to price changes—that dictate the exact change that will occur.

The key points to note are that a permanent decision to reduce meat consumption (1) does ultimately reduce the number of animals on the farm and the amount of meat produced (2), but it has *less* than a 1-to-1 effect on the amount of meat produced. The opposite is also true. By eating more of a food item one increases the demand for that food, but this also raises the price for other consumers, indirectly causing them to reduce their purchases. Increasing one's chicken consumption by 5 lbs causes more chickens to be raised, but the increase in chicken consumption will be less than 5 lbs. To clarify this idea, let us replace one consumer with a million consumers. If a million consumers increase meat purchases by 500,000 lbs, they will cause the demand and hence price of meat to rise. This price increase deters purchases by other people, and though meat production may rise, the increase will be less than 500,000 lbs.

Using estimates of the elasticities of supply and demand for different animal products we can determine how total production of a food item will respond to changes in a person's consumption patterns. Table 8.1 shows the impact of a decision to reduce consumption for six animal food products. Forgoing one lb of beef reduces total beef consumption and production by 0.68 lbs, while shunning one lb of milk reduces total milk consumption and production by 0.56 lbs. The reason for the difference has to do with the differences in the elasticities of supply and demand for milk and beef. Eggs seem to be most responsive to changes in the diet. A decision to consume one fewer egg results in 0.91 fewer eggs being produced, because egg production is very sensitive to changes in egg prices. A 1 percent decrease in egg prices will decrease egg production by a larger percentage than a 1 percent change in beef prices will reduce beef production.

Allow us to provide some intuition to the results in Table 8.1. Consuming one egg leads to a rather large decrease in total egg production for two reasons. First,

Table 8.1 Long-Run Effects of Reducing Consumption of Six Animal Food Products

If you give up...	Total production eventually falls by	*Per capita consumption of food item*
One pound of beef	0.68 lbs	65.20 lbs
One pound of chicken	0.76 lbs	85.10 lbs
One pound of milk	0.56 lbs	600.00 lbs
One pound of veal	0.69 lbs	0.50 lbs
One pound of pork	0.74 lbs	50.80 lbs
One egg	0.91 lbs	250.00 eggs

Note: All products are assumed to be conventionally raised and sold as a generic animal food product. If pork is produced in a crate-free (confinement-pen) system the number 0.74 can be replaced by 0.71, and if the pork is raised in a shelter-pasture system the number can be replaced with 0.53. More details on these alternative numbers can be found in the footnotes of Table 8.A1.

egg production can easily be increased or decreased by producers in response to changing egg demand. Second, consumers tend to purchase roughly the same number of eggs regardless of the price. If one segment of consumers refrains from their normal egg purchases, producers will decrease prices only a little because they can easily decrease production, and even if prices did fall, other consumers will not respond by consuming many more eggs. Put simply, changes in one person's egg purchases have little impact on the purchases of other people.

The opposite is true for beef and milk. Because it takes a year between the time a cow is bred and the time her calf is born, and then it also takes a long period before that cow can be transformed into beef or produce milk, it is difficult for beef and dairy producers to alter production according to changes in consumer preferences. Moreover, consumers are highly sensitive to changes in milk prices. If prices fall because one segment of consumers no long purchases milk, other consumers will quickly take up the slack. The point is that consumers respond differently to changes in the prices of different goods, and the production methods used differ among these goods. These differences have a significant impact on how changes in one person's consumption of animal food products affect total consumption. When it comes to the well-being of animals, it is the *total* consumption of the food products that matters, not *any one person's* particular choices.

The results in Table 8.1 also hold if we consider increases in consumption. A decision to eat an additional 1 lb of pork will ultimately lead to 0.74 lbs of additional pork being produced. The reason is that when you increase pork consumption you effectively raise the bid price of pork, which causes other consumers to consume slightly less pork.

How Happy is a Farm Animal?

The Human Society of the United States (HSUS) encourages eaters to follow the three Rs: *Reduce, Replace, and Refine.* We are encouraged to *reduce* consumption of all animal food products and *replace* meat products with vegan alternatives. Moreover, the HSUS encourages the *refining* our diet by shunning products produced by animals under miserable conditions in favor of food products derived from more humane production settings.

The HSUS brochure on the three Rs specifically states, "the chicken, egg, turkey, and pork industries tend to be far more abusive to animals than the beef industry," but they noticeably stop short of actually advocating that the reader eat this meat.[3] Given the make-up of the HSUS leaders (virtually all are vegan, although the members of the board of directors are not) and given the organization's mission to reduce animal suffering, the HSUS is unlikely to encourage consumption of beef even if an animal received extraordinary care, attention, and enrichment. Somehow, intuitively it seems reasonable to abstain from animal food consumption for ethical reasons, and it is therefore unintuitive to increase animal food consumption for ethical reasons. But this chapter may cause us to question your intuitively reached conclusion.

It is their pragmatism that leads HSUS to realize that some people will continue eating animal foods regardless of how the animals are raised, and as such they encourage eating foods which have been produced in a way that causes the least suffering. HSUS leaders applaud any action that reduces animal suffering, whether it be a conversion to veganism or abstaining from pork.

However, some of us may believe that if an animal has had an overall pleasant life—if it has experienced many positive emotions and very few negative emotions—some of us may believe it is then ethical to raise the animal for food. Whether we are vegan or compassionate omnivores, we must come to terms with what we believe about animal suffering in different production systems. This requires some difficult choices. In Chapter 5 we provided detailed descriptions of how farm animals are raised in different production systems, in part, to clarify how animal well-being varies across farm settings. To encourage contemplation on the relationship between our dinner plates and livestock, we might consider each individual animal used to produce that food on our plates and assign it a score between −10 and 10 related to our perception of the animal's well-being (see Figure 8.1). The first step is to decide whether we

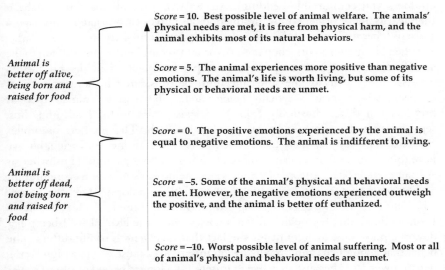

Animal is better off alive, being born and raised for food

Animal is better off dead, not being born and raised for food

Score = 10. **Best possible level of animal welfare. The animals'** physical needs are met, it is free from physical harm, and the animal exhibits most of its natural behaviors.

Score = 5. The animal experiences more positive than negative emotions. The animal's life is worth living, but some of its physical or behavioral needs are unmet.

Score = 0. The positive emotions experienced by the animal is equal to negative emotions. The animal is indifferent to living.

Score = –5. Some of the animal's physical and behavioral needs are met. However, the negative emotions experienced outweigh the positive, and the animal is better off euthanized.

Score =–10. Worst possible level of animal suffering. Most or all of animal's physical and behavioral needs are unmet.

The score for any animal is between –10 and 10.

Figure 8.1 Developing a personalized welfare score for specific livestock
The Score for any animal is between -10 and 10.

believe the animal is better off alive and raised for food, or if the animal would
have been better off if it had never existed. If we believe the animal was better
off for having lived, the animal welfare score should be a positive number
(between 0 and 10). Conversely, if we believe the animal suffers so much that it
would have been better off not to have been born the score should be a negative
number (between −10 and 0). The better off we believe an animal to be, the
higher the score it should receive; the more positive emotions we believe the
animal experiences relative to negative emotions, the higher the score.

The purpose of assigning animal well-being scores, as will be fully developed
later in the chapter, is to help guide our choices about which animal products are
most consistent with our beliefs and values. Consider an example. Adrian
believes that veal calves suffer greatly throughout their lives, so much that all
living veal calves should be euthanized and no additional veal calves should be
allowed to be born. He gives the veal calves a score of −7. Adrian then considers
cage egg production, and decides that the hens are also better off dead, but that
each hen does not suffer as much as a veal calf. Consequently, Adrian's score for
a laying hen equals −4. After learning about cage-free egg production, Adrian
believes hens suffer less in a cage-free system than a cage system, but not to the
extent that their lives are worth living. The score for cage-free egg hens equals
−1. Finally, Adrian learns about milk production and believes that milk cows
do not suffer, but that many of their needs are unmet (e.g., little access to
outdoors and comfortable bedding). Adrian scores a milk cow at 2. Adrian
then learns about organic milk production, and gives each organic milk cow a
score of 6 because most of the animals' needs and desires are met.

When biting into a pork chop it is important to consider that, in addition to the
actual animal you are eating, there were parents involved in the production
process. The quality of the sow's life must be considered in tandem with its
offspring whose meat is actually being eaten. The lives of animals used as
breeders can differ drastically from those raised strictly for food, and thus,
separate animal welfare scales are needed for both. The mothers of broiler
chickens, for example, likely live a much worse life than their offspring, but we
have to remember that we cannot have offspring without parents. Thus, whereas
Adrian gives broilers a score of +2, he gives the broiler's mother a score of −5.

The length of an animal's life may need to be addressed in the scores.
Chickens raised to lay eggs live longer than chickens raised for meat (broilers).
If one believes that both chicken types possess the same level of well-being per
day, one may want to account for the fact that their length of life differs. The
belief that laying hens and broilers live in misery may cause one to assign laying
hens a lower score because they live in misery for a longer period of time. Or, the
belief that both chicken types are content may suggest a higher score for laying
hens, due to the fact they experience this contentment for a longer period. How
the duration of an animal's life interacts with their average daily well-being is
not a question science can answer; this interaction thus depends on the readers'

perception, and consequently, must be reflected in the welfare scores the consumers assign.

Our respective scores may differ from that of Adrian. Or, we may be too
overwhelmed to assign scores at all. Farm animal welfare is complex and
difficult. Nevertheless, when we decide to buy cage-free instead of cage eggs,
to become vegetarian, or that organic milk is too expensive, we have *implicitly*
assigned such welfare scores by our actions. As authors, we suggest that it is
necessary for consumers to attempt to make these scores explicit so that they can
logically think through the consequences of their actions.

To help the consumer to decide, it is possible to simply use results from the
SOWEL, FOWEL, and COWEL models discussed in Chapter 5 to create
welfare scores for hogs, laying hens, and dairy cows. These models use data
from scientific studies to translate animal well-being in different production
systems into a 0 to 10 score—note those scales are in the 0 to 10 continuum,
whereas our scale is on the −10 to 10 continuum. The SOWEL model does not
seek to determine whether sows live a life worth living, only if a sow is better off
in one system or another. Although these models are useful tools for guiding
decisions about the *relative* well-being of animals in different systems, they have
nothing to say about whether an animal is better off dead or alive, a key issue in
deciding whether or not to eat a given animal product. It has to be said that it is
difficult to imagine that any amount of scientific evidence could definitively
answer the question as to whether an animal would have been better off had it
never been born.

Bailey's View

Because assigning welfare scores to animals in different systems can be a
challenge, it is worth walking through how one of us (Bailey) went about the
task. Bailey's scores, although informed by years of careful research and
thought, are not meant to be iron-clad assessments of animal well-being with
which one should concur. Indeed, the two authors are not in complete agreement about the scores. Although Jayson and Bailey tend to agree on issues such
as whether hens are better off in cage or cage-free systems and whether beef
cows are better off than hogs, Jayson is more likely than Bailey to believe that
chickens and hogs in conventional systems are better off alive than dead.
Fortunately, Jayson and Bailey can make their own choices on what to eat, as
can every consumer.

With these provisos behind us, let us journey with Bailey as he expresses his
opinions about the quality of life experienced by different farm animals (see
Table 8.2). Beef is given a high rating for all the reasons discussed in Chapter 5.
Mother cows are given a high score of 8 because they spend most their lives in
pasture. By contrast, the market (non-breeding) animal spends the last 100 to

200 days of its life in a feedlot. There are desirable qualities about feedlots. For example, the animals receive a type of food that they enjoy eating, but they do not receive the room and grassy areas provided by a pasture. Thus, market steers and heifers get a score of 6. Breeder chickens are given a very low score (−7, better off dead); Bailey believes they suffer greatly from the feed restrictions. The market chickens are presumed to have a life worth living, but because of leg problems and tight space allocations that come about later in life, broilers are not given as high a score as beef cattle. Breeder animals are irrelevant for dairy because it is the dairy cows that are being bred. Breeder animals are not considered for veal production either, because the milk cows which produce the veal calves would experience the same life regardless of whether their calves were raised for beef, raised for veal, or killed at birth. Even if veal production was banned, the life of dairy cows would be unaltered.

Pork receives a low score because Bailey believes hogs to be creatures in need of mental stimulation, which is not provided in the barren environments of conventional production systems. Because the breeder animals (sows) cannot even turn around in their cages they receive a lower score (−7) than the market animals (−2). The most likely alternative to the confinement-crate system is the confinement-pen system described in Chapter 5, so in this chapter, crate-free pork is pork produced from a confinement-pen system. Whether sows receive any benefit from the switch is debatable, but Bailey interprets the scientific literature to imply that hogs in group-pen systems are better off than in gestation crates, but not by much. Thus, the score for breeding hogs improves somewhat (−5), but remains negative. When switching from conventional pork to a shelter-pasture system, the move has important consequences. In Bailey's opinion, hogs under conventional pork production systems would be better off dead, whereas life is not only worth living in the shelter-pasture system, but it is pleasant. The reader may wish to refer to Figure 5.15 in Chapter 5 for a reminder of the differences across hog production systems.

Most pork produced under shelter-pasture systems could likely be sold using the Animal Welfare Approved (AWA) label. Developed by the Animal Welfare Institute, in Bailey's opinion, the AWA label is associated with the highest welfare standards possible. Most food sold under the AWA label not only refers to animals raised humanely, but animals that were most likely happy—scientists are perturbed when we use the word "happy," but we, the authors, suggest that animal happiness is what this book is about.

Finally, Bailey gives a low score to hens raised in cage systems (−8) for all the reasons discussed in Chapter 5. It may seem odd that Bailey gives laying hens a slightly lower score than sows. The reason is that laying hens are confined to small spaces with other hens, who sometimes cannibalize one another; hens are de-beaked, which may result in chronic pain; but, most importantly, because the sow has more variation in its life. Hens stay in the same setting for their entire lives, whereas sows at least periodically are moved to a different cage and have

Table 8.2 Bailey's Personalized Welfare Scores for Various Livestock

Welfare of animals raised to produce (raised conventionally unless otherwise noted)	Welfare score of one breeder animal	Welfare score of one market (non-breeder) animal
	(−10 score for worst state, 10 score for best state)	
Beef	8	6
Chicken meat	−4	3
Milk	not relevant	4
Veal	not relevant	−8
Pork	−7	−2
Crate-free pork[1]	−5	−2
Shelter-pasture pork[2]	4	4
Egg from cage system	3	−8
Egg from cage-free system	3	2

Notes: [1]Pork produced in the absence of gestation crates; farrowing crates are still permissible. This system is referred to as a confinement-pen system in Chapter 5.
[2]This system is defined in Chapter 5. It provides pigs with both shelter and pasture, and in many cases could be sold under the private label Animal Welfare Approved, developed by the Animal Welfare Institute.

the opportunity to nurse offspring. The breeder birds receive a higher score because the breeders must be raised in a cage-free setting with nests (this is true for the parents of cage and cage-free hens). Hens laying eggs in cage-free systems are better off alive (score = +2), but only marginally so, in Bailey's view.

Of Mice and Men

An important concept is required before we proceed: even veganism is murder. In order to harvest the grains and beans crucial for vegan (and non-vegan) diets, combines are sent into the fields which insects, mice, rabbits, and other animals call home. Many of these wild animals succumb to a cruel death-by-combine. Vegan diets still require the sacrifice of animal lives. It has even been argued that some diets which include meat could be more humane than some vegan diets. For example, meat-eaters who only eat beef from cattle raised on a pasture for their entire lives may kill fewer animals than a vegan dining on the wheat, soybeans, and corn grown in fields which require harvesting equipment. Of course, the opposite may be true, depending on the relative mortality rates and well-being of the animals residing on harvested versus foraged acres.[4]

Ideally, food choices made to influence animal well-being would account for the lives of non-farm animals living on farms, but unfortunately, the data is insufficient to properly account for these animal lives. However, a few observations might help put non-farm animals in perspective. One might suspect that beef production would support a larger number of non-farm animals than other livestock production. Pastures are used extensively with cattle, and pastures provide a habitat for wildlife; however, the issue is complicated. For example, although cattle may feed on pastureland for many months, they need hay during the winter. Hay production requires the same type of harvesting that goes on with grain production. Pastures may be harvested two to four times each year, while cropland rarely exceeds two harvests per year. However, cropland is often plowed and tilled while pastures are not.

Moreover, the quality of life of non-farm animals is not clear. Animal advocates have a tendency to idealize the lives of animals in the wild, in our view, but even nature is cruel. Who is happier, a rabbit living in a wheat field constantly searching for food and being pursued by snakes and foxes, or a cow grazing in a pasture with every need tended to by the rancher? Animal rights thinkers have been so busy opposing the use of domestic animals that they have not articulated their views on wild animals,[5] but any careful analysis would almost certainly recognize that life as a wild animal is fraught with misery.

The lack of good data and the complexities involved lead us to generally ignore the effects of food consumption on the lives and well-being of non-farm animals, but one could partially account for these animals by adjusting their welfare scores for livestock or by including welfare scores for grain consumption. There is little information we can provide to help account for wild animals, except for the following. The production of animal food is generally thought to require more cropland than does a vegan diet. The suggested caloric intake for the average adult is about 2000 calories per day. At this level of energy intake, one acre of corn can feed 370 people for one day—if they eat the corn directly.[6] If, instead, the 370 people obtained their calories by consuming eggs, it would require 2.63 acres of corn to feed the hens that lay the eggs that provide each of the 370 people 2000 calories.[7] Or, if the 370 people ate chicken meat to meet all their daily caloric needs, they would need to harvest 4.92 acres of corn to feed the broilers.[8]

Based on these numbers, it appears that consumption of animal foods requires more cropland, and more cropland displaces natural habitats and kills wild animals. How this translates into an overall measure of well-being then depends on your personal preferences. To make the problem even more complex, the calculations above assume that vegans consume very bland diets, consisting mostly of corn. In reality, vegans consume large amounts of vegetables and nuts as well. While we think these more realistic vegan diets still require less cropland than do animal products, we have no data or reference to verify this claim.

Society is at a point in history where it is (seriously) considering the happiness of livestock. There is still much to be learned about farm animal welfare, and many people are still making up their mind about the issue.

Animal Numbers

Imagine a scenario where one day, far into the future, a mad scientist might breed a chicken that can lay 76 billion eggs a year—enough to supply every American consumer's current annual rate of consumption. This chicken, however, is terribly unhappy and must be housed in a barren cage, undergoing painful medical procedures, physical pain from the cage floor, and frustration at its inability to exhibit natural behaviors. Although the chicken leads a miserable existence, one might very well argue that it is still ethical to eat her eggs. The reason is that the poor chicken is just one suffering hen, and when her suffering is divided by 76 billion eggs, suffering per egg is basically zero.

The extra happiness that many millions of people receive from eating eggs might far outweigh the sadness of that poor chicken. We may undertake advocacy campaigns to improve her well-being, but the fact remains that she is only one and her suffering bears goods that benefit many. We are faced with the reality that the *number* of animals affected by our food choices matter. Misery and merriment are emotions that occur within an animal brain. It is not the pounds of meat that matter in considerations of animal welfare but rather the number of animal brains. The relationship between pounds of meat and numbers of brains is paramount in determining the effect of our dietary choices on animal well-being.

Shortly, we will calculate the number of animals associated with a certain amount of meat, dairy, or egg products, but first it is useful to observe the total amount of animals currently harvested in the US. Shown in Table 8.3, by far the livestock category with the most animals slaughtered is the broiler. The large number of birds harvested is due to the fact that per-capita consumption of chicken is higher than that of beef and pork, and the fact that a single broiler contains only about 6 lbs of meat. Both beef and pork are consumed in similar amounts by the average American, but many more chickens must be harvested than hogs and many more hogs than cattle must be harvested for the same quantity of meat, due to the great variation in the respective animals' sizes.

If you care about animal well-being and if you believe the hens suffer, if you learn that it takes two rather than one caged hen to supply the eggs in your morning omelet, you should perhaps consider switching to oatmeal. However, we should not allow our emotions to override our logic, and logic dictates that a situation where there are two suffering hens is worse than one where there is only one suffering hen. In fact, if some animals are going to suffer in the process of supplying our food, total suffering can be minimized by putting fewer, more

Table 8.3 Number of Livestock Harvested in the US in 2008

Livestock Type	Number harvested in 2008
Beef cattle (mother cows and bulls)	4,173,900
Beef cattle (not bred)	27,040,000
Veal calves	942,000
Hogs (sows)	3,960,000
Hogs (not bred)	111,460,700
Dairy cows	2,591,200
Broilers	9,031,035,000
Turkeys	264,969,000
Laying hens (inventory)	340,000,000

Notes: Broiler and turkey numbers refer to 1997. Laying hens inventory refers to the average number of laying hens in the US actively laying eggs. All numbers refer to livestock slaughtered under federal inspection.
Sources: NASS (*2008, 2009a, 2009b*).

productive animals in more miserable conditions than many, less productive animals in better conditions.

Because the logic of this outcome is not necessarily obvious, the decision-making framework being developed in this chapter is intended to help us avoid irrational choices. Based on production data and our knowledge of livestock production, we have calculated the number of breeder and non-breeder animals that are affected when someone eats 1 lb of beef, pork, chicken, veal, and milk, or eats one cage (or cage-free) egg. To avoid breaking up the flow of the chapter, we have placed the calculations in the Appendix.

To understand the calculations, it is useful to walk through a brief example. How many laying hens are affected when you eat 1 egg? During its 2.2 years of life a laying hen in a cage system will produce 509 eggs. Consuming 1 egg implies that you are responsible for $1/509$ of a hen's total output. There is also the breeder bird to consider. For every laying hen there are about 0.01399 breeder birds, implying that 1 egg is not only relevant to $1/509$ laying hens but $(1 / 509) \times (0.01399)$ breeder birds as well.

Now consider cage-free eggs, which typically come from a different breed of bird that is less productive and has a shorter life span. Even if the same bird breeds were used in cage and cage-free egg production, cage-free production would remain less efficient because mortality rates are higher in a cage-free system. Instead of producing 509 eggs throughout its life, a cage-free hen only produces 314 eggs. Consequently, 1 cage-free egg is associated with $(1/314)$ laying hens and $(1/314) \times (0.01399)$ breeding hens. Eating a cage-free egg

Table 8.4 Number of Animals Associated With Production of Select Food Products

Food item (raised conventionally unless otherwise noted)	Number of breeder animals associated with		Number of non-breeder animals associated with	
One pound of beef	0.000600601	Cows	0.001201201	Cattle
One pound of chicken	0.001804675	Chickens	0.259740260	Broilers
One pound of milk	———	———	0.000017429	Cows
One pound of veal	———	———	0.006849315	Calves
One pound of pork	0.000167336	Sows	0.007195450	Hogs
One pound of crate-free pork	0.000167336	Sows	0.007195450	Hogs
One pound of shelter-pasture pork	0.000175193	sows	0.007182901	Hogs
One cage egg	0.000027491	Chickens	0.003184713	Laying hens
One cage-free egg	0.000044554	Chickens	0.003184713	Laying hens

Notes: Crate-free pork refers to pork raised in a confinement-pen system as described in Chapter 5. Shelter-pasture pork is also described in Chapter 5 as a system that provides comfortable shelter and access to pasture for most of the year.

affects more birds than eating a cage egg. If you believe birds in both cage and cage-free systems suffer, the fact that one cage-free egg requires more birds to suffer should be a salient feature underlying the logic behind our eating ethics.

Our final calculations for several different farm animals are shown in Table 8.4. The calculations indicate the number of breeder and non-breeder animals required to produce 1 lb of meat or milk (or one egg).[9] The statistics provide a direct link between our dinner plates and the daily lives of farm animals. The numbers demonstrate that drinking 1 lb of milk affects far fewer animals than 1 egg or 1 lb of meat. This is because one cow can produce an extraordinarily large amount of milk. Conversely, 1 lb of chicken meat is associated with a large number of birds, relative to the other food products. These numbers are useful for relating your perceptions about the well-being of a single farm animal into a number dictating how your actions affect the lives of animals in general. If you believe dairy cows and broilers suffer greatly, shunning 1 lb of chicken will have a far greater impact than shunning 1 lb of milk. Conversely, if you believe both dairy cows and broilers to experience pleasant lives, consuming more chicken meat will lead to more happy chickens than consuming more milk will lead to more happy dairy cows.

There are a few caveats about the numbers in Table 8.4. One is that they do not reflect the different lifespan of various animals. Breeder chickens live longer than broilers, and cows live longer than laying hens. The lifespan of an animal may impact the degree of misery or merriment it experiences. Sows are constrained in a manner similar to laying hens, but a single sow experiences that environment for a longer period of time. Does it create additional suffering when a sow is confined to a crate for three years instead of two, or does she become more accustomed to the environment the longer she lives, reducing the total amount suffered? These are questions we cannot answer, and so we leave it to each consumer's personal adjustment to account for different lifespan however he or she deems appropriate, through the welfare scores each person assigns to animals.

It should also be mentioned that the linkage which exists between the dairy industry and the veal industry, as well as the dairy industry and the beef industry, is not reflected in the model. The dairy–veal linkage is actually quite small. If dairy farmers did not have the possibility of selling calves to the beef industry they would simply kill them at birth. Veal calves bring in little money, anyway. However, while the dairy–veal linkage is small, this is not the case for the beef–dairy linkage. As Table 8.3 shows, a significant number of dairy cows are slaughtered for beef. Should we then assume that a hamburger affects animals in both the beef and the dairy industry? We could, but prefer not to. The purpose of this chapter is to utilize what we have learned about the different livestock industries and articulate the linkage between our dinner plates and animal well-being in the relevant sectors. When we eat beef, it is preferable to document how that beef consumption translates into animal welfare within the beef industry. It should then be easy to utilize the data provided in this chapter and modify the results to reflect the linkage between dairy and beef production.

An Ethical Eating Assessment Tool

All the components are now in place to determine the consequence of a dietary change on animal well-being. Our previous discussions of supply and demand, animal well-being scores, and animal numbers can now be put to use in something we call the Ethical Eating Assessment Tool (EEAT). The EEAT works as follows. Consider an action, such as forgoing all egg consumption, decreasing veal consumption by 1 lb per year, or increasing pork consumption by 1 lb per year. EEAT calculates the ultimate consequence of the action on farm animal well-being. The EEAT score articulates how changes in market consumption alter the population of breeder and non-breeder animals of the species, and weighs these changes against the welfare scores assigned to the breeder and non-breeder animals, where a higher EEAT score number is preferred to a lower EEAT score. The EEAT score reveals to us the farm animal welfare consequences of an action, and it can be mathematically calculated as:

EEAT Score = (change in total consumption of food item due to change in diet) × [(number of breeder animals affected per consumption unit) × (welfare score of one breeder animal) + (number of non-breeder animals affected per consumption unit) × (welfare score of one non-breeder animal)].

Let us put the tool to work. Suppose Bailey is thinking about giving up all conventional pork, and he, like the average American, currently consumes 50.8 lbs of pork per year. What is the effect of his decision on the well-being of pigs? We can calculate the EEAT score in several steps.

Step 1. Determine how total consumption will change. We know that giving up 50.8 lbs of pork will not reduce pork production by 50.8 lbs. From Table 8.1, we can calculate the change in total pork consumption that will result from Bailey's decision. His decision to give up pork will ultimately result in 50.8 × 0.74 = 37.59 fewer pounds of pork being produced.

Step 2. Determine animal well-being. We need to know something about Bailey's perception of hog well-being. We can see from Table 8.2 that Bailey gave a welfare score −7 to one breeder animal (a sow) and −2 to one market hog.

Step 3. Determine how many animals are affected. The number of animals affected by the decision (on a per-pound basis) is provided in Table 8.4. These statistics tell us that for each pound of pork that is forgone, there will be 0.000167336 fewer sows and 0.007195450 fewer market hogs.

Step 4. Put it all together. Plugging all our statistics into the EEAT model: EEAT score = *EEAT score for Bailey forgoing 50.8 lbs of regular pork = (−50.8 × 0.74 = -37.59 lbs) × [{0.000167336 sows per lb} × {welfare score for sows = −7} + {0.007195450 market hogs per lb} × {welfare score for hogs = −2}] = +0.59.*

Step 5. Interpret the result. The EEAT score provides a single number indicating the consequence of a dietary decision. If the number is positive, then animal welfare is improved, but if the EEAT score is negative then animals are made worse off by the consumption change. Bailey's decision to give up 50.8 lbs of pork per year yields an EEAT score of 0.59. The score is positive, which means that hog well-being is improved by his decision. The actual units are not meaningful, but their values can be compared relative to one another. For instance, if instead of giving up pork suppose that Bailey (still assumed to be the average American) gave up cage eggs. The average American consumes 250 eggs per year, and the EEAT model suggests that reducing one's annual egg consumption by 250 eggs changes animal welfare by: EEAT score = *EEAT*

score for Bailey forgoing 250 cage eggs each year = (−250 × 0.91 = −227.5 eggs) ×
[{0.000027491 breeder birds per egg} × {welfare score for breeder birds = −3} +
{0.001964637 layers per egg} × {welfare score for layers = −8}] = +3.60. The EEAT
score of 3.60 for giving up eggs is larger than the respective score for giving up
pork, implying that if Bailey had to choose between giving up eggs or pork and
the only factor considered is the impact of his choice on animal welfare, Bailey
should give up eggs.

In many ways, Bailey's decision is not surprising. His welfare scores indicate
that he believes sows and hogs suffer, and when he eats less pork, animal
suffering falls. That result is intuitive, but the result that giving up eggs relieves
animal suffering more than giving up pork is not immediately apparent without
crunching the numbers. Let us consider another question. What if Bailey gives
up conventional pork and replaces it pound-for-pound with crate-free pork?
Following Steps 1 through 4 above tells us that: EEAT score for Bailey eating
50.8 lbs of crate-free pork = *(50.8 * 0.71 = 36.07 lbs) × [{0.000167336 sows per*
lb} × {welfare scale for sows = −5} + {0.007195450 market hogs per lb} × {welfare
scale for hogs = −2}] = −0.55.[10]

Because Bailey believes hogs suffer in crate-free systems, deciding to eat 50.8
lbs of crate-free pork produces a negative EEAT score: −0.55. But our question
was not whether Bailey should eat crate-free pork, but whether he should
replace his consumption of conventional pork with crate-free pork. By comparing our two EEAT scores, we can evaluate the overall impact of Bailey's decision
on farm animals. By forgoing consumption of conventional pork Bailey
increased welfare by +0.59. By increasing his consumption of crate-free pork
by the same amount he reduced animal welfare by −0.55. Replacing conventional pork with crate-free pork results in a net change of farm animal welfare
equal to (0.59 − 0.55 =) 0.04. Bailey's decision improved the lives of sows and
hogs. Of course, hog well-being would be even higher if Bailey decided not to
eat either crate-free or conventional pork.

Bailey is a great consumer of bacon. Is there any way for Bailey to continue
eating pork while improving animal well-being, given his beliefs about how
hogs feel in different production systems? Consider the consequence of Bailey
giving up all conventional pork and replacing it pound-for-pound with shelter-
pasture pork, which could likely be sold under the private label: Animal
Welfare Approved (AWA) pork. Following Steps 1 through 4 above tells us
that: EEAT score for Bailey eating 50.8 lbs of shelter-pasture pork = 0.79. The
action of Bailey deciding to eat shelter-pasture pork *alone* increases animal
welfare by 0.79. Bailey's decision also included a choice to decrease conventional
pork consumption, which increases animal welfare by 0.59. Thus, total animal
welfare increases by 0.79 + 0.59 = 1.38. Given Bailey's beliefs about hog well-
being in different systems and all the assumptions incorporated into the calculations in EEAT, Bailey finds that replacing his consumption of conventional pork

with shelter-pasture pork will improve the welfare of sows and hogs. Hogs that would have been raised in miserable conditions are no longer brought into existence; hogs that have the potential to lead happy lives are now brought into existence. This change in diet has only positive consequences.

The EEAT scores offer a key lesson regarding the animal *rights* compared to animal *welfare* debate. The debate often centers on the question of whether animal welfare should be improved by encouraging consumers to purchase more humane products or by encouraging vegan diets. Our calculations suggest that if Bailey gives up consumption of conventional pork then the EEAT score is 0.59. However, if he replaces conventional pork with shelter-pasture pork, the EEAT score is 1.38. In this case, the *welfare* approach wins over the *rights* approach. The rights (or vegetarian) approach simply eliminates unhappy animals, while the welfare approach replaces sad animals with happy animals. The rights approach, in this context, never stood a chance.

Animal rights activists are unlikely to be persuaded by the EEAT scores. First, they are unlikely to agree with Bailey's welfare scores and would probably give negative scores to all methods of raising farm animals. They tend to believe that it is impossible to simultaneously raise an animal for money and provide the animal with a pleasant life. But more importantly, animal rightists are unlikely to engage in the logic embodied in the tool for personal and professional reasons. Such people, often called *animal abolitionists*, believe that the only animals that should exist are those in the wild. This is a philosophical position about animal ownership or a preference that is *outside* the logic of the tool; *even if* the EEAT score is higher consuming shelter-pasture pork than no pork at all, an abolitionist would still want to eschew pork production because they do not believe that humans should be allowed to own animals.

An animal abolitionist's philosophy about animal ownership can lead them to ignore outcomes of the *EEAT* no matter how high or low the scores. Animal abolitionists agree with the logic that the existence of more sad animals is bad, but will not concur with the logic that more happy animals is good. A meat-eater is often willing to bring a sad animal into existence to make him-or herself happier, but a vegan is unwilling to bring a happy animal into existence because it would make the vegan sadder. Who, then, is unethical? The meat-eater and abolitionist both inject their personal interests into the diets they consume and advocate. Meat-eaters tend to relish in the enjoyment of meat and are less squeamish at the thought of slaughter, while vegans tend to dislike meat and are more squeamish at the sight of blood. Though they both may be concerned about farm animals they will utilize different forms of logic when arguing over the morality of their diet, and this logic is swayed by their personal interests.

It is also likely that the animal abolitionist movement also succumbs to the same desire for market share as firms in the private sector. People receive both monetary and personal rewards for their participation in the farm animal welfare debate. Authors do make at least some money from publishing books.

More importantly, some individuals extract intense personal meaning from being recognized as, say, animal rights leaders. Every social movement consists of two struggles: (1) a struggle between the advocates and opponents of the movement; and (2) a struggle within the movement for social recognition. Economists are convinced that personal incentives such as these influence the views and arguments individuals offer in ethical debates. Personal incentives sway the logic individuals use. The purpose of constructing the EEAT is to provide the consumer with a tool for evaluating the morality of choice in diet that is logically consistent and impervious to personal interests (save for how the welfare scores are assigned, of course).

Cage or Cage-Free?

At the center of the farm animal welfare debate is the question of whether eggs should be produced in cage or cage-free systems (see Table 8.5). Let us use EEAT to try and parse the various consequences of our egg consumption decisions. If your opinions are like Bailey's (see Table 8.2), the issue is straight-forward. Bailey believes hens in a cage system are better off dead, while those in a cage-free system are better off alive. Thus, for Bailey it is preferable for eggs to be raised using cage-free methods—if animal welfare was the only factor in the decision. Some people, however, believe that hens suffer greatly in both systems. Recall that the HSUS does not just encourage replacing cage eggs with cage-free eggs, but also reducing total consumption of eggs.

Suppose a person named Paul believes that hens in cage *and* cage-free system live miserable lives, but because he loves omelets he remains undeterred from

Table 8.5 Does Consuming Cage or Cage-Free Eggs Lead to Higher Animal Well-Being?

Egg type	Change in egg consumption	Welfare score of one breeder bird	Welfare score of one laying hen	Change in animal welfare
Paul's beliefs imply he should prefer cage eggs				
Cage	−12 eggs	+3	−8	0.17056
Cage-Free	+12 eggs	+3	−6	−0.18975
		Total change in welfare = 0.17056 − 0.18975 = -0.0192		
Margaret's beliefs imply she should prefer cage-free eggs				
Cage	−12 eggs	+3	−8	0.17056
Cage-Free	+12 eggs	+3	−4	−0.12605
		Total change in welfare = 0.17056 −0.18975=0.0455		

purchasing eggs. Although Paul is an egg-lover, however, he is not heartless, and he is willing to change the types of eggs he purchases based on the recommendations of the EEAT. Paul assigns breeder birds in both systems a welfare score of +3, layers a score of −8 in the cage system, and layers a score of −6 in the cage-free system. Paul's choice is far from obvious. Although he believes layers suffer less in a cage-free system, more hens are required to produce the same amount of eggs, so more hens suffer. When Paul calculates the EEAT score associated with the decision of replacing the purchase of a dozen cage eggs with the purchase of a dozen cage-free eggs, it turns out that this choice *reduces* total welfare (EEAT score = −0.0192). The cage system inflicts more suffering on each bird, but the higher level of productivity in a cage system means that fewer hens are needed to produce each egg. If Paul wants to improve animal well-being (conditional on his choice to continue eating eggs), he should consume cage eggs.

Paul's decision nicely illustrates the usefulness of a logical welfare model like the EEAT. Paul might intuitively believe that it is more ethical to purchase cage-free eggs, but his intuition would ignore the fact that it requires more hens to produce an egg in a cage-free facility. Without the logic provided by EEAT, Paul's attempt to improve the lives of laying hens would instead produce more suffering. Of course, not everyone has the same preferences and beliefs as Paul. Consider Margaret, who differs from Paul in only one respect. The welfare score she assigns to hens in a cage-free facility is −4 rather than −6. As illustrated in Table 8.3, cage-free eggs are the preferred choice for Margaret (assuming she does not mind paying the higher price for cage-free eggs). For Margaret, the fact that fewer hens suffer in the cage system does not outweigh the fact that each hen suffers more in the cage system, and cage-free eggs are logically her preferred choice.

The EEAT can help clarify the conditions under which purchasing cage-free is preferred to purchasing cage eggs. Ignoring the effects on breeder animals, who are very few in number anyway, the EEAT indicates that purchasing cage-free eggs rather than cage eggs produces higher levels of layer welfare when:

1. The welfare score for cage-free hens and cage hens are positive, but the score for cage-free hens is larger than cage hens;
2. The welfare score for cage-free hens is positive, while the score for cage hens is negative; or
3. The welfare score for cage-free and cage hens is negative, and the ratio (*score for cage-free market animals*) / (*score for cage market animals*) is less than 0.62. The ratio for Paul is 0.75, which is not less than 0.62, and thus he prefers cage eggs; the ratio for Margaret is 0.5, which is less than 0.62, and thus she prefers cage-free eggs. For another example, if a person named Jackson assigns score of −1 for cage-free hens and −3 for cage hens, the ratio 1/3 is less than 0.67, so Jackson prefers cage-free eggs.

The Selective Carnivore

People make the choice to become vegetarian for many various motivations. Some people have concluded that the lives of any farm animal are filled with such misery that it is better the animals are never born. Others believe it is "not right" to raise an animal for food, and others embrace vegetarianism for lifestyle or health reasons. Vegetarianism is a personal choice, and we neither promote nor oppose the diet. However, we argue that vegetarians should seriously consider the fact that their diet may preclude the existence of a farm animal whose life, from the animal's point of view, is worth living. Our research suggests that the vast majority of people are neither vegetarians nor indifferent about farm animal well-being. Most of us can be characterized as *selective carnivores*.[11] Selective carnivores eat meat, dairy, and egg products, but are willing to consider the impact of their choices on the misery and merriment of farm animals. The EEAT can provide guidance for the selective carnivore.

Making small changes in consumption is unlikely to be expensive or have much influence on eating satisfaction. Thus, we ask, what if we decreased consumption of any *one* animal food product by 1 lb (in the case of eggs, by 1 egg), while keeping the remainder of our diet unchanged? The EEAT can be employed to judge the desirability of such actions. Given Bailey's beliefs about animal well-being, the EEAT scores for the one unit reduction of six animal products are shown in Table 8.6.

First consider beef, which has an EEAT score of −0.00819. This number is negative, indicating that decreasing beef consumption by one pound will lower the overall level of farm animal welfare. The number −0.008 has no tangible

Table 8.6 Impact of Decreasing Consumption of Animal Food Product for Bailey

If Bailey gives up ... (item is conventionally produced)	The change in animal welfare is ...
One pound of beef	−0.00819
One pound of chicken	−0.58485
One pound of milk	−0.00004
One pound of veal	+0.03779
One pound of pork	+0.01153
One egg	+0.01421

Notes: Values refer to changes in the value of the E.E.A.T. Readers who calculate these numbers by hand may obtain slightly different results; this is due to rounding.

meaning other than its sign and its respective value for other food items. Beef, according to the EEAT and Bailey's welfare scores, is a humane food source. The EEAT score for chicken, −0.587, is also negative but even more so than beef. The negative number indicates that eating less chicken meat results in fewer chickens existing. Because Bailey believes chickens are overall happy animals, fewer chickens means less happiness in the world. The EEAT score for chicken is much larger than beef because chickens are much smaller than cattle. One could eat one chicken during a single meal, but the average person must eat for over eight years before they will consume one entire cow. Misery and merriment are experienced in one brain, not 1 lb of meat. There are far more brains per pound of chicken than there are brains per pound of beef. Thus, Bailey's beliefs coupled with EEAT lead him to conclude that eating chicken is humane and is more so than eating beef, even though he believes cows lead overall better lives than broilers.

The EEAT scores in Table 8.6 for veal, pork, and eggs are all positive, indicating that reducing the consumption of such products leads to higher level of well-being. Veal has the highest EEAT score because Bailey assigns veal a low welfare score (−8) and because one veal calf produces a relatively small amount of food. Reducing consumption of eggs generates slightly more well-being than reducing consumption of pork.

For the selective carnivore, the EEAT scores in Table 8.5 are not just there to satisfy idle curiosity; rather, they can be used to make small dietary changes to improve the lives of farm animals. Bailey, for example, loves to eat ham. But, as his EEAT scores in Table 8.5 suggest, he believes hogs would be better off if he were to eat less ham. Moreover, beef is a more ethical choice for Bailey. EEAT has led Bailey to replace the ham bologna in his daily lunch sandwich with beef bologna. Likewise, EEAT has given Bailey a clearer conscience about eating fried chicken with a glass of milk. While it is true that some broilers and dairy cows lead poor lives, it is important that food choices be dictated not by the exception, but the rule.

The EEAT can help you make modifications to your diet to improve the well-being of farm animals. It should be noted, however, that we eat food for a host of reasons and the goal in life is not necessarily to maximize animal well-being. We want food that is cheap, tasty, safe, and nutritious, and our desires for these food characteristics can often conflict with our desire to improve farm animal well-being. What the results in Table 8.6 show is that some small changes can influence animal well-being. Eating chicken rather than pork, or oatmeal rather than an omelet is probably not going to make a meal much less tasty or any more expensive, and yet the change might just result in happier farm animals.

A Repugnant Conclusion

There is something odd about the results in Table 8.6, something that might trouble us if we were to take the EEAT results to their logical extreme. Chickens raised for meat have lives that are only mildly pleasurable, according to Bailey's views. The animals experience more happiness than sadness, but not much more. Cattle, on the other hand, experience very pleasurable lives. Bailey should then prefer a pasture filled with cattle to a factory farm filled with chickens.

This is not the world Table 8.6 suggests Bailey should seek. The numbers in the figure assert that Bailey should consume as much chicken as possible in order to encourage a world where millions of little chickens live a meager existence, at the expense of cattle who would be largely content. Which is better, a world with one billion chickens possessing a life barely worth living, or a million cattle that experience very little discomfort?

This conflict is what moral philosophers refer to as the *Repugnant Conclusion*; it is the conflict utilitarianism philosophers confront when they seek to infer their ideal world based on a few principles. Economist Tyler Cowen describes the Repugnant Conclusion as one which "postulates a society with a large amount of total utility obtained by having very many persons living at near-zero levels of utility."[12] The Repugnant Conclusion suggests that it is difficult to sketch out an ideal world from a few basic principles that will not—when taken to its logical extreme—seem repugnant.

It is for these reasons that the EEAT is not a panacea to ethical eating dilemmas. It is simply a tool. It helps us make connections between the physical world, the economic world, our beliefs, and our choices. If there is one thing we have learned about the ethics of eating it is this: you cannot reason your way out of ethical eating dilemmas. Every logical sequence you attach to your assumptions will eventually carry you to a place you will feel uncomfortable. It is at these uncomfortable places where you must incorporate your intuition—and emotion. In fact, to some degree, you have no choice.

Your Turn

We have introduced you to Bailey, Adrian, Paul, Margaret, and Jackson, but chances are some would disagree with some of their animal welfare scores. It would be surprising for it to be otherwise. We all have different tastes and preferences; some people could not imagine giving up hamburgers, but others make the sacrifice easily. If you want to get serious about your eating choices and determine how they might influence animal well-being, you too can put EEAT to work. To provide a little guidance, we have prepared the worksheet shown in

Table 8.7.[13] An electronic version is available at the book's website: <http://asp. okstate.edu/baileynorwood/Survey4/Default.aspx>.

Column A, titled "Your Change in Consumption," is the location to enter the change in diet you are considering. For example, if you are thinking of consuming an additional pound of beef, place +*1* in row marked *Beef* underneath Column A. If you are contemplating giving up a dozen cage eggs, place a −*12* in the row marked *Cage Eggs*. Columns B and C are where you can enter your beliefs about the well-being of breeder and non-breeder animals. In columns B and C enter your animal welfare scores from −10 to +10 for each food item being considered. Columns E through F simply report the statistics we have previously discussed. Finally, the EEAT score for each food change can be calculated in the last column. The total change in animal well-being that will result from your planned dietary change is the sum of the EEAT scores for all food items considered. Go ahead, give it a try!

Public Views

If you made an attempt at completing the worksheet, you no doubt found it difficult to assign welfare scores for animals in different systems. You can rely on expert opinion, but, frankly, experts are often funded by and aligned with special interests groups. What would a group of impartial citizens think about the issue? To answer this question, we have spent countless hours conducting phone, internet, and face-to-face interviews with people all over the US.

Some of the answers received provide some insight into how the average person might complete the animal welfare scores. Consider, for example, a nationwide telephone survey we conducted with over 1000 randomly selected US households in 2007. We asked respondents several questions regarding the practices they thought were important in determining animal well-being, including questions about how important animal well-being is compared to other factors, such as the price of food. Based on people's answers, we found that most people could generally be categorized into one of three groups.[14] The first group, constituting 14 percent of the population, were comprised of people we refer to as *price-seekers*. These individuals might have some preferences for how farm animals are raised, but when shopping they mainly focus on price. Farm animal welfare is simply not much of a concern for price-seekers.

A larger group, constituting 40 percent of the population, is comprised of people we call *basic-welfarists*. These individuals are concerned about the well-being of farm animals but are primarily concerned with making sure the animals have adequate feed, water, and health care. Basic-welfarists might have a problem with the modicum of food broiler breeders receive and the nutritional deficiencies of veal feed. However, broiler breeders make up only a

Table 8.7 Ethical Eating Assessment Tool (EEAT) Worksheet

Food item	A. Your change in consumption	B. Your welfare scores for breeder animals	C. Your welfare scores for non-breeder animals	D. Consumption factor	E. Number of breeder animals	F. Number of non-breeder animals	EEAT score = A×D×(B×E + C×F)
Beef	___ lbs	___	___	0.68	0.000600601	0.001201201	___
Chicken meat	___ lbs	___	___	0.76	0.001804675	0.259740260	___
Milk	___ lbs	___	___	0.56	0	0.000017429	___
Veal	___ lbs	___	___	0.69	0	0.006849315	___
Conventional pork	___ lbs	___	___	0.74	0.000167336	0.007195450	___
Crate-free pork	___ lbs	___	___	0.71	0.000167336	0.007195450	___
Shelter-pasture pork	___ lbs	___	___	0.53	0.000175193	0.007182901	___
Coge eggs	___ lbs	___	___	0.91	0.000027491	0.001964637	___
Cage-free eggs	___ lbs	___	___	0.83	0.000044554	0.003184713	___

small percentage of all chickens within the broiler industry, and the problems with veal diets seem to have improved substantially.

Overall, both price-seekers and basic-welfarists, which represent a combined 54 percent of the population, should be generally happy with modern livestock production. The factory farms so demonized by animal advocacy groups provide animals with exactly what these 54 percent of Americans desire. Factory farms are efficient, thus resulting in the low prices that bargain-seekers want. They also provide the basic needs of the animal, meeting the requirements of the basic-welfarist.

A third and final group of people believe that it is important that farm animals perform and exhibit natural behaviors: access to the outdoors and opportunities for dust bathing, for example, are important for this group of people. Because of their interest in natural behaviors, we refer to the third group of citizens as *naturalists*, which comprise 46 percent of the population (these percentages are rough estimates; the true percentage could be 50 percent or a little more). The price of food is less important for naturalists, and as such they are likely to be willing to pay higher prices so that farm animals can exhibit natural behaviors. The naturalist will likely take issue with factory farms because they desire that meat, eggs, and dairy be from animals raised outside of cages and concrete floors. It is the vocal members of this group who publicly oppose factory farming.

Naturalists are likely to generally agree with Bailey's welfare scores, although there will certainly be some disagreement among members of this group relating to individual species. Most naturalists are likely to agree that veal and egg production are among the most inhumane industries, and that sows in pork production are treated inhumanely. Once aware of how different farm animals are raised, naturalists might modify their eating habits to accommodate more beef, dairy, and chicken, and attempt to forgo veal, eggs, and pork. Naturalists are more likely to seek alternative food retailers who provide cage-free eggs and shelter-pasture pork.

To summarize, roughly half the population are comprised of bargain-seekers and basic-welfarists and they seem to generally approve of all the food products in Table 8.7. Though they are likely to agree with the naturalists that the egg and pork industries are among the least humane, the welfare scores for all products are likely to be positive. Naturalists are more likely to assign scores throughout the −10 to 10 range of possible welfare scores, indicating beef, broiler, dairy, cage-free egg, and shelter-pasture pork deserve positive welfare scores; the remainder would receive negative scores. It should be noted that just like grouping voters into Democrat and Republic party supporters would mask the true variability in political beliefs, a significant portion of individuals belonging to one group yet share some views with other groups. Finally, these three groupings, related to animal welfare, cannot capture more extreme beliefs. Just as there are some people who believe animal feelings deserve no attention, there are some who would assign negative welfare scores to all the foods in Table 8.7, based on the ideology of opposition to the ownership of animal lives.

A Line in the Sand

One of the key decisions in setting the animal welfare score is whether one believes an animal is better off dead than alive. There is a *line in the sand* separating miserable animals which are better off dead from those which experience more positive than negative emotions. In this context, it might be useful to know where ordinary people do draw the line. Do animals "suffer" under certain circumstances, and are farm animals thought to be "happy and content" in other?

To answer this question it is more relevant to ask such questions of *informed consumers*. Many consumers do not know much about how farm animals are raised. Asking a passerby on any street in America whether they believe chickens "suffer" without giving the person any information or context about how chickens are raised is likely to yield a spurious answer. For example, although 95 percent of all egg-laying hens in the US are raised in cage systems, our research shows that most people *think* only 40 percent are raised in this way (they believe the remainder reside in cage-free facilities).

The lack of knowledge about modern farming poses a challenge for those of us who wish to better understand consumers' views on how farm animals should be raised. To overcome this obstacle, we visited three US cities and hired marketing firms to recruit representative samples of individuals in each location. We visited about 300 people, half of whom discussed pork and half of whom discussed eggs. During the visit we provided detailed information about how farm animals are raised, including information on factors affecting farm animal welfare and a discussion of different systems for raising farm animals. After the information and question/answer session, people participated in a number of activities, one of which included some questions on people's views on whether laying hens and hogs "suffer" in some circumstances and are "happy and content" in others.

When respondents were asked whether they agreed or disagreed with the statement, *"most laying hens in a cage system lead a happy and content life,"* two-thirds disagreed. Similarly, when asked whether they agreed or disagreed that, *"most laying hens in a cage system regularly experience pain and discomfort,"* 52 percent agreed (only 25 percent disagreed and the rest neither agreed nor disagreed with the statement). The findings suggested that informed Americans believe that hens in a cage system lead miserable lives. The answers suggest that hens in cage systems are believed better off dead than alive.

When asked about hens in a free-range system, however, we found that people jumped over the *line in the sand*. A majority of these same respondents, 71 percent to be precise, believed that hens in the free-range system "*lead happy and content lives*" and only 18 percent thought the free-range hens "*regularly experience pain and discomfort*." The responses suggest that most people believe the welfare scores for hens in a cage system are negative (meaning that the animals experience more negative than positive emotions throughout their life), but that the welfare scores for hens on a free-range farm are positive. As discussed earlier in the book, a "free-range" system is defined in this book, and was defined for the subjects, as an aviary, cage-free system with access to outdoors that includes shelter and predator protection. Free-range is not defined as a system that allows hens large pastures and forests to wander. The outdoor access is always fenced-in with chicken wire fitted from below the ground (to prevent predators from digging a hole into the area) and overhead (to prevent hawks, for example, from swooping down and grabbing chickens).

A similar set of questions were asked about hogs. About 70 percent of respondents did not believe hogs in crate systems live a "*happy and content life*," but surprisingly, only 50 percent thought hogs in crate systems "*regularly experience pain and discomfort*" (of the remaining, 25 percent did not think crate hogs regularly experience pain and discomfort and 25 percent were neutral). The answers suggest that people are mixed in their beliefs about whether hogs in a crate system suffer, but are in wide agreement that the life is not "*happy and content*." The answers suggest a slightly negative or near-zero welfare score for crate hogs. By contrast, 67 percent of respondents agreed that the shelter-pasture system led to hogs that were thought to live a happy and content life and only 18 percent of respondents thought shelter-pasture hogs lived with regular pain or discomfort.

As you may have noticed, there was some disagreement between the 300 individuals we interviewed face-to-face and the 1000 people we surveyed over the phone. On the phone, roughly half of the individuals seemed to accept the methods used in the conventional egg, pork, and veal industry, whereas the 300 people we met with in person seemed opposed to those same methods. The difference is due to more than just measurement error. The salient difference between the two groups is that the 300 people we met with were *educated* about livestock agriculture. Consequently, the difference in consumers' preference between the two samples imply that the more consumers are educated about egg, pork, and veal production, the more inhumane they deem the industries to be.

From Animal Well-Being to Dollars

People do not always choose the animal products that produce the highest EEAT scores. One might say that Bailey is a hypocrite because he continues to eat eggs from cage systems with full knowledge that such a choice results in negative EEAT scores. Bailey is quick to admit his hypocrisy, and Bailey did not offer his own personal welfare scores in the pretence he is an ethical person. In fact, Bailey readily asserts that *if* he were an ethical person, he would donate much more to his favorite charity too. At no place in this book do we argue that we set an ethical example for others to follow, because this book is not about us and our beliefs. This book is concerned with educating consumers about the farm animal welfare debate.

Moreover, what a moralist calls a hypocrite, an economist calls a rational consumer who makes trade-offs. The goal in life is not to maximize the EEAT score. An EEAT score is but one factor affecting food consumption decisions. The EEAT summarizes our beliefs about what will happen to animal well-being when a consumption decision is made, but to know what action to take, it should now be apparent that we need more than beliefs; we need to know something about people's *preferences* for improving an EEAT score relative to other things we want out of life. Even if we can all agree on our beliefs and have similar EEAT scores, we are still likely to witness very different behaviors across different people because people often have different preferences.

Bailey, we learned, believes pork from pigs raised on conventional farms is the result of low welfare but shelter-pasture pork is the result of high welfare. Yet, Bailey still purchases conventional pork (though infrequently) and only occasionally purchases shelter-pasture pork. Now consider a fellow named John who has the exact same beliefs as Bailey, and thus the exact same EEAT scores. John, however, only purchases shelter-pasture pork. The reason for these differences in behavior must be that Bailey and John have different preferences. Compared to Bailey, John cares relatively more about animal welfare than all the other things he might consume. It is not that Bailey is indifferent to the well-being of farm animals, only that he is relatively more interested in all the other things he could buy. It is thus important to measure both a person's beliefs and their relative preferences; and as economists, we accomplish this task with the combined measure of *WTP*.

In economics, the words "happiness" and "satisfaction" are all summarized with the word *utility*. Think of your utility score as your happiness score at any point in time. Utility is a number indicating the well-being of a person, and it is a number that is presumed to drive our behavior. If you choose to watch an hour of television instead of reading *War and Peace*, we would say you did so because the utility of the former is greater than the latter. Your choice reveals that you expected to have more utility watching TV than reading. Let us return to the decision of whether to substitute shelter-pasture pork for conventional pork. Suppose that Bailey and John consider replacing the consumption of 1 lb of conventional pork with 1 lb of shelter-pasture pork. Because Bailey and John have the same beliefs about how farm practices affect animal welfare, we know that they have the same EEAT sores:

Farm animal welfare consequence of Bailey & John forgoing 1 lb of regular pork +
Farm animal welfare consequence of Bailey & John eating 1 lb of shelter-pasture
pork

$$= (-1 \times 0.74) \ [\{0.000167336\} \ \{-7\} + \{0.007195450\} \ \{-2\}] + (1 \times 0.53)$$
$$[\{0.000175193\} \ \{4\} + \{0.007182901\} \ \{4\}]$$
$$= 0.01152 + 0.01560$$
$$= 0.0271$$

The EEAT score of 0.0271 indicates Bailey and John's shared beliefs of how hog well-being is impacted by the food choice. However, Bailey and John experience different levels of *utility* from knowing their actions helped farm animals. Not only do they differ in their utilities for farm animal well-being, Bailey and John differ in their willingness to part with a dollar. John may be wealthier than Bailey and hence find it a smaller sacrifice (i.e., loses less utility) to pay more for food. Or, Bailey may place a greater importance on saving money than John, which leaves Bailey less money to spend on food. Or, Bailey may place far more emphasis on giving to human charities than to animal charities. For a thousand different reasons, an improvement in the life of an animal means different things to different people.

Spending a dollar creates negative utility or *disutility* because we do not like to give up our money. Giving up a dollar to buy one good prohibits a person from using that dollar on some other good. Paying more for better animal treatment requires giving up other things. The disutility represents the utility a person would have gained by spending a dollar on something other than the good at hand. By combining the perceived change in animal welfare, the utility from improving animal welfare, and the disutility of one dollar, we have a mathematical relationship showing the maximum amount an individual will pay for more humane food over a less humane food.

Table 8.7 shows the formula used for calculating WTP. In the numerator is the perceived change in animal welfare (i.e., the EEAT score) multiplied by the extra utility a person gets from improving animal welfare. The numerator is the benefit of the change or the bang for your buck. In the denominator is the disutility from spending \$1, which constitutes the cost, or what one gives up, to improve farm animal welfare. Notice that the price of pork is not a consideration in calculating WTP. WTP is the *maximum* amount someone would pay for more humane food, not the *market* price; if someone's WTP premium is greater than the actual price premium, we would expect them to buy the more humane food (see Figure 8.2).

Recall that both Bailey and John agree that animal welfare is improved when they purchase shelter-pasture instead of regular pork by a factor of 0.027. John, however, receives more personal satisfaction from helping the hogs. The extra utility John gets for every one unit increase in EEAT equals 500, compared to Bailey's utility of only 50. Bailey also experiences more disutility of parting with \$1 than John (10 vs 3). All things considered, the maximum premium Bailey is

Willingness-to-pay premium for the more humane food over the less humane food

=

(perceived change in animal welfare)(utility from improving animal welfare)
───
(disutility from spending \$1)

Example A—Bailey replacing 1 lb of regular pork for 1 lb of shelter-pasture pork

$$\textit{WTP premium for shelter-pasture pork} = \frac{(0.0270)*(50)}{(10)} = \$0.14$$

Example B—John replacing 1 lb of regular pork for 1 lb of shelter-pasture pork

$$\textit{WTP premium for shelter-pasture pork} = \frac{(0.0270)*(500)}{(3)} = \$4.50$$

Figure 8.2 Willingness-to-pay (WTP) for more humane food

willing to pay for shelter-pasture pork over conventional pork is \$0.14 per lb, while John is willing to pay up to \$4.50 more per pound.[15] Intuitively, WTP calculates how much sadness you would have to experience (or disutility) by giving up dollars to exactly offset the extra happiness (or utility) you would receive by improving animal welfare. If the price of shelter-pasture pork is \$2 per lb higher than conventional pork, Bailey would not purchase the shelter-pasture option because the cost (\$2) is higher than his WTP (\$0.14). John would readily make the upgrade to shelter-pasture pork because his maximum WTP (\$4.50) is greater than the actual cost of \$2.

The example helps illustrate how WTP is a concept that can translate people's beliefs about the well-being of farm animals and their relative preferences for farm animal well-being into a single metric that can be directly compared to costs. The next chapter reports on our attempt to measure Americans' WTP for better animal treatment. The WTP framework also provides clear insight into why people make different food choices. The calculation for WTP shown in Table 8.7 indicates that there are three factors explaining differences in peoples' shopping behaviors. People's WTP for more humane food will differ as:

1. People have different beliefs or perceptions about the well-being of farm animals in different production systems;
2. People differ in the personal pleasure they receive from helping farm animals; and
3. People differ in their willingness and ability to part with a dollar.

Consider each of these three factors in turn. On the first point, this chapter has already provided ample evidence that people's beliefs differ about the well-being of animals in different systems (recall the naturalists and the basic-welfarists).

Indeed, although hens producing cage-free eggs are generally thought to have higher levels of well-being, the folks at Peaceful Prairie Sanctuary have a very different set of beliefs.[16] Second, even if two individuals agree on how farm animals are affected by changes in how they are raised, their desire to provide animals with a more pleasant life will vary in intensity. Our research shows, for example, that women generally care more about helping farm animals than men, which might be one reason why animal rights groups tend to be dominated by female members.[17] Finally, there is a good reason why shelter-pasture pork is sold in Whole Foods but not Wal-Mart, and one of the reasons Whole Foods shoppers tend to be wealthier and to experience less pain paying a few extra dollars than those normally shopping at Wal-Mart.

How the three factors interact to form people's WTP is ultimately an empirical matter—an *important* empirical matter. Retailers need to know consumer WTP before they offer new "animal compassionate" products, farmers need to know if they can recoup the costs of cage-free systems, and regulators need to know whether the benefits of animal welfare policies exceed the costs. The only way to really know the factors needed to calculate a person's WTP (their EEAT score, their utility from improved animal well-being, and their disutility from a dollar) is to ask. In the next chapter we discuss our interviews with over 300 Americans about their willingness to pay more money to reduce the misery and enhance the levels of well-being on livestock farms.

Appendix—Relationship Between Individual Food Consumption and Total Consumption

In this chapter, we have argued that if you chose to stop eating 1 lb of beef, the overall amount of beef that would continue to be produced and consumed would likely fall by something *less* than 1 lb. In this Appendix, we present our calculations used in determining the amount by which consumption would actually fall. The market for animal food products is modeled using the standard model of supply and demand. For further details of the economic model, see chapter 3 of our textbook, *Agricultural Marketing and Price Analysis*.

Let X be the per capita consumption of a food item and N be the population, making NX total consumption of the good. Suppose that one individual decreases their consumption of the good by δ. This represents a demand shock, in the sense that if price did not change, the percentage change in quantity demanded would be

$$\%\Delta QD = -\delta/NX.$$

However, price will change. Price will fall, inducing some consumers to increase their consumption and inducing suppliers to produce less. To predict these changes we must utilize the concepts of supply and demand elasticities. A supply elasticity predicts the percentage change in production of a good by firms in the market if the price of the good rises by 1 percent. For example, if the supply elasticity equals a value of 2, then a 1 percent increase in prices causes firms to increase production of the good by (2)(1 percent) = 2 percent. Similarly, if the price of the good rises 2 percent, the predicted response by firms is to increase production of the good by (2)(2 percent) = 4 percent, and if prices fell 5 percent firms would react by decreasing supply by (2)(−5 percent) = −10 percent (the change is −10 percent, the "decrease" is 10 percent—semantics).

Demand elasticities are the consumer corollary to the supply elasticity. If prices rise by 1 percent and the demand elasticity is −1.5, consumers respond to the 1 percent increase in price by changing their purchases by (−1.5)(1 percent) = −1.5 percent. Similarly, if prices fell by 20 percent consumers would increase their purchases by (−1.5)(−20) = 30 percent.

Both supply and demand elasticities depend on the sensitivity of firms and consumers to price changes. Using price and production/consumption data, economists can estimate elasticities, though such statistical estimation can be difficult. In many cases common knowledge of the industry structure and consumer habits allow one to develop reasonable elasticities without data. Economists assume that at any point in time the quantity supplied will equal quantity demanded, meaning firms will sell all they produce. If something changes in the market, such as a decrease in consumer demand for the product, the percentage change in the quantity supplied must equal the percentage change in quantity demanded. Otherwise, the amount produced by firms would not equal the amount purchased by firms.

Let E_D and E_S be the demand and supply elasticities, respectively, and $\%\mathit{\Delta}P$ be the new equilibrium price after the person decreases their consumption by $-\delta$. The change in quantity supplied (QS) and demanded (QD) are

$$\%\mathit{\Delta}QS = E_S(\%\mathit{\Delta}P)$$
$$\%\mathit{\Delta}QD = E_D(\%\mathit{\Delta}P) - \delta/NX.$$

Setting the two equations equal to one another, we solve for the change in price.

$$\%\mathit{\Delta}QS = \%\,\mathit{\Delta}QD$$
$$E_S(\%\mathit{\Delta}P) = E_D(\%\mathit{\Delta}P) - \delta/NX$$
$$(\%\mathit{\Delta}P) = [-\delta/NX]/[E_S - E_D]$$

Next, plugging the $\%\mathit{\Delta}P$ back into the supply and demand equations, we can solve for how one person's decrease in consumption by an amount $-\delta$ changes total production and consumption of the good.

Table 8.A1 Elasticity Assumptions Behind Ethical Eating Assessment Tool (EEAT)

Food item (raised conventionally unless otherwise noted)	Demand elasticity	Supply elasticity
Beef	−0.35	0.75
Chicken meat	−0.64	2.00
Milk	−0.80	1.00
Veal	−0.90	2.00
Conventional pork	−0.70	2.00
Crate-free pork	−0.80[a]	2.00
Shelter-pasture pork	−0.90[a]	1.00
Cage eggs	−0.20	2.00
Cage-free eggs	−0.40[b]	2.00

Notes: The milk category includes cheese, where one pound of cheese is equivalent to ten pounds of milk. Demand elasticities are taken from Huang and Lin (2000), Huang (1985), and Sumner *et. Al*. (2009). Supply elasticities and the veal demand elasticities are based on authors' research and assessment. The elasticity of pork supply is taken from Norwood and Lusk (2008). The supply elasticity of beef is informed by Marsh (1994). The milk supply elasticity is informed by Ahn and Sumner (2006). The total population in the US is assumed to equal 303,824,640. There are undoubtedly elasticities of substitution between goods, especially cage and cage-free eggs (see notes below). These substitution elasticities are accounted for indirectly in how the user chages his/her food choices. For example, if purchases of cage and cage-free eggs are negatively correlated, the user can decrease cage-egg purchases by one egg each times—her cage-free egg purchases rise one egg.

[a] If the pork was produced in a crate-free system or shelter-pasture system, demand elasticities may differ due to the fact that the product is differentiated, making the good somewhat novel. It is impossible to empirically estimate these elasticities, though we do know they should be more elastic than their generic countrerpart. Thus, if the good in question is crate-free pork it is assigned a demand elasticity of -0.8, and if a shelter-pasture product the demand elasticity is -0.9.

[b] (See previous note) If the eggs are cage-free the demand elasticity is assumed to equal -0.4.

[c] Producing shelter-pasture pork has greater land requirements than pork produced in confinement, and thus the supply elasticity is halved.

$$\%\Delta QS = E_S([-\delta/NX]/[E_S - E_D])$$
$$\%\Delta QD = E_D([-\delta/NX]/[E_S - E_D]) - \delta/NX = E_S([-\delta/NX]/[E_S - E_D])$$
The total decline in consumption of the good is
$$\Delta QS = E_S([-\delta/NX]/[E_S - E_D])(NX) = -\delta(E_S/[E_S - E_D]).$$

Thus, by identifying the population, per capita consumption, and elasticities, the change in quantity in reaction to one person decreasing their consumption of an animal food product can be calculated (Table 8.A1).

Number of Breeder and Non-breeder Animals in Food Item

The purpose of this section is to describe the assumptions used to determine the number of breeder and non-breeder animals associated with a given amount of food. For beef, pork, and chicken the calculation refers to the number of breeder animals and animals raised exclusively for meat (never bred, or non-breeder animal) associated with a single pound of meat. For eggs, the calculations refer to the number of laying hens and breeder birds required for one egg, and for dairy the numbers describe the number of milk cows producing the milk.

We would like to draw to your attention a few qualifications to this model composition. First, the animal numbers do not account for differences in the lifespan of an animal. The numbers relating to chicken meat and pork refer to the number of chickens and hogs used to produce 1 lb of meat, with no adjustments made for the fact that broilers live a much shorter life than the hog. How these lifespan differences impact on welfare must be left to the opinions of the person using the model, and reflected in the welfare scores they apply to each food item. It is assumed that beef comes only from beef cows, despite the fact that a significant portion of cows slaughtered are dairy cows. When you eat a hamburger, there is a very good chance that the beef was derived from a dairy cow. However, we separate the link between dairy production and beef in the model so that we can focus on the different lives of the beef and dairy cows we eat. However, those interested in establishing the link between dairy and beef can do so using the directions followed under the dairy section. At no point are the breeding males accounted for in the beef, dairy, and pork industry because they comprise such a small portion of animal food production.

Cage Eggs

Cage eggs are assumed to be white eggs from a White Leghorn type breed. This hen will produce 509 eggs throughout its 2.21 years of life. So for every one egg eaten, $(1/509) = 0.001964637$ birds are affected. It is assumed that the number of breeder birds needed to produce chicks is the same for laying hen and broiler breeds. However, since all male chicks are killed in a laying chicken hatchery, a laying chicken operation will need twice as many breeders to obtain 100 hens than a broiler operation needs to obtain 100 chickens (male or female). So, we assume that of all the birds used in layer production, $0.69 \times 2 = 1.38$ percent are breeders (see the broiler section for more details).

If X denotes all layers, and XB denotes those used for eggs, $XB = (1 - 0.0138)$ X and XR is breeder birds, $XR = 0.0138(X)$. For every layer used for eggs, $XR = 0.0138(X) = 0.0138(X = XB / (1 - 0.0138)) = 0.01399$ birds are used in breeding.

Thus, one egg is associated with $(1/509)$ = 0.001964637 layers and $(1/509)$ (0.01399) = 0.000027491 breeder birds.[18]

Cage-free Eggs

Cage-free eggs are assumed to be produced using brown birds. One laying hen will produce 314 eggs throughout its 1.54 years of life. So for every egg eaten, $(1/314)$ = 0.003184713 birds are affected. Assumptions about breeder birds are identical to those under the cage system. Thus, one egg is associated with $(1/314)$ = 0.003184713 layers and $(1/314)(0.01399)$ = 0.000044554 breeder birds.[19]

Beef

One steer/heifer weighs 1200 lbs and produces 0.45×1200 = 540 lbs retail beef; 1 cull cow weighs 1300 lbs and provides 0.45×1300 = 585 lbs retail meat. Cows tend to produce three offspring, after which they are harvested at 5 years of age. Let X be the number of breeding cows in the herd. Every year, $(1/3)X$ cows are culled and $(1/3)X$ replacement heifers are added to the breeding herd. So the rate of heifers (as a percentage of breeding cows) reserved for breeding (not sent to slaughter) each year is R: $R(0.5)X = (1/3)X$; $R = (1/3)/0.5 = 2/3$; $R = 2/3$. Or, $(2/3)(1/2)$ = $1/3$ of every X is reserved for cow replacement.

One-third of the offspring (two-thirds of the heifers) are reserved for breeding. Consequently, one cow is slaughtered for every $(3)(0.5 + 0.5(1/3))$ = 2 market cattle. This group of one cow / 2 market cattle produces $585 + 2 \times 540$ = 1665 lbs of meat. Thus, for every lb of beef you eat, you are responsible for $(1665)^{-1}(1)$ = 0.0006006006 cows and $(1665)^{-1}(2)$ = 0.0012012012 market cattle.

Confinement-crate and Confinement-pen Pork

Sows in conventional systems tend to be slaughtered at 3 years of age and breed for 2 of those years, producing 22 babies per year. Thus, each sow produces 44 babies, though one of those will be reserved for breeding to replace the sow when she is slaughtered. One 250 lb hog produces 0.535×250 = 134 lbs retail pork and a 400 lb sow produces 0.535×400 = 214 lbs retail pork. Thus, when the sow is slaughtered she provides 214 lbs of retail pork, and she was responsible for producing 134×43 =5762 lbs of retail pork indirectly through her offspring. The sow and her offspring together then constitute $5762 + 214$ = 5976 lbs of pork. Thus, eating 1 lb of pork makes one economically responsible for the lives of $(5976)^{-1}(1)$ = 0.000167336 sows and $(5976)^{-1}(43)$ = 0.00719545 hogs. To see why this calculation is valid, suppose that you consumed 5976 lbs of pork.

On average, those lbs would have been derived from one sow and 43 market hogs.

There is some debate as to whether sows are more or less productive in a confinement-crate or confinement-pen system, but no compelling evidence suggests the number of market animals produced by each sow is different in the two systems.[20] Thus, the numbers relating retail pork to animal numbers are the same for regular pork and crate-free pork.

Shelter-pasture Pork

Pork raised in a shelter-pasture system will have different parameters due to the fact that the breeding animals are not as productive. The number of market hogs produced by each sow is 14, compared to 22 in the confinement system.[21] Utilizing sows that are good mothers is paramount, so farmers will likely keep good sows on for longer than they would in a confinement-pen or confinement-crate system. We assume sows are harvested at 4 years of age, on average, producing offspring in 3 of those years. After these changes, we make the same calculations as in regular pork.

Each sow will produce $14 \times 3 = 42$ farrows during her life, but one out of those 42 offspring will be reserved for breeding to replace the sow when she is slaughtered. Thus, hogs tend to slaughtered in a manner where one sow is slaughtered for every 41 hogs. This group of 41 hogs and 1 sow produces $214 + 134 \times 41 = 5708$, and consuming one pound of shelter-pasture pork makes one economically responsible for $(5708)^{-1}(1) = 0.00017519271$ sow and $(5708)^{-1}(41) = 0.0071829012$ hogs.

Broilers

The average broiler today weighs 5.5 lbs, 70 percent of which is retail meat. Of all the birds involved in broiler production, 0.69 percent are raised in breeding facilities. If X is the total number of birds, XB regards those made into meat $(XB = [1 - 0.0069]X)$, and XR is breeder birds, $(XR = [0.0069]X)$. For every bird eaten, $XR = 0.0069(X) = 0.0069(X = XB / (1 - 0.0069)) = 0.006948$ birds are used in breeding. So for every 1 lb of chicken meat eaten, $1 / (5.5 \times 0.7) = 0.2597$ broilers are affected and $(1 / (5.5 \times 0.7)) \times 0.006948 = 0.001804675$ breeder birds are affected. It is assumed that meat from breeder birds is not processed for human consumption, but relegated to rendering instead.

Dairy

One dairy cow produces about 7.5 gallons of milk each day. Each cow only milks 10 months in each year, so the annual production for each cow is about 2250 gallons each year. Those who enjoy cheese may be interested in the fact that it takes 10 lbs of milk to make 1 lb of cheese. Also, a gallon of milk weighs 8.5 lbs. Assume that one dairy cow lives 4 years, with 3 of those years being spent providing milk. Consuming 1 lb of milk thus corresponds to $1 / (2250 \times 8.5 \times 3)$ = 0.0000174292 milk cows.[22] One serving of milk is typically thought to refer to 8 oz of milk. For those interested, one 8 oz serving of milk corresponds to (8 oz / (128 oz per gallon))(8.5 lbs per gallon) = 0.53 lbs of milk. Thus, 1 lb is milk is roughly two servings of milk.

Breeder animals are not considered because milk comes from the breeder cow herself, and the use of artificial insemination makes the number of bulls employed incredibly small. One important question is how we relate the interactions between the dairy, beef, and veal industry. It is true that male calves born on dairy farms are subsequently raised for beef or veal, but does that mean one is responsible for veal because they drink milk? If the world drank milk but consumed no beef or veal, dairy products would simply euthanize male calves just as the hatcheries (for layer breeds) kill male chicks soon after birth. Drinking milk creates profits for dairy products, which acts to subsidize beef and veal. Though milk production subsidizes veal production, the linkage is insignificant, and thus ignored in the model.

A significant number of dairy cows are processed into beef though, and a regular beef eater is also a regular consumer of dairy cow muscle. Because the purpose of the model is to distinguish between the lives of beef and dairy cows, the linkage between dairy and beef is ignored. Consuming milk does subsidize the production of beef, because dairy producers will sell the male dairy cows at almost any price. The extent to which this subsidizes beef production is unclear though. Because this linkage is so uncertain, and likely insignificant, it is ignored.

Veal

Veal calves are harvested at 325 lbs[23] and produce $0.45 \times 400 = 146$ lbs of retail meat each.[24] These calves will be born regardless of how many dairy cows are produced. As previously stated, being able to sell them for a price subsidizes milk and vice versa, but the link is small. So we simply say that one lb of veal meat affects $1/146 = 0.0069$ veal calves.

CHAPTER 9

Consumer Expressions

The Willingness of Consumers to Pay Higher Food Prices in Return for Improved Animal Care

> "*Improvements to farm animal welfare can only come about within the context of the forces that drive the free market. In essence, consumers need to afford a greater extrinsic value to farm animals ... The responsibility is therefore on the consumer to convert an expressed desire for higher welfare standards into an effective demand.*"
>
> A. J. F. Webster in *The Veterinary Journal* (2002)

Silent Majority

The farm animal welfare debate has primarily been held between special interest groups. The Humane Society of the United States (HSUS) purports to serve the interest of animals and the National Pork Producers Council (NPPC) purports to serve the interests of pork producers. These are just two of the many groups and individual advocates that have traded punches over the past decade. The vast majority of the population, whom we collectively refer to as "consumers," is not involved with livestock production or animal advocacy though. Who will represent this large consumer group; who is undeniably most affected by animal welfare laws? We the consumers are—and this is why we have included this chapter.

There is a tendency on the part of advocates for and against modern livestock farming to discount the need to know the thoughts of the average consumer—unless a poll on consumer thoughts happens to support their interests. In our experience, farm groups claim to make production decisions based on "sound science," with the implicit (sometimes even explicit) argument being that consumers are too ignorant or gullible to make decisions about the raising of animals. These farm groups often make these claims without being aware of a substantial number of studies that oppose their claims. Farm groups should remember who buys their products, and recognize that consumers are under no obligation to spend their scarce income on animal products. Animal advocates can also be dismissive of consumer concerns, deeming the consumers to be misinformed or uneducated. This is in many ways understandable, as their responsibility is towards their donors, most of whom donate no money towards their group. Producers and activists are correct; the average American has very little knowledge of the modern livestock farm.

In their rush to improve animal living conditions, animal advocacy groups can be less than forthright about the costs of the activities planned, sometimes even arguing that improving animal welfare would decrease costs—alas, if only that were the case.

Similarly, in their rush to fend off attacks, livestock groups sometimes claim that modern agriculture could not possibly improve animal care even if it wanted to; this claim can be proved false with even a modicum of research.

The reality is that improvements in animal welfare can be easily achieved, but this will require consumers to pay higher food prices. Are consumers willing to pay this higher price? This is a question special interest groups on either side wish to avoid asking, not because they are unaware of the question's importance, but because they fear the answer may be contrary to their cause.

It would intuitively seem obvious that food retailers would be interested in understanding consumers' willingness to pay (WTP) for better animal care, but this does not seem to be the case. One reason is that they may prefer to ignore the issue, because products on sale claiming better animal care may seem to degrade other food products on sale in the store. In some ways, does not the presence of cage-free eggs suggest conventional egg production is inhumane? If so, would this not question the humaneness of beef, pork, and dairy production? It is also our observation that food retailers tend to believe that their intuitive understanding is more enlightened than marketing research conducted by others, and tend to disregard marketing research for the simple fact that it was conducted by other individuals. At the same time, they often spend large amounts of money on marketing research.

What about politicians? Should they show interest in their constituents' opinions about farm animal welfare? Presumably policy-makers in a republic are elected to office to serve the public's interest. However, farm animal welfare is a relatively minor issue compared to abortion, health care, and wars. Only when politicians are forced to confront the farm animal welfare issue are they interested in constituent views, and only recently has this taken place.

One of the primary motivations for undertaking writing this book is to give a voice to the large, but disregarded group that has yet to make its voice heard in the farm animal welfare debate. In this chapter, we report on our attempt to measure the intensity—in dollars—with which consumers desire to improve the lives of farm animals.

Measuring and communicating consumer preferences is difficult. Much of the skepticism expressed about the use of research from consumer surveys is well-founded. People can say whatever they want in telephone polls with no consequence to themselves. Indeed, the evidence is quite clear that when people are asked hypothetical questions like, "How much more would you be willing to pay for cage-free eggs as compared to cage eggs?" the typical answer is often two to three times higher than what people would pay if really faced with a purchase decision.[1] We once conducted a research study where we asked some consumers in our hometown if they would be willing to purchase locally grown organic ground beef, and based on what they *said* they would purchase, we predicted that a local store should have sold about 360 lbs of the organic ground beef. When we actually put the new ground beef product for sale in the store, we sold only 12 lbs![2] When asked if they would pay more for a local, organic product, subjects readily agreed, but when it came time to actually pay money their enthusiasm waned. Our goal is to avoid these kinds of gaps between stated preference and real action, and we do this by asking consumers to "put their money where their mouth is." Every result reported in this chapter is based on how consumers *actually* behave, not how they *said* they would behave. We accomplish this by conducting *economic experiments*—we construct markets in which people must exchange real food (or real animals) for real money.

Compassion in the Grocery Store

If consumer research is best done by observing what people actually do when shopping, one might ask why surveys or economic experiments are needed at all. The answer is simple. With the exception of eggs, many consumers do not have the choice of buying pork, milk, and chicken meat from animals raised under alternative production systems; these options are not readily available in most stores. Although specialty grocery stores, such as Whole Foods, sell products like Animal Welfare Approved pork, it would be inappropriate to use such information to make statements about general consumer preferences because shoppers at Whole Foods tend to be wealthier and more concerned about food safety, animal welfare, and the environment than the typical consumer. Thus, to measure the willingness of consumers to pay more of their real money for higher standards of care, we must construct our own markets where real purchases take place by a representative sample of Americans.

As we mentioned, the exception to this argument is eggs. Walk into almost any grocery store and you will find a variety of eggs that differ in color, size, and, importantly for our purposes, production systems. Many grocery stores closely track sales on all items scanned and sold, and we were able to acquire information regarding the sales of over 108 billion eggs occurring in major grocery chains all across the US from 2004 to 2008.[3] Observing purchase patterns for regular, cage-free, and organic eggs provides insights into consumer desire for better animal care. Although hens in organic systems are required to be cage-free and have access to outdoors, we should recognize that people might buy organic eggs for reasons other than animal welfare (e.g., concerns over their own health or the environment).

Figure 9.1 illustrates the market share for cage-free and organic eggs over time relative to all eggs.[4] Two striking features stand out. First, the market share for cage-free and organic eggs has doubled in four years; apparently these varieties are much more popular than they once were. Second, and more importantly, despite the doubling of market share, the market share of cage-free and organic eggs are both less than 2 percent. Eggs which provide higher hen welfare represent a very small percentage of total eggs sold, and this evidence would seem to suggest that a majority of consumers care little for providing laying hens with a better life.

However, it is important to ask to what extent the relatively small market shares are the result of high prices, as opposed to a lack of desire for better animal care. Figure 9.2 shows the average price of a dozen cage, cage-free, and

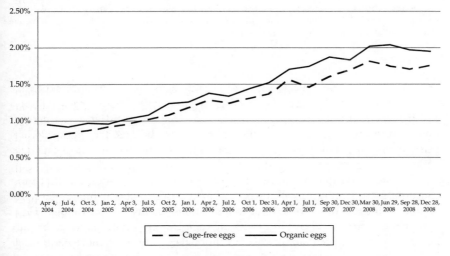

Figure 9.1 Market share of cage-free and organic eggs

Note: These data are obtained from grocery store scanner data from stores across the US and include sales information on 126 varieties of organic eggs, 51 varieties of cage-free eggs, and 1637 varieties of regular eggs. Each point refers to the market share for the 13-week period prior to the date indicated on the x-axis. Market shares were calculated as the volume of sales for one product divided by the total sales of cage, cage-free, and organic eggs. Sales of eggs labeled natural, Omega-3, pasteurized, etc., were disregarded as they comprise a very small portion of total sales, and because they do not provide higher levels of animal welfare than traditional eggs.

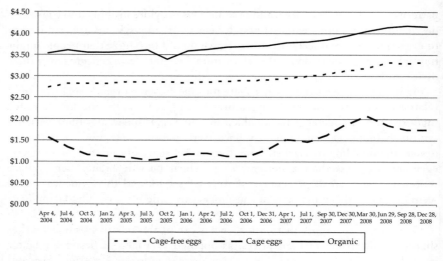

Figure 9.2 Egg prices ($ per dozen)

Note: These data are obtained from grocery store scanner data from stores across the US. The price data include 126 varieties of organic eggs, 51 varieties of cage-free eggs, and 1637 varieties of regular eggs. Each point refers to the weighted average price of each variety (weighted by volume of sales) for the 13-week period prior to the date indicated on the x-axis.

organic eggs over time. Cage-free and organic eggs are *much* more expensive than conventional cage eggs. On average from 2007–8, organic eggs sold at a 175 percent premium to cage eggs, and cage-free eggs sold at a 120 percent premium to cage eggs. On a per-dozen-eggs basis, the cage-free eggs sold at a price $1.57 more than cage eggs. One reason that the market shares of cage-free and organic eggs are so low is that their relative prices are so high.

One conclusion that might be drawn from Figures 9.1 and 9.2 is that most consumers are simply unwilling pay what it costs to obtain higher levels of animal welfare. There are several qualifications to this verdict though. First, note that the farm-level cost of producing cage-free eggs is estimated to be only about 20 percent more than that of conventional eggs, or $0.35 more per dozen.[5] It is unclear exactly why the estimated farm-level costs differences between cage and cage-free are only 20 percent whereas the retail price differences we observe are around 120 percent.

Fortunately, we are in possession of another data source that can help us answer this question. A personal contact provided us with cost, price, and sales information for a chain of grocery stores owned and operated under the same general management. With these data we can calculate the additional cost the grocery store pays for cage-free eggs and also the price premium they charge for cage-free eggs. This chain of grocery stores is well-known and has stores located in many different states. The location of these stores probably does not place them in the same vicinity as the high-end grocery stores discussed in the preceding paragraphs. The data were provided under the condition of anonymity, using data from a trusted provider.

This grocery store chain charges a $0.97-per-dozen markup (markup is defined here as the difference between the retail price and the cost of a specific item to the store)[6] for brown, cage-free eggs and a $0.69-per-dozen markup on regular white eggs. Consequently, the chain does make more money for each dozen of cage-free eggs sold.

Although the store may have higher per-dozen costs managing cage-free eggs,[7] it seems probable that the store still makes higher profits from each unit of cage-free eggs sold. These data alone do not determine whether farmers make higher profits on their cage-free egg sales, but they do suggest that the large discrepancy between price premiums at the grocery store and cost premiums at the farm are partly, and perhaps largely, attributable to higher profits for the food retailer. The discrepancy is certainly not due to cost differences alone.

If food retailers make higher per-unit profits from cage-free egg sales, should they not sell cage-free eggs only? The answer is no, because while there are some consumers willing to pay higher prices for better animal care, many consumers are not. Cage-free eggs are a relatively novel item. As the market matures and if more farmers begin producing cage-free eggs, the cage-free price premium may drop. After years of market adjustments the price premium at the retail level may equal the cost premium at the farm level. However, it is also possible that cage-free prices need to be high for consumer appeal. Consumers may be suspicious of cage-free eggs that sell at relatively low prices, just as they would be suspicious of, say, cheap wine.

These two data sources suggest that the observed retail price differences between cage-free and cage eggs do *not* reflect differences in costs alone. More importantly, it is possible that many more consumers might be more willing to pay the true cost of cage-free production than the market share estimates in Figure 9.1 suggests—if food retailers lowered their prices. This observation also points to another problem with using data from grocery stores to infer people's values for improved animal well-being: it is difficult for us as researchers to identify exactly what causes price and/or market share to be high or low. By moving to an experimental setting we can more conclusively identify consumer demand and WTP. Moreover, producers and retailers might be interested in knowing how consumer purchases of cage-free and organic eggs might change as prices fall, but the grocery store scanner data have so little variation in egg prices that it is difficult to measure how consumer purchases vary with price. For example, if every store charged the same high price premium for cage-free eggs, it is impossible to gauge the change in sales if the premium were lower.

There is yet another reason to move to an experimental setting to study preferences for improved animal well-being. Some people do not buy eggs and still care about hen well-being. Their preferences will not be reflected within grocery store data. Additionally, even egg consumers who care about hen well-being might not buy cage-free eggs because of something called the *free-rider* problem (this issue will be studied in more detail later). If someone derives satisfaction of knowing that someone else buys cage-free eggs without having to purchase the more expensive eggs themselves, they are a free-rider. These observations indicate that there is a *public-good* dimension to the value of animal welfare that should be measured in addition to the *private-good* value, which is what we normally expect to see reflected in grocery store purchases.

If we have not yet convinced you of the need for a more robust investigation into consumer preferences for farm animal welfare, one more example might help make our case. The need for an experimental approach is nicely illustrated by what might be called the California egg paradox.

The California Egg Paradox

As shown in Figure 9.1, sales of cage-free and organic eggs are only about 5 percent of all egg sales. Yet in November 2008, 63.5 percent of voters in California decided to ban cage eggs by passing Proposition 2 (hereafter, Prop 2). Why are consumers seemingly so compassionate in the voting booth and yet so indifferent in the grocery store? Several factors are at play. First, the profile of people who purchase eggs and the profile of people who vote are not the same. Almost all voters buy eggs, but not all egg consumers vote. The average voter is likely to be older, better educated, and more concerned with social issues than the average food shopper. A second factor relates to the fact that Californians may be different from people in other parts of the US. California is known as a state with a penchant for greater government influence in the daily lives of its citizenry.

Though the data indicate that Californians are indeed different from their eastern counterparts, they are not that different. For example, retail scanner data of egg sales for the San Francisco, Oakland market indicates that expenditures on organic and cage-free eggs represent only about 10 percent of all egg expenditures. While this is higher than the nationwide 5 percent share, it is nowhere close to the 63.5 percent of Californians voting in favor of Prop 2.

Another reason contributing to the paradox is something we have already mentioned: free-riding. Although a shopper can free-ride on the effects of another shopper's purchases of cage-free eggs without having to pay the cost themselves, the Prop 2 referendum mitigates the free-rider problem by making everyone purchase cage-free eggs. When casting a vote to ban cage eggs, the voter attempts to force him- or herself and all other Californians to purchase cage-free eggs. The incentives in the grocery store are different. Although free-riding may explain part of the Californian egg paradox, in our opinion it does not tell the whole story.

A major reason for this discrepancy has to do with human psychology and the framing of any decision. Social scientists argue that people display different preferences in different settings; they act as *consumers* in the grocery store and as *citizens* in the voting booth. When we wear our citizen hat, we tend to think more about ethical outcomes than we would in the market-place. When we wear our consumer hat, we tend to think less about what is ethical and more about what is practical and in our self-interest. The manner in which we perceive costs differs in the grocery store and voting book context. If we choose to spend more on cage-free eggs, this immediately translates into less money to spend on other things, because shopping requires us to think specifically about what items cost and the amount of money we possess. As citizens in the voting booth, the lack of direct feedback on the monetary consequences of our actions causes us to dwell more on the non-financial aspects of the decision. Put simply, you do not have to immediately pay higher prices to vote in favor of cage-free eggs, and thus our willingness to pay higher prices is greater.

Individuals vary their desire to improve animal welfare depending on whether they wear their citizen or consumer hat. This begs the question: which hat should consumers wear when preference measurements are taken? In our assessment, the best strategy is to pursue a mixture of both. Consumer preferences should be measured where they think in terms of ethics, but also when they express those ethics using their own, real money. Consumers express themselves best when they act as both citizens and consumers.

A final issue is related to consumer information. Many consumers simply do not understand modern agriculture, and tend to have an idealized notion of modern farms. One of the consequences of placing Prop 2 on the ballot in California is that it prompted media attention. Many consumers saw for the first time the reality of modern egg production. A more accurate depiction of modern farms altered social norms regarding food purchases. In fact, our analysis of the issue suggests that the mere fact of putting Prop 2 on the ballot increased the demand for cage-free and organic eggs by 180 percent and 20 percent, respectively.[8]

Consumers' lack of information about farm animal welfare makes it difficult for them to express their preferences for eggs, pork, beef, and milk produced under alternative farm conditions. If one only possesses data on consumer purchases of eggs in a grocery store, it is difficult to know whether the small market share for cage-free eggs is due to a lack of concern for farm animals or incorrect beliefs about how farm animals are actually raised. If we really want to know the value consumers place on improved animal care, a better approach is to recruit a representative cross-sample of consumers, educate them about farm production issues, and then document their WTP for improved animal care in a setting involving real economic consequences. That is exactly what we did.

Empathy in the Laboratory—Private Values From Three Hundred Human Guinea Pigs

During the summer of 2008, we went in search of consumers in Wilmington, NC, Chicago, IL, and Dallas, TX. There is nothing particularly special about these locations except that they are geographically disparate and offer diversity in culture, demographics, and food consumption habits. We hired marketing research companies to recruit 100 people in each city such that the mix of people participating in our research sessions would be diverse and roughly representative of the respective locations. As you might imagine, it is not easy to convince people to give up hours of their time to help two economists they do not know, so we offered people between $65 and $85 (depending on the location) to attend a 90-minute food preference study. We did not tell participants that we would be asking about animal welfare issues because we did not want to recruit only those people who had very strong opinions about the issue. To ensure that we studied a diverse mix of people, the research sessions were limited to 25 people, each at various times during the day: in the morning, in evenings, on weekdays, and on weekends. As a result, our guinea pigs included men and women, old and young, rich and poor, some without high school educations, and a few with PhDs. In each location, half of the participants were asked questions about pork and the other half were asked about eggs. We limited our time to questions about eggs and pork because these production systems are among the most controversial. Although veal is equally controversial, far fewer animals are raised for veal than for pork and eggs.

Given the controversial nature of the topic, we started our research sessions with a few introductions. We told participants that we were going around the country studying a variety of food issues, in addition to animal welfare so that we did not bias people toward them thinking that we, as researchers, viewed this topic as more or less important as other food issues. We also emphasized that as researchers, our only interest was that the participants tell us what they really thought; we were not working for farm production groups or animal activists groups. Because some of our questions related to real-money purchases, we also wanted to make sure that people did not feel obliged to spend money a particular way. Thus, we told participants that the money that had been promised during the telephone recruitment for showing up was their money to keep, and was given in compensation for their time. They could choose to spend some of that money during the research session if they so desired, but there was no expectation or obligation to do so.

The participants' lack of knowledge about livestock production practices required us to spend about 30 minutes educating them about pork or egg production systems. This information session began with a presentation (including pictures) regarding the factors affecting hen or pig welfare, including a discussion of how different animals are raised.[9] For example, in the case of pork we discussed factors such as space per sow, space per growing pig, group sizes, minor surgeries (e.g., tail docking), bedding availability, access to outdoors, and the mortality rate of piglets. We discussed how and why these factors are important for animal well-being and indicated why producers use certain practices. Participants were encouraged to ask questions, but they were also asked not to offer opinions on the issue. We did not want one particular individual with strong views to offer a polemic to the group, which might alter the subjects' preferences during the research session.

What is Farm Animal Welfare Worth to You?

An Overview of the Value Expression Tool

Imagine that you are someone who knows very little about livestock production. You have just attended a 30-minute information session and are now asked to articulate the additional amount of money you would be willing to pay to provide farm animals with a better life. This question would undoubtedly be difficult for you. To assist the participants in arriving at a WTP number, we created a computer program that walks them through a series of simple decisions, and then aids the person in synthesizing all of their thoughts into a single WTP number.

The key strategy of the program is to help the participant answer a complex question through a series of simple question. To illustrate the nature of the program, imagine that you are shopping for a house, and are currently deliberating over the value you place on a particular home. Consulting a friend, they ask you questions such as: "How important is the square footage of the house?" "Do you prefer the privacy of the country or do you prefer a dense neighborhood?" and "How much money are you willing to spend on a house that meets all your specifications?" After answering these and other questions, your friend suggests that you should be willing to pay at least $220,000 more for a particular house. Your friend's suggested price may seem reasonable, or you may decide that the price seems a little too much. Given this feedback, your friend may then help you resolve your thoughts, perhaps helping you realize that you do not value a larger house as much as you first believed.

Individuals in our study were given such a friend in the form of a computer program. This chapter will describe this program (and interested readers can see the exact survey at the book website, and they can download academic papers describing the mathematics behind the software;[10] see also Figure 9.3). After learning about livestock production, each person in the survey was given a laptop computer and was asked to answer a series of questions. The program first required the individual to answer a series of short, simple questions; for example, it asked participants to rate, on a scale of 1 to 10, the desirability of providing only 14 square feet per sow. Then the program asked the individual to indicate whether space-per-sow is more or less important than some other issues, such as whether their tails are docked. Based on these answers, the computer predicted the person's WTP for a pork chop from a pig raised in five different farm settings. The person observed this prediction and was asked to say whether it was too high or too low. The person could then adjust the WTP numbers by returning to the questions and answering them differently.

As we argued earlier, ordinary Americans know little about livestock production, and even after an information session our participants could not easily form an opinion regarding the additional amounts of money they might be willing to pay for better animal care. This tool assisted the individuals in discovering these values in a rational and systematic way. Ultimately, the tool

> **Background information**: Research participants are educated about livestock farm systems. Each farm is described by a series of attributes, such as barn space per animal and access to outdoors, and each attribute is described by various levels. For example, for the attribute "barn space per animal," the attribute levels are the precise square inches per animal.

> **(Step 1) Intra-attribute ratings**: For each farm attribute, participants rate the desirability of each attribute level on a scale of 1 to 10, where 10 means highly desirable.

> **(Step 2) Inter-attribute rankings**: Participants rank each attribute according to the most important attribute and the least important attribute. Each attribute is assigned a number indicating its importance where the numbers across all attributes sum to 100.

> **(Step 3) Product descriptions**: Five egg products or five pork chop products are presented. Each product is described by the attributes of the farm on which the animal was raised.

> **(Step 4) Auction bids**: Software generates auction bids for each food item, based on their answers to the questions in Steps 1 & 2.

> **(Steps 5 and 6) Calibrate auction bids and holds auction**: Participants alter their answers in Step 1 and 2 to alter the auction bids in Step 4 to their desired level. These auaction bids are then entered in a real auction.

Figure 9.3 Overview of value expression tool

provided us with numbers indicating the value these people placed on better animal treatment. To illustrate how the tool worked, it is perhaps easiest to look at the steps in the experiment.[11]

The Value Expression Tool: Step 1—Rate the Attribute Levels

There are numerous factors or attributes affecting farm animal welfare, such as barn space per hen and group size. For each attribute there can be many attribute levels. In looking at eggs and the connected production process, the attribute of *group size* could take the values of 7, 2000, or 3000 or more hens per group. For the attribute *barn space per hen*, the attribute levels could assume the values of 48, 69, 110, 171, 252, or 353 or more square inches. The expression tool first asks the research participant to indicate the desirability of each attribute level, within a single attribute. For example, the respondent may indicate that a *barn space per hen* of 48 square inches received a desirability rating of 1, while 110 square inches received a rating of 5.

To aid the participants, objective information was provided about farm processes to accompany each attribute question. For example, when answering the questions about space allotment, participants were informed that hens need 67 square inches to stand and lie down comfortably, and 300 square inches to flap their wings. In addition to the attribute *barn space per hen*, the computer program asked similar questions related to eight other attributes; including *egg price, floor space per hen, whether beaks were trimmed, foraging room, nest availability, access to outdoors, group size*, and the *type of feed*. To illustrate the questions asked about some of these attributes, let us consider *nest availability*. The levels for nest availability included no nest, group nest with no bedding, group nest with bedding, individual nest with no bedding, and individual nest with bedding. The attribute *type of feed* was unrelated to animal welfare, but was included to allow consumers to express preferences for organic and antibiotic, and hormone-free. This helped ensure our estimated preferences for animal welfare were not confused with people's concerns for these factors.[12] Participants in the pork sessions answered similar questions relevant to the pigs' welfare.

The Value Expression Tool: Step 2—Rate the Relative Importance of Attributes

Step 1 aimed to identify how important people thought it was for, say, a hen to have 48 sq inches of barn space rather than 69; however, these answers do not provide any information about how important barn space is relative to another attribute, say, outdoor access. Some people might believe outdoor access is essential for high animal well-being, while others might think it is irrelevant. Some people might believe animals enjoy outdoor access, but personally might care very little about the extra happiness the animal experiences from going outside when they are shopping for pork chops. Step 2 was designed to capture this kind of information.

The second step in our tool requested participants to assign weights to each of the production attributes. In particular, Step 2 of the computer program asked about the relative importance participants placed on each attribute when buying eggs or pork. We began the process first by asking people to simply rate each attribute on a scale of 1 to 7, where 1 was completely unimportant and 7 was very important. Let us refer to this stage as Step 2.A.

Unfortunately, we know people sometimes answer survey questions in the easiest way possible, and not necessarily reflective of their true beliefs. We expected that many people want to assign high importance scores to every attribute—effectively saying, *"everything matters to me."* What we were really after, however, was information about relative importance of each attribute. How important is price relative to barn space? How important is barn space relative to the provision of nests? We paused before individuals were allowed to answer Step 2, and stressed that they should avoid selecting the same score for each attribute. Participants were encouraged to think hard about which attributes mattered the most to them, assigning those attributes a higher importance score than the other attributes. Then, in the second portion of Step 2, which we refer to as Step 2.B, the computer took the participants' scores of 1 to 7 and created weights which summed to 100.[13] Here, participants could tell us that, say, barn space was more important than a lower price when buying eggs, but they could only communicate this fact by taking some of the weight assigned to *price* and giving it to *barn space*, such that the total remained 100.

Before describing the next steps in the value expression tool, it is worth mentioning the benefits of asking people to answer the questions described in the first two steps. First, the questions help participants remember and reflect on the information given regarding all the attributes in egg or pork production. Second, the questioning helps participants discover the degree to which they might like different production systems by weighing the pros and cons along with the participant's intensity of the likes and dislikes. A participant might not know how they want eggs to be produced until they sum up all the scores. Third, the computer program helps the participant keep track of all the farm practices that affect animal welfare, as well as each person's likes and dislikes of the practices. As we will describe momentarily, the computer uses this information to arrive at a final score for different food items—a score that is expressed in dollars. The tool is essentially a score-keeping device used to assist the thought process of each person.

There is an additional advantage of the tool that might be unrecognized to the user. Because we know people's ratings of different attributes and their relative attribute weights, we know why certain foods are given a higher overall score, and thus a higher monetary value than others. The tool not only helps participants determine which egg or pork product is most valuable to them, but allows us to understand what aspects of the farm process contributed to a low or high WTP. It allows us to understand which farm systems consumers prefer *and* why.

The Value Expression Tool: Step 3—More Information

Once the participants had explored and revealed their preference for various attributes of a farm process, we went on to introduce food products from different farms. At this point, we introduced five cartons of eggs (each consisting of a dozen brown eggs) in the egg session and five packages of pork chops (each consisting of 2 lbs of pork) in the pork sessions. Each of the egg or pork products came from a production system that was systematically related to the animal welfare factors discussed in the previous steps. Participants in the egg sessions learned about five egg production systems (cage, barn, aviary, aviary w/free-range, and organic) and those in the hog research program learned about five pork systems (confinement-crate, confinement-pen, confinement-enhanced, shelter-pasture, and organic). We included an organic food item to allow people the opportunity to express interest in pesticide-free food, the absence of commercial fertilizers, and the like. However, in terms of animal welfare, the organic pork and organic egg items were identical to one of the other products. For example, the organic pork provided identical levels of animal welfare as the shelter-pasture pork. Consequently, we will not dwell on the participants' preferences for organic food, and focus instead on the topic of this book: farm animal welfare.

The different farm systems were chosen to represent the range of productions systems providing the lowest to the highest levels of farm animal welfare. These systems resemble those we discussed previously in Chapter 5. Cage eggs constitute conventional eggs from hens housed permanently in cages, while barn and aviary systems represented different types of cage-free eggs. Aviary w/free-range is a cage-free system with outdoor access. Confinement-crate pork is conventional pork with pregnant sows living in gestation crates. Confinement-pen system pork is pork from farms that use group pens instead of gestation crates. Confinement-enhanced systems and shelter-pasture systems raise hogs with larger space allotments, bedding, natural nesting and mothering by the sow, and outdoor access. The outdoor access in the confinement-enhanced system is one where there is a barren concrete pad, while the pasture system provides pasture with dirt and grass. Cage eggs and crate pork constitute the baseline egg and pork products, and represent the type of factory farms that animal advocacy groups oppose with fervor.

Our descriptions of the different production systems were not vague; in addition to providing pictures, we detailed exactly the amount of space in each system and went into detail on how each system related to factors affecting farm animal welfare. Table 9.1 shows exactly how each of the production systems related to various attributes affecting animal well-being. For example, participants were told that hens in a cage system had 69 square inches of space, had their beaks trimmed, had no room for scratching, foraging, or dust-bathing,

Table 9.1 Detailed Description of Five Egg and Pork Systems

Eggs		Production system			
	Cage	Cage-free (barn)	Cage-free (aviary)	Cage-free (aviary with free-range)	Organic
Barn space per hen (sq. inches)	69	155	186	186	186
Barn floor space per hen (sq. inches)	69	155	97	97	97
Beak trimming	trimmed <10 days	trimmed <10 days	trimmed <10 days	trimmed <10 days	trimmed <10 days
Room for scratching, foraging, and dust-bathing (sq. feet per hen)	0	1.35	1.35	1.35	1.35
Nest availability	no nests	group nests without bedding	group nests without bedding	group nests without bedding	group nests without bedding
Free-range	no free-range	no free-range	no free-range	free-range with shelter and predator protection	free-range with shelter and predator protection
Group size	<7	>3000	>3000 with perches	>3000 with perches	>3000 with perches
Type of feed	non-organic	non-organic	non-organic	non-organic	organic
Pork	**Confinement-crate**	**Confinement-pen**	**Confinement-enhanced**	**Shelter-pasture**	**Organic**
Space per gestating sow (sq. feet)	14	24	90	90	90

Space per nursing sow (sq. feet)	14	14	90	90	90
Space per growing pig (sq. feet)	8	8	32	32	32
Nesting provisions	with privacy/no straw	with privacy/no straw	with privacy/with straw	with privacy/with straw	with privacy/with straw
Survival rate of farrows	90%	90%	80%	70%	70%
Minor surgeries	performed < 7 days	performed < 7 days	none	none	none
Free-range	no free-range	no free-range	free-range with shelter and no pasture	free-range with shelter and pasture	free-range with shelter and pasture
Group size (number of sows)	1	5	20	20	20
Provision of dry straw (inches)	0	0	12	12	12
Type of feed	non-organic	non-organic	non-organic	non-organic	organic

they had no nests, no outdoor access, were in group sizes less than 7 hens, and were fed non-organic feed. Other products were similarly described as shown in Table 9.1. It should be noted that participants were not shown Table 9.1, but rather the information was presented in a more user-friendly way over the course of several slides.

As will become apparent later, the expression tool does not limit us to valuing the five egg and pork products shown in Figure 9.1. The tool collects data on the person's preferences for each attribute and attribute level, revealing why they prefer one particular product by a certain amount. One can say we "got into the head" of the participant. More will be said about this shortly. Consequently, the tool allows us to study preferences for farms other than those specifically described in Table 9.1. For example, although the barn space allocation for eggs only took the values 69, 155, and 186 square inches per hen, we can study any space allocation between 69 and 186 square inches. Just as we will obtain a WTP number for a farm with 155 square inches, we can obtain a corresponding number for 154, 150, and 71 square inches.

The Value Expression Tool: Step 4—Auction Bid Generator

The ultimate question of our research concerns the additional amount of money each person is willing to spend for egg or pork products from different production systems. We are not interested in how much they say they will pay, but how much money they truly forgo. We sought to use the information people gave us in Steps 1 and 2 to calculate what people would be willing to pay, and use this value in an actual market setting. Achieving this outcome required some creativity and computer programming.

Step 4, which might be called the auction bid generator, is where we calculated each person's predicted WTP for each of the given food products, based on the answers they previously provided. An example might best illustrate how the auction bids are generated. Imagine a very simplified situation with only two attributes: *space per hen* and *price*. Suppose a person named Jackson assigned an importance score of 2 for the attribute *space per hen*, and a score of 6 for the attribute *price*. As we previously mentioned, the computer turned these scores into "importance scores." So, in our example, the importance score for *space per hen* equals $(2 / (6 + 2)) \times 100 = 25\%$ and the percentage for *price* equals $(6 / (6 + 2)) \times 100 = 75\%$. These weights are obtained from Step 2, and along with the answers obtained in Step 1, we can calculate someone's WTP for a carton of eggs with a given space per hen.

Imagine only two cartons of eggs: eggs from System A, giving hens 69 square inches per bird, and eggs from System B, giving hens 110 square inches per bird. Suppose that in Step 1, Jackson gave the attribute level 69 square inches a rating of 2 and the level 110 square inches a rating value of 8. In general, the score Jackson would give to eggs from any system is:

(rating of square inches in system) × (importance score for space) − (price of eggs from system) × (importance score for price).

Now, we can calculate Jackson's scores for cartons of eggs from the two systems:
Overall score for System A = 2 × 0.25 − (Price of carton A) × 0.75
Overall score for System B = 8 × 0.25 − (Price of carton B) × 0.75.

All things equal, we can see that at the same price level, Jackson would prefer eggs from System B because it gives more room to hens—something Jackson likes. All things are not likely to be equal though; the price of Carton B is likely to be higher than Carton A. How much higher does the price of Carton B have to be before Jackson no longer prefers it to Carton A? The answer to this question concerns what we call Jackson's WTP, and is the ultimate item of interest in this chapter. WTP in this context is the price difference between eggs (or pork) from two systems that would cause the overall score (i.e., the desirability) of the carton of eggs to be the same. A bit of algebra leads us to believe that Jackson is willing to pay (8 × 0.25 + 2 × 0.25) / 0.75 = $2.00 more for eggs from System B than System A. When the price of System B eggs are $2.00 more per dozen, Jackson is indifferent between the egg items, and if the premium is anything less (more) than $2.00, he prefers eggs from System B (A).

In our study, we did not want our participants to have to make all the calculations necessary to determine their WTP amounts (the amounts that would subsequently be entered into an auction). The equations would be very long and tedious. The computer did this task for them. When the participants first encountered Step 4, the computer program provided what might be called an intelligent assessment of their WTP for the five different egg or pork products based on the previous answers given in Steps 1 and 2. Remember that by this point the computer has collected detailed information about how the individuals believe the welfare of the laying hen is affected under different conditions (e.g., different space allotments), and how important those conditions are relative to the price they are willing to pay. Someone who indicated that *price* was more important and thus, *space per hen* was less important, would have a lower bid than someone with the opposite ratings.

The Value Expression Tool: Step 5—Wash, Rinse, Repeat

Although people's answers in Steps 1 and 2 were used in Step 4 to calculate an intelligent guess regarding each person's WTP for the five products, consumers were given the opportunity to modify the values in whatever way they desired, *but only by changing their answers given in Steps 1 or 2*. This requirement forced people's WTP values to be precisely linked to their underlying values for the animal welfare attributes such that we could know exactly *why* peoples' values are what they are.

Participants may have increased all their bids by decreasing the importance score assigned to *price* (indicating *price* was less important to them), or decreased all the bids by increasing the importance score assigned to price. A participant may have decided that he or she wanted to increase the bid for the shelter-pasture pork, but leave the bids for all other four pork products (not counting organic pork) unchanged. He or he had to accomplish this by indicating a stronger choice for outdoor access with pasture and shelter, as only shelter-pasture provided outdoor access with pasture. This could be achieved by increasing the rating of the level *outdoor access with shelter and pasture* for the attribute *outdoor access*. Participants who changed their mind and now believed the space provided per hen was less important than previously indicated could indicate this change by decreasing the importance score assigned to *barn space per hen*.

We gave participants ample time to complete this exercise, and we encouraged them to move some of the attribute importance scores to see how their WTP amounts would change. We also assisted people who needed help by asking questions like, "What do you want to bid on product X?" "How does product X differ from product Y in terms of the attributes?" and "What will happen if you said attribute Z is more important?" Ultimately, everyone was able to make the bid amounts correspond to what they wanted.

The Value Expression Tool: Step 6—Going Once, Going Twice...Sold

The WTP amounts generated in Steps 4 and 5 were not just abstract dollar values created by a computer program. Indeed, between Steps 3 and 4, we paused for about 15 minutes and told participants that they would participate in a *real* auction to buy five types of egg or pork products. When people were completing Step 5, they knew that the bid amounts being generated would be the values that they would enter into an auction.

Most people are likely to be familiar with how auctions work, and many will have participated in one through eBay. However, as our auction was a little different we spent 15 minutes describing how the auction would be administered and explaining the optimal bidding strategy, this being to submit a bid for each product equal to that maximum WTP for the product. This latter point is important because incentives in many auctions, such as that on eBay, are different. Conventional auctions have a few simple rules: the highest bid wins the auction, the person with the highest bid receives the good, and the winner pays an amount equal to their bid. Participating in these conventional markets causes you to weigh two competing factors when forming a bid: bidding too low could reduce your chances of winning, but bidding too high means you give up more money. The rules of the conventional auction are such that your highest bid is something *less* than your maximum WTP. The reason is that if you bid your maximum WTP immediately, you win the product but give up an amount of money exactly equal to the good's worth; this leaves you no better or worse off than if you never purchased the product. If we had sold the food items using a conventional auction, the participants' bids would not reveal the bidders' maximum WTP, but this maximum WTP is exactly what we were seeking!

To measure *exactly* the maximum people will pay for improved animal well-being, we turn to insights from auction theory. If we can separate what people *say* they will pay (i.e., what they bid) from amount they finally *pay*, then we can often induce people to bid an amount equal to their maximum WTP. These incentives are not necessarily obvious, and thus we spent some time discussing our auction and how it works. A few pretend and non-pretend auctions are even held with candy bars, to reinforce understanding of the auction. We will not dwell on the training exercises we conducted, and instead will simply describe the auction we used. In brief, though, the participants knew their WTP amounts would be entered in an auction, and they knew exactly how the auction would work, and they knew that they had an incentive to submit bids exactly equal to their maximum WTP.

The egg and pork auctions proceeded as follows. After calibrating their answers to Steps 1 and 2 such that the bid for each product equaled their maximum WTP, they clicked a *submit bids* button. At this point they waited on other people to finish their bids. Then we randomly picked *one* of the five cartons of eggs on which people bid (or one of the five packages of pork chops) by drawing a number from a hat. It could have been the eggs from the cage system, eggs from the barn system, or the eggs from the organic system. The random draw determined which egg or pork type was actually auctioned. We did this so that no one would be worried about potentially buying five dozen eggs or 10 lbs of pork. If the auction was won, it was known that the individual would carry home only one dozen eggs or one package of pork chops. This ensured that respondents submitted bids revealing their maximum WTP for each single product if they were purchasing only that randomly chosen product. The next step entailed randomly picking one of the participants in the room by drawing another number from the hat. At this point there was one randomly selected food item and one randomly selected person. The auction was then held, for that one person and for that one food item.

Rather than comparing the person's bid to all the other bids, as one might do in more conventional auctions, we compared his or her bid against something we called a *secret price*, which itself was randomly determined. If the person's bid for the item was greater than the secret price, they won the auction and received the item. The amount they paid for the product equaled the secret price. If the person's bid for the item was less than the secret price, they paid nothing and won nothing.

However bizarre the rules may sound, the procedures were designed such that individuals would submit a bid equal to their maximum WTP. Because people knew that any of the five egg or pork products might be randomly chosen, they had to carefully consider their bids for each item (even though only one would actually be sold) and because there was a chance that they might be the randomly selected person, participants were forced to take the auction seriously. Consequently, the individuals submitted bids as if they knew they were going to be the person chosen for the auction.

By using the secret price to determine whether someone won and what they paid, we satisfied our rule from auction theory that we separated what people say they will pay from what they are actually prepared to pay. The incentive to bid an amount equal to the maximum WTP is explained as follows. A person would never want to bid *more* than they were really willing to pay because the secret price might be higher than their bid, and thus the person would pay more for the good than it was worth to them. Consequently, one would never want to bid more than their true WTP. If someone bid the true maximum value and won the auctioned item, they would gain from the transaction because they would always pay a price less than their bid. Conversely, a person would never want to bid less than they were really willing to pay, because the secret price might preclude them from purchasing the good at a gain. For these reasons, no one would ever want to bid less than their maximum WTP. So long as the research participants understood the auction (and we conducted numerous practice trials to ensure that they would), each person would submit a bid precisely equal to their maximum WTP for the good.

The bids were not made public to the group. It was our desire to keep the actual bids confidential so that a person would not inflate or deflate their bids to avoid the judgment of others. Before the auction bids were submitted the participants were told that if they were randomly chosen for the auction they would meet with us in private to hold the auction. Their bids, therefore, were submitted knowing they would not become public knowledge to their peers.

Assistance in Being Rational

In addition to helping the research participants form their auction bids, the computer program has some additional beneficial features. Recall that this research session requires participants to process a large amount of information about a topic they have rarely pondered, consider five food items that differ according to a large number of attributes, and then to submit auction bids for each item. This could be a lot to ask for some participants, and in cases where participants face a significant cognitive burden they often behave in two undesirable ways. Participants may simply submit a bid having given it little thought. Even if they want to give the researchers an honest answer, the process might seem too complicated, and they might be scared to ask questions. So they might submit a number—any number—with no real basis for their choice. For example, the subjects might have submitted bids of zero, even though deep down they did care about farm animals and would have been willing to pay more for better animal treatment. Hopefully, this type of behavior was averted in our research because the participants are paid a fair sum of money (most receiving $75), which they were under no obligation to spend; the software was designed to be easy to use; subjects were given ample time to complete the questions; and individual assistance was provided to anyone. Some of the sessions contained older participants who were uncomfortable with computers. These individuals were given special, individual attention. In cases where we felt the survey had not been treated seriously, for example those who finished the survey before most people had got started and in one case a person fell asleep, these individuals' answers are not included in the results. The rooms contained two-way mirrors so that we were able to watch how the respondents behaved from all angles.

A second type of behavior is *irrational behavior*. Economists often ask too much from consumers and design survey questions that confuse the respondents—inducing decisions that appear irrational. For example, preference reversals have been noted, where a person indicates they prefer Good A to Good B, and then later expresses the opposite preference, out of confusion. Our software program specifically prohibited this type of behavior: the auction bids had to be formed through the software interface, and the software generated bids based on logical, mathematical manipulations of each person's survey answers, making preference reversals impossible.

Suppose that a respondent named Audrey is willing to pay more for hogs that are provided outdoor access. When she views the auction bids on her computer screen she sees a pork product named *shelter-pasture pork*. This product obviously comes from animals that were allowed outdoor access, so Audrey makes sure it has a higher bid. Then she sees the product called *enhanced-confinement pork*, the name of which does not indicate whether outdoor access was provided. She wonders whether this product does indeed have outdoor access and whether it should be given a high or low bid. If Audrey were asked to simply write down a bid for this item, without the aid of the computer, she might not give a higher bid for this pork product, not being able to recall whether it was from a process that did allow outdoor access. However, in our experiment the computer code *forces* Audrey to bid more for *Enhanced-Confinement Pork*, because she previously indicated that she believed outdoor access was important. If Audrey bid a higher amount for *shelter-pasture pork* because it is the result of a process that provides the animals with outdoor access, she must bid higher for *enhanced-confinement pork* as well—it is only logical. The software helps the participants behave in a logical manner, which they sometimes would not do when asked to process large amounts of information.

Looking Inside the Consumer's Head

Our research could have been conducted differently. The participants could have been given the same pre-auction information session, followed by the same auction, but instead of using our computer program to formulate bids they could have just been asked to write five bids on a piece of paper. The participants would have made their own calculations in their heads to determine their bids, but we, the researchers, would not have had any idea of the mental processes they had used to calculate their bids—what arguments they had used to formulate their decisions or what thought processes they had gone through. We would have seen the bids, but not understood why some bids were higher than others.

Ideally, however, we would like to see inside the consumer's head, and download the neural network controlling his or her thought processes into our computer. Then we would be able to study how the thought process works to better understand consumer views. For example, you may bid a higher price for cage-free eggs because you place a high priority on allowing hens to lay eggs in nests. The greater space allotment also given to hens in cage-free egg processes might be of less concern to you. If subjects only wrote their bids on a sheet of paper we would see the higher bid for cage-free eggs but not understand why the bid was higher. However, if we had access to the participants' neural networks we would see which issues are most important for each participant. Based on this information we could construct a new farm which would be different from any of the four egg products. A hypothetical farm could be created which maintains hens in cages but provides them with a private nest for egg laying, and so on. You value the product, in the above case, more because of the nest the hen has been given, meaning that to please consumers the farmer does not have to convert to the expensive cage-free system. Accessing the neural network in your mind would allow us to not only understand what factors influence your product bids, but also to predict how you might bid on eggs from a farm system different from the four included in the auction.

The case of eggs illustrates the point well. As described in Chapter 5, there is a system called *enriched cages*, where the hens remain in cages but those cages contain features such as nests, perches, and areas to dust-bathe. Enriched cages are receiving much interest from animal scientists and the egg industry as a possible alternative to cage-free egg processes. It is thus important to measure the value consumers place on eggs from enriched cages, but that your product is not included in the auction. If only we could download that neural network (containing all the perceptions and logic used to form your bids) we could predict how you would bid in the auction for enriched-cage eggs.

Using the computer program we can predict bids for enriched-cage eggs. The reason is that the computer program serves as a surrogate, or a companion for the neural network in your head. We do not need to look inside of participants' head because we have all the information needed in the program. Because we required participants to use the software to generate their auction bids, we were then able to use the information from the program to construct possible bids for (almost) any hypothetical egg product, including enriched-cage eggs.

On the Interpretation of Auction Bids

Before we delve into the results, a few points regarding the interpretation of auctions bids are warranted. The bids are *consumer expressions* of their beliefs regarding the well-being of an animal on a farm and the extent to which each person cares about that well-being. Also, bids are a *one-time expression*. It would be nice to conduct the auction repeatedly every month for two years. Such information would indicate whether enthusiasm for improving the lives of farm animals wanes over time. Alas, our research budget was far too meager for such an extensive project. Although some survey methods, such as an internet survey, would allow us to track people's values over time, they would rely on hypothetical stated intentions of what people say they will do, rather than true economic values revealed in a market setting.

1. The auction provides a platform for an *informed expression*. Data from our experiment is not necessarily reflective of the average grocery shopper. Our participants were informed about the farm animal welfare issue, whereas most consumers know little about the issue. Participants knew the exact conditions in which hens and pigs associated with each product were raised. Conversely, grocery store labels like "cage-free" and "free-range" provide suggestions on how the animals were treated, but not in the precise details like the products in our auction.
2. The expressions were obtained in a laboratory setting, not a grocery store. The fact that our research subjects knew that their responses were being studied may have induced them to act more as a *citizen* than a *consumer*, which potentially means they might have paid more attention to ethical considerations than they would have done in a grocery store. This is not necessarily a weakness.
3. WTP is an *expression of intensity*. Participants did not simply tell us whether they liked cage-free or cage eggs, they expressed the *intensity* of preference.
4. The *intensity* is *expressed* in *dollars*. Because we measure consumers' intensity of preference in dollars we can directly compare the amounts to the *cost* of a product to determine if people would truly prefer the product, given the costs of production.
5. WTP is a *non-hypothetical expression*. These experiments did not involve pretend money. When people submitted their bids, they knew there was a real chance their bids would have an economic consequence. The result

was that people could not simply *say* they wanted hens to have more space, but rather, they were required to "put their money where their mouth was."

6. The *expression* is unusually *rational*. By recording, processing, and showing the consequences of consumers' inputs, the value expression tool helped participants make choices in a rational manner. It is very difficult for the individuals to make a mistake or contradict themselves, because the computer helped each person remember the many details associated with the food products, and helped them perform numerical calculations that ordinary people cannot perform alone.

7. The *expressions* can be deconstructed into *expressions for individual farm attribute levels*. Participants were required to use the software program to formulate their bids, and the program could deconstruct the value of a product into the value of the product attributes and attribute levels. This allowed us to determine the extent to which a single attribute level contributes toward the overall value of the product. Simply put, the program allowed us to determine why participants valued a product a certain amount.

Egg Auction Results

How did our specimens actually behave? Recall that we assigned half our study participants to research sessions involving eggs. Some of these people failed to adequately answer all the questions or fell asleep or clearly did not understand what was being asked of them. Limiting our attention to the remaining participants yielded a usable sample of bids from 120 participants. We focused our addition on bid differences: the difference between a particular egg product and the bid for cage eggs. We focused our attention on the bid differences (or premiums) because these values were the most informative. The participants were not expecting to buy eggs when they entered the research session and they may not have been returning home straight after the session.[14] This would imply a lower overall WTP in the research sessions as compared to when they were in the grocery store. When submitting bids, the overall level of bids represented a participant's interest in *any* egg product at the time, but the *difference* between bids represented the value they placed on animal welfare. We would not expect individuals' concern for animal welfare to differ greatly in a research session than in a grocery store, so the differences (or percentage differences) between bids were reflective of their preferences for how laying hens are treated. This suggests that while the overall level of bids was likely lower than it would have been when grocery shopping, we should have confidence that the bid differences represented something useful about consumer preferences.

Approximately 10 percent of the participants submitted bids equal to $0.00 for all five egg products. Consequently, the computed WTP premiums for higher welfare standards for these individuals all equaled zero. However, one cannot assume the actual premiums for these people actually were zero. Bids equaling zero for all egg products indicated that a person did not want to purchase eggs at the time of our experiment. If the participant were interested in purchasing eggs, they might have been willing to pay a high premium for animal care, or they may not have been. The point is that zero bids for all egg products only indicate a lack of demand for *any* egg product, not a lack of interest in animal care *when eggs are purchased*. Consequently, individuals who submitted bids equal to zero for all egg products were not included in the analysis. Thus, we studied preferences for different types of eggs *among active egg purchasers*. That is, we studied WTP for animal welfare attributes conditional on someone being "in the market" for eggs.

The average and median auction bid premiums for four egg systems are shown in Table 9.2.[15] The figures represent the extra *premium* consumers are willing to pay for one dozen eggs from a particular system, over what they are willing to pay for cage eggs. For this reason, there are no premiums to report for cage eggs. The reader should note that all egg products were brown eggs, so the differences in premiums are not attributable to differences in egg color, or any other intrinsic property of eggs. The description of cage-free facilities in Table 9.2 is slightly different from those used in the auction. At the time the experiments were being planned and administered we used cage-free system descriptions that were published in the scientific literature. It later became apparent that the suggested standards for cage-free facilities had evolved since those publications, and we altered the definition of cage-free farms to be consistent with the cage-free standards used by the United Egg Producers. For example, the square inches provided to each bird in a cage-free (barn) system is 200 square inches in Table 9.2, whereas the same system in the consumer experiments allowed only 155 square inches. The exact farm description for each system in Table 9.2 is provided in the footnotes to Table 9.2. The beauty of our preference measurement tool is that it easily accommodates these types of alterations—we make these changes to ensure our results are relevant to current policy debates.

The average premium for cage-free versus cage eggs is $0.55/dozen if the cage-free system is a barn and $0.47/dozen if an aviary. The major differences between the barn and aviary systems is that the aviary system has more total space for the bird (total space including perches and suspended walkways) but less floor space. Most cage-free eggs in the US are produced in a barn system, although the popularity of aviary systems is rising. The median premium suggests that if *all* consumers were informed about egg production, and if cage-free eggs were placed on sale at a $0.44/dozen premium, approximately one-half of all consumers would purchase the cage-free barn eggs. Recall that

Table 9.2 Willingness-to-Pay Premiums Relative to Eggs from Cage System

Egg system	Average premium	Median premium	Average percentage premium	Median percentage premium
	$ Per dozen		As a percentage of cage egg bids	
Cage eggs				
Cage-free eggs (barn system)	$0.55	$0.44	78%	49%
Cage-free eggs (aviary system)	$0.47	$0.35	70%	39%
Cage-free eggs (aviary system w/free-range)	$1.06	$0.85	141%	85%
Enriched cages (projected)	$0.41	$0.28	51%	32%

Notes: The last two columns then denote the average and median of these percentage premiums. These premiums represent the preferences of 120 participants from Wilmington, NC, Dallas, TX, and Chicago, IL, and does not include the 10% of participants who submitted a bid of zero for all egg products. It does, however, include participants who submitted a non-zero bid for at least one egg product. The cage system is assumed to provide hens with 67 square inches of room and a flock size less than 7. The barn system allows 200 square inches per hen, plus scratching room, plus group nests without bedding, and a very large flock size. The aviary system allows 240 sq. inches per hen but 144 sq. inches of floor space per hen, scratching room, group nests without bedding, perches, and a very large flock size. The aviary w/free-range system has the same specifications as aviary system except that it includes outdoor access with shelter and predator protection. The enriched cage system specifications are: 116 sq. inches per bird in cages of less than 7 hens, scratching area, group nests with no bedding, and perches. For cage-free systems the scratching area was defined as 1.35 square foot per bird, at times when the bird seeks scratching area. Scratching area for enriched cages equals 0.675 square feet per bird, when the bird seeks scratching area. In all cases, birds are assumed to be beak trimmed at less than 10 days of age and are given non-organic grain. Percentage premiums are calculated by dividing the premium for egg variety by the bid submitted for cage eggs. If cage egg bid was zero the percentage premium cannot be included in the calculation. This excluded 10 individuals.

these bids were obtained from informed consumers, and as such they do not necessarily predict sales for a grocery store, as most consumers are *un*informed about egg production. It is worth nothing that the data we previously showed in Figure 9.2 on actual retail egg prices suggests that cage-free eggs often sell at a premium of $1.50/dozen or more to cage eggs. Even most of our *informed* consumers are not willing to pay such a high premium.

The numbers in Table 9.2 only describe the central tendencies in the participants' bids, and they mask significant variation in values. To describe the variation in premiums for cage-free eggs, consult the histogram presented in Figure 9.4. The x-axis indicates a range of premiums for cage-free eggs produced in a barn system relative to cage eggs. The y-axis shows the percent of the

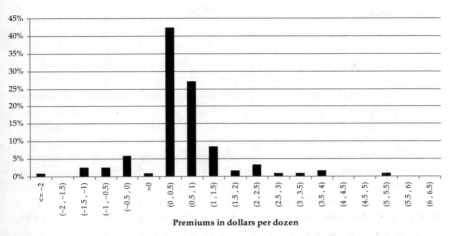

Figure 9.4 Histogram of premiums for brown cage-free (barn) egg premiums over brown cage eggs

Notes: This histogram shows the percentage of individuals in the consumer experiments whose premium values for cage-free eggs (relative to cage eggs) reside within a given interval.

WTP premiums that fall within the respective category. As expected, a majority of participants submitted premiums that were positive but less than $1.00 per dozen. A small percentage of individuals were willing to pay several dollars in premium, and a very small percentage were prepared to pay up to $5 more per dozen for cage-free eggs. More interesting are the negative premiums. Not everyone believed that cage-free eggs are more humane, and we believe this is not attributable to a lack of understanding about egg systems. During the information sessions, participants learned that feather pecking and cannibalism can occur in large flocks, such as the ones in cage-free systems. These individuals did prefer that birds be given greater space allotments, shown by their ratings of various space allotments in Step 1 of the expression tool. However, the individuals placed a greater importance on maintaining very small flock sizes of less than six birds. Consequently, these individuals preferred the cage system and the protection it provides the birds. Because they preferred the cage system, the premiums they placed on cage-free eggs were negative. These individuals would only purchase cage-free eggs if they were sold at a lower price than cage eggs.

Although some individuals preferred cage eggs, a majority preferred cage-free eggs, and both the average and the median premiums for cage-free eggs were positive. The price data in Figure 9.2 indicates that from 2004 to 2008, cage-free eggs sold at an average premium of $1.57/dozen to cage eggs. During this period, cage-free eggs sales only represented about 1.3 percent of combined cage-free and cage egg sales. Let us conduct a thought experiment and ask: if the subjects in our experiment had the opportunity to purchase cage-free eggs instead of cage eggs for $1.57 more per dozen, what percentage would we predict to purchase the cage-free eggs? This question is easily answered by calculating the percentage of participants with an auction bid premium for cage-free eggs (from a barn system) that exceeds $1.57/dozen. The answer is 9 percent. Again, the difference in our calculation and the actual market data might be explained by the fact that we had only a sample of informed consumers. Still, the data suggest that even if all consumers were perfectly informed the market share of cage-free eggs would still be low at prevailing market prices. The low market share for cage-free eggs is not due to a lack of interest in cage-free eggs, but because cage-free prices are so high they outweigh consumer concerns for laying hens.

It has been estimated that the cost of producing cage-free eggs is about $0.35/dozen more than cage eggs.[16] Comparing the $0.35/dozen cost to the $0.44/dozen median benefit (under barn system), this simple cost–benefit analysis suggests that a majority of the participants in the research session prefer cage-free eggs even after accounting for the additional costs of production. What drives the estimated $0.55 average premium consumers place on cage-free eggs from a barn system? Fortunately, our value expression tool allows us to parse the WTP premium into people's values for each farm attribute. To illustrate, suppose we begin with a cage system and then slowly change the attributes of the system to convert it to a cage-free (barn) system, and evaluate how each change along the way affects the average projected bids. First, starting with a traditional cage system, begin the conversion by opening all the cage doors but do not provide any additional amenities. The result is misery and violence. The hens are still in a cramped, barren environment, but instead of residing in cages with small flocks to protect the birds from hen aggression they live in a huge flock of maybe 100,000 birds, piled on top of one another and pecking one another. Our value expression tool reveals that this change *reduces* the average value consumers place on eggs by $0.46/dozen, and represents the value consumers place on keeping hens in small flocks to protect them from hen aggression. To reduce hen aggression, suppose that we now increase the space per hen from 67 to 200 square inches per bird. This change *increases* the value of eggs to consumers by $0.57/dozen, resulting in a total change in value of (−$0.46 + $0.57) = $0.11 relative to cage eggs. Next, suppose hens are given group nests with no bedding. This is an improvement over the cage system which has no nests at all, and thus people's WTP increase by $0.10/dozen. Finally, suppose we now covered the barn floor with dirt or straw for scratching and foraging—a change which increases the value consumers place on eggs by $0.34/dozen. We now have moved from a cage to a barn system. All changes considered, the average value consumers place on this change equals (−$0.46 + $0.57 + $0.1 + $0.34) = $0.55, which is precisely the average amount reported in Figure 9.4.

Consumers place a value on all the features distinguishing cage-free from cage systems. The drawbacks of a larger flock size are taken into consideration; consumers are not naive about the extent to which hens will peck at each other. Our research participants felt that providing chickens with an area to scratch and a greater amount of space would contribute greatly to the well-being of the hen; providing the hen with a nest was important, though to a lesser extent.

Our approach allowed us to assign a value to small changes in any one production process—not just entire changes in systems. Consider, for example, the availability of nests. Most cage-free facilities provide nests that leave much to be desired. The nest is simply a small room with three walls and a curtain, where a few hens can separate themselves from the large flock to lay. The nest does not contain straw, but rather it is made of a foam or rubber-like substance. Imagine instead of this group nest with no bedding, that there was a cage-free facility that provided hens with an individual nest with bedding. Our results indicated that consumers would value eggs from such a system at $0.50/dozen more than eggs from a group nest without bedding. Consequently, the results do not indicate that consumers feel nests are of little importance, but that the specific nests used in cage-free production could use improvement.

The results suggested that a more hen-friendly farm would increase the value of eggs to the consumer. Increases in animal welfare are also costly. The data we have collected allow us ask questions like: *which changes can be made to would benefit the hen at a cost that consumers are willing to pay?* Consider how the premium for cage-free aviary eggs changes with and without free-range aspects. Free-range was described as outdoor access with shelter and predator protection. The presence of the free-range aspect raised WTP by $0.59/dozen. The value of the change in this single attribute was larger than the entire value participants placed on moving from cage to cage-free egg production! We have little doubt that the cost of providing outdoor access would exceed this large premium, but we could be wrong. The cost of a small, enclosed area outside the barn should not be large. The largest cost might be increased risk of disease. Mice, fleas, and insects can carry disease and viruses into the free-range portion of a farm, and if one bird gets sick a thousand birds will get sick. Avian influenza is a contagious illness that can strike humans and birds alike.

A new system that has been a topic of much recent discussion is the *enriched-cage* (sometimes called the furnished-cage) production system, in which each cage has a nest, scratching area, and a perch. Although we did not ask people specifically about their values for eggs from an enriched-cage system, our value expression tool can *project* what people's bid might have been, if they had been asked. As shown in the last row of Table 9.2, the data indicated that participants would have been willing to pay $0.41/dozen more for eggs from an enriched-cage facility, relative to traditional cage systems. What about the cost? Our conversations with industry experts suggest that moving from traditional cages to enriched cages is likely to increase farm costs by about 12 percent. Assuming cage eggs cost $0.75/dozen to produce,[17] the cost of producing eggs from an enriched facility would be only $0.09 more per dozen than cage eggs. This estimate is consistent with the experience of a Canadian producer, who suggests enriched-cage eggs can be produced at an additional cost of about $0.07/dozen (in US dollars).[18] Although consumers are willing to pay $0.41/dozen on average for eggs from enriched cage system, the costs are less than $0.10/dozen—a four-fold difference between WTP and cost. If the egg market consisted of only educated consumers, the farmers would enthusiastically produce enriched-cage eggs. Our cost data for enriched-cage eggs is not as reliable as the data for cage-free eggs, but it is the only data available. Our intuition says that the true costs of an enriched-cage system would be larger than the two sources suggest, but we have no data to support that intuition. We would need better data before we could advocate the implementations of the enriched-cage system. Thus, the claim that the value-added of enriched-cage eggs exceeds the costs by a magnitude of four provides more *motivation* for further study of the system—but it does not provide *evidence* that enriched-cages are a socially desirable system.

Pork Auction Results

Half the participants in our study bid on different types of pork chops rather than eggs. One product was pork produced in the *confinement-crate* (or *crate*) system. We use the crate pork chops as the baseline against which we calculated participants' WTP premiums. They also bid on pork from a *confinement-pen* (or *crate-free*) system, which resembles the crate system except that small group-pens replace the gestation crates. When people employ the term "factory farm," they may be talking about the confinement-crate or the confinement-pen system.

Bids were also collected for pork from a *confinement-enhanced* system. Pork produced from the confinement-enhanced system is increasingly sold at more expensive grocery stores aimed at the more affluent customer, and is receiving greater interest as producers do not want to operate a crate system, but do not want to return to the antiquated systems of previous generations. There are a number of variations of the confinement-enhanced system. Producers are experimenting with different styles, but there is one particular system we have studied the properties of which are described in the footnotes of Table 9.3. Generally, this system has everything that a pig needs for a good level of well-being, except access to pasture. Though the animals have access to the outside, this outside is a barren concrete lot. The fourth product is pork produced in a *shelter-pasture* system, which largely satisfies all of the welfare standards to be sold as Animal Welfare Approved pork. Except for the lower survival rate of baby pigs, the shelter-pasture system received higher welfare scores than the other three systems in every aspect.

As with the eggs, the overall level of the bids is not of interest. Instead, we were interested in the difference between an individual's bid for confinement-pen pork, confinement-enhanced pork, and shelter-pasture pork, relative to the confinement-crate pork. Individuals who submitted zero bids for all pork products were not included, which left us with premium data for 109 individuals.

Table 9.3 reports the mean and median bids for the different systems, as well as the premium percentage. The premium percentage was calculated as the value of the premium divided by the bid for the confinement-crate pork. The results indicated that participants were willing to pay a premium of $0.16/lb, on average, for pork chops from a confinement-pen (crate-free) system as compared to those from the confinement-crate system. The median premium for crate-free pork was only $0.02/lb. This meant that about half the people cared very little about banning gestation crates. Recall that banning gestation crates is at the center of the HSUS's campaign against factory farming. It is this crate which voters in California elected to ban. The result seems somewhat counter-intuitive—do most people not want hogs to have room to turn around? Again, one must keep in mind that our consumers were *informed*. It was not that the participants liked gestation crates, it was just that they considered the crate system as being only slightly inferior to the alternative—which was a very small and cramped group-pen. Our participants were informed that moving from a crate to a pen would be beneficial because the hogs would get slightly more room to move, but participants would have also learned that the sow would then need to compete with other hogs for feed, and in the case where pen membership was not stable, they would fight for their place in the group hierarchy.

Some of our colleagues at Michigan State University have also sought to measure the value consumers place on banning gestation crates. They find a much higher estimate of $2.11 per pound![19] Why is their value for a gestation crate ban so much higher than ours? There are two reasons. One reason is that the Michigan researchers asked consumers hypothetical questions in mail surveys, whereas we asked our participants to submit bids in a real auction. Research strongly shows that people answering surveys tend to *say* they are willing to pay more than they are truly willing to pay when real money is used. Another reason for the discrepancy is that the participants in the Michigan study were only informed about the role of gestation crates in farm animal welfare, whereas our subjects were informed about every feature of a farm that affects animal welfare. The respondents in the mail survey might have inferred that the *only* issue affecting animal welfare is gestation crates. Supposing that the simple act of banning these crates would provide a big boost to animal welfare, they were willing to pay a large amount. Conversely, our respondents understood that gestation crates are only one part of the farm, and that there are many other changes to the standard factory farm that would improve the welfare of the sow. The amount our participants would pay for only banning the gestation crate was low, and it should be low, as it has only a small impact on the welfare of the sow.

Table 9.3 Willingness-to-Pay Premiums Relative to Pork Chops from a Confinement-Crate System

Farm system producing pork chops	Average premium	Median premium	Average percentage premium	Median percentage premium
	$ per pound		As a percentage of crate pork bids	
Confinement-crate system	————	————	————	————
Confinement-pen system (crate-free)	$0.16	$0.02	16%	3%
Confinement-pen system (crate-free)	$1.09	$0.71	111%	76%
Shelter-pasture system*	$1.17	$0.73	112%	80%

Notes: These bids represent the preferences of 109 participants from Wilmington, NC, Dallas, TX, and Chicago, IL. The crate system is assumed to provide sows with 14 sq. feet of space for sows in the gestation and farrowing phase, 8 sq. feet for market hogs at their largest size, no nesting provisions (but nesting privacy), a 90% survival rate of farrows born alive, minor surgeries performed less than 7 days of age, no outdoor access, a sow group size of one, and no straw. The crate-free system is the same as the crate system, except that sows get 24 square feet of space at gestation in a group size of 5. The confinement-enhanced system provides 90 sq. feet per sow in the gestation and farrowing phase, 32 sq. feet per growing pig (at their largest size). The nesting provisions include straw and privacy, and the farrow survival rate is 80%. There are no minor surgeries performed, and all hogs have access to the outdoors, with shelter but no pasture. The sow group size is 20, and all hogs have 12 inches of dry straw at all times. The shelter-pasture standards include 64 sq. feet of space for sows at the gestation stage and 96 sq. feet at the farrowing phase, 25 sq. feet for market hogs at their largest weight, access to nesting provisions and ability to build individual nests at farrowing, 70% survival rate of farrows born alive, no minor surgeries, outdoor access with pasture and shelter, 12 inches of dry straw at all times, and a sow group size of 20. Percentage premiums are calculated by dividing the per pound premium by the bid for pork from the crate system.

* Pork raised largely following the AWA standards, but not including the hormone and antibiotic restrictions contained in AWA standards. Thus, this premium reflects the value of the animal welfare components of the AWA standards only.

Recall that our bids concern pork chops, and as such we have to be careful when trying to extrapolate the value consumers might place on pork sausage or ham. It is reasonable to conclude that if consumers will pay more for pork chops from pigs raised in different systems, they will do the same for sausage and ham. One way to extrapolate our results for other pork products is to apply the *percentage* increase in value we observe for pork chops to all pork products. Table 9.7 indicates that consumers are willing to pay 16 percent more, on average, for pork produced without gestation crates (using group pens instead). The average price of retail pork, using a weighted average of all pork products, is approximately \$2.80/lb.[20] Transferring this 16 percent premium to all pork, banning gestation crates increases the value of retail pork by \$0.49/lb. Using the median percentage-premium of 3 percent, the retail pork increase is only \$0.084/lb.

The difference between the median and the average percentage-premiums has to do with outliers. The median 3 percent premium is calculated as the premium for which half the participants had a higher percentage-premium and half had a lower percentage-premium. That 3 percent would be the same if the highest percentage-premium was 4 percent or 40,000,000 percent. The average does change with outliers, and responds to especially high values by increasing its value. The presence of a few people with unusually high bids will cause the average to rise but will not change the value of the median.

Which is better: the median or the mean? It depends on the question. If one is interested in the total or sum of all benefits across all Americans it is the mean that is the most useful. By multiplying the mean premium by the total number of participants, the calculation yields the total amount of money the group is willing to pay for a welfare improvement. If the demographic profile of the respondents reflects well that of the nation, the average premium can be multiplied by the US population to measure the nationwide WTP for a welfare improvement.

An alternative question is whether a majority of people value a welfare improvement more than the improvement costs. The median premium is that which half of the premiums are smaller and half are larger in value. When the median premium of a welfare improvement is greater then the cost, it is consequently true that a majority of participants approve of the improvement.

It matters little whether the median or the average is used in the case of banning gestation crates. The additional cost of using group-pens instead of crates for gestating sows is about \$0.065/lb.[21] Converting hog farms from the confinement-crate to confinement-pen systems would thus create a net average benefit of \$0.49 − \$0.065 = \$0.425/lb using the average percentage-premium and \$0.084 − \$0.05 = \$0.019 using the median. The magnitude of the numbers differs greatly, but they both indicate that most consumers prefer sows to be raised in group-pens instead of gestation crates.

Consumers appear to approve replacing gestation crates with group-pens. Let us now explore whether consumers desire more improvements by studying their preferences for pork from the confinement-enhanced system. This system houses gestating sows inside barns with large space allotments, 20 sows per group, and plenty of straw for bedding. When giving birth and nursing each sow is provided a farrowing room with ample space, ability to build a nest, and protection for her baby piglets, and the sow is allowed to come and go as she pleases. Growing pigs are also given considerably more space than the traditional factory farm, with deep straw bedding. All animals have access to the outdoors, which is a barren concrete lot. Because of the large space allotments, no tail docking or teeth clipping are needed.

Our informed consumers respond to this confinement-enhanced system with enthusiasm, and are willing to pay an average premium of 111 percent compared to the confinement-crate system. They will pay roughly twice the price of regular pork. With an average pork price of $2.80 per lb, this amounts to an average premium of $5.91 per retail pound—a huge premium! The premium becomes even more compelling when we compare it to the additional cost of producing pork in a confinement-enhanced facility. It costs approximately $0.20 more to produce live-hogs in this improved environment, which translates into a $0.37 increase at the retail level.[22] The average additional value created by the confinement-enhanced facility is $5.91 per retail lb and the additional cost is $0.37 per retail lb. Even if a reader criticizes our use of the percentage-premiums and use the raw per lb premiums in the lab, one would find the value of the confinement-enhanced facility to exceed the costs by a large margin. Informed consumers have a clear preference that hogs be raised in confinement enhanced facility instead of the traditional factory farm.

Now consider the premium placed on pork raised in a shelter-pasture system. This system closely mimics the welfare guidelines set forth by the Animal Welfare Institute, and could likely be sold under their label Animal Welfare Approved. The sows under this system have large space allotments where they can walk and turn around within the barn, and have access to bedding within the barn. When they give birth they can build their own nest and are not confined to a farrowing crate. The market hogs also have more room and access to bedding. Because the animals have large space allotments there is no need to dock tails, and all animals have access to the outdoors with pasture. The downside of this system is that more baby pigs die from crushing, but the results indicate that most consumers are willing to accept this in return for the many benefits of an shelter-pasture system.

The average premium for shelter-pasture pork equals $1.17/lb, with the median equaling $0.73—very large. When these premiums are divided by the total bid for the confinement-crate pork, the percentage-premiums are 80 percent using the median and 112 percent using the average. These premiums are thus similar to those under the confinement-enhanced system, but slightly larger. The bids suggest a substantial improvement in perceived hog welfare when adopting the Animal Welfare Institute standards. Perhaps surprisingly, research indicates that the cost of shelter-pasture pork is less than pork from the confinement-enhanced pork. Estimates are that the additional cost of producing shelter-pasture pork relative to pork from a crate system is only $0.11 per retail lb of pork.[23] With larger premiums and smaller costs compared to the confinement-enhanced system, shelter-pasture pork appears to provide the most benefits of any other animal welfare improvement. Remember that shelter-pasture pork largely meets the standards for Animal Welfare Approved pork, which is available for sale at specialty food retailers. Perhaps you will now seek this pork during your next shopping excursion.

As with the egg results, the value expression tool can also be used to determine people's values for the underlying attributes of pork production systems. For example, while people value an increase in group size from one to five sows at about $0.03 per lb (all the values in this paragraph are average bids), they actually dislike increasing group size from one to ten sows (WTP of −$0.10 per lb). This finding is a logical reaction to animal well-being, with people valuing socialization as hogs are moved from isolation to some level of companionship, but recognizing the decline in welfare that can result when group size becomes large and inter-fighting ensues. People place more value on providing space to gestating sows than nursing sows ($0.18/lb vs. $0.13/lb for in increase from 14 to 60 square feet), which makes sense as sows spend a longer period of their life in the gestation phase as compared to the nursing phase. People also value survival rate of baby pigs with a change from 60 percent to 90 percent being valued at $0.23/lb. These findings illustrate how complicated an issue like a gestation and farrowing crate ban can be, and that the overall value people might derive from such a crate ban is comprised of many factors—all of which may not be apparent to a person unless they have participated in an experiment like ours. For example, banning gestation and farrowing crates would increase space per sow—something people value; however, it could result in group sizes larger than 10 and could also result in lower survival rates. If you simply asked someone, "*Should we ban gestation and farrowing crates?*" they will probably reply in the affirmative. That is because they know so little about the

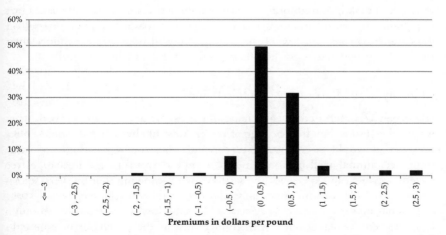

Figure 9.5 Histogram of pasture-shelter pork chop premiums over regular (confinement-crate) pork chops

Notes: This histogram shows the percentage of individuals in the consumer experiments whose premium values for pasture-shelter pork (relative to confinement-crate pork) reside within a given interval.

purpose of the crates and the unintended consequences of eliminating their use. The expression tool provides a useful way of both informing the consumer of the issues involved and providing them with a nifty process to express their preferences for animal care.

As a finale to the section regarding pork, let us survey the variability in perceptions and preferences for pig care. Figure 9.5 provides a histogram of premiums for shelter-pasture pork, relative to the traditional pork raised on factory farms. The first salient feature is the fact that some consumers prefer the factory farm, and this is not by mistake. These individuals perceive that the protection provided to the sow by gestation crates and the high survival rates of piglets outweigh any of the disadvantages. A large majority of the Americans we studied were willing to pay some amount, but less than $1, per pound for shelter-pasture pork. Others were willing to pay premiums of 200 percent or more.

Welfare as a Public Good

Informed Americans generally support improving the lives of laying hens and hogs, and are willing to pay the associated cost of doing so. The numbers used to justify this assertion have imperfections. However, this study goes further than any marketing study we know of to ensure that the consumer preferences measured are based on accurate, objective information, reflecting the consumers' WTP real money, and are free of many of the biases that infect marketing research. It took hundreds of thousands of dollars to collect these data, and years of planning before the money arrived. If this study does not convince the reader that informed consumers want the laying hens and hogs which produce their food to be provided a better life, then we doubt that anything would.

Empathy in the Laboratory—Public Values

If a vegan was recruited to participate in the experiments just described, what would they bid to buy the package of pork? Most likely their bid would be $0; however, it would be a mistake to conclude that the vegan does not value improved animal well-being. If a hen is moved from a cage to a cage-free system, anyone can experience the satisfaction of knowing that the hen experiences a more pleasant life, regardless of whether they purchase eggs.

Purchasing eggs from a cage system is like smoking in a restaurant. Although a smoker enjoys a few puffs with dinner (just as the purchaser of cage eggs enjoys paying a lower price), they make the non-smokers' meals less enjoyable (just as the purchase of cage eggs makes hens and animal lovers less happy). When we want to know the value of a restaurant choosing to go smoke-free, we have to calculate the effects on smokers *and* non-smokers. Likewise, when we want to know the value of a farm going cage-free, we have to calculate the effects on those who do and those who do not directly consume the hens' eggs. In similar spirit, we should account for the preferences of people who are not vegans but are selective omnivores. Some individuals use eggs only sparingly for animal welfare reasons and benefit when others decrease their egg consumption. There is even the prospect that a veracious meat-eater receives benefits when others refine their diet towards more humane foods. This is just another way of saying that farm animal welfare is a *public good*.

The value expression tool described in the last section was designed to measure the private-good values people place on improved animal well-being in the eggs and pork they *personally consume*. Here, we describe another economic experiment designed to measure the public-good values people place on improved animal well-being in the eggs and pork they will *never* consume. If you thought the real-money, real-food auctions we conducted to measure private values were a bit unusual, get ready for an even more bizarre—though appropriate—experiment.

Another Unusual Auction

After our participants submitted bids in the private-good auction for eggs or pork, we gave them a chance to bid in a second auction. This auction did not sell food though. We sold the ability to change where a group of hens or hogs lived. People bid for the opportunity to decide whether a group of animals would live out their lives on a farm the participant believed (as indicated through their previous questions) to generate higher levels of animal welfare, instead of a farm with lower levels of welfare. If the person won the auction, they paid the auction price, and the animals went to the "happy" farm. If the person lost the auction, they paid nothing, and the animals went to the "sad" farm. We clearly communicated the fact that the research participants could not consume any of the food products produced by these animals, regardless of whether they won or lost the auction. The auction bids represent the economic value our participants placed on higher welfare standards for farm animals irrespective of who eats the food. The bids represent the participants' *public values* for farm animal welfare.

Participants in the research sessions related to egg-laying hens were asked to consider two production systems: a cage system and cage-free (aviary) system with free-range. This second system is referred to simply as a free-range system, but we ask readers to keep in mind that the birds in this system have access to both a cage-free barn and an outdoor area with shelter and predator protection. Participants were told that a group of baby layer hens would soon be born and used for egg production, and that the decision of whether the hens will live in a cage system or free-range system depends on the outcome of an auction. After reminding participants about the rules and mechanics of the auction, they entered bids to have 1 hen, 100 hens, or 1000 hens live in a free-range system as opposed to the cage system.[24] We also asked people to (hypothetically) indicate how much they would pay to move *all* hens in use to a free-range system.

The rules of the auction were very similar to that used in private-good auctions. Each participant submitted three bids (one to move 1 hen, one to move 100 hens, and one to move 1000 hens), after which we picked a number from a hat to decide which auction would be the auction that actually took place. Then, we randomly picked one person in the session by picking another number from a hat. The randomly selected person was taken aside in a separate room, and we then took a look at their bid on the randomly selected item (either the 1-, 100-, or 1000- hen auction). This bid was compared to a randomly drawn "secret price." If the bid was greater than the secret price the participant paid the secret price and the hens were raised in the free-range system. Conversely, if the bid was less than the secret price, the participant paid nothing and the hens were raised in the cage system.

This unusual type of auction with the secret price was used for the same reasons described in the previous section—it provides people with an incentive to enter bids that equal their true WTP to provide the animals a better life. Moreover, this was a *real* auction and we went to great lengths to convince the participants that it was real. Participants were repeatedly told that if they won the auction they would really have to pay the secret price. Participants were told that we would accept cash, checks, or credit cards. Moreover, we also described the protocol established for ensuring the correct number of hens is truly raised on the correct farm. To enhance our credibility, participants were given our contact information and were told that if they desired, we would send videos demonstrating the movement of real animals to real farm systems. Finally, we stressed that it would be impossible for the auction winner to ever eat food produced by these animals, as the animals would be harvested and consumed at a University far from the participants' homes.

We asked people to submit bids for 1, 10, and 1000 hens for several reasons. First, we wanted to check whether people's bids were logical; the value of moving 100 hens should be at least as large as that to move 1 hen. If participants gave us bids to the contrary, the computer program prompted them with a warning and asked if this was really how the participant wanted to bid. The goal of this is to help participants discover their true values and express them in a logical way, and helping them increase their bids with the number of animals affected contributes towards logical bids. Secondly, you may remember from our discussion in Chapter 7 that humans often respond in irrational ways when valuing a change in the well-being of one animal. A single animal tends to evoke strong emotions, while many animals lead to participants thinking only in terms of numbers. There is ample evidence to believe that an individual may express a higher value to benefit 1 hen than 100 or maybe even 1000 hens. To mitigate the emotional appeal of helping a single hen, and to elicit values that were rational in the sense that the benefit of an action increased as the number of animals benefited increase, we asked participants to simultaneously submit bids for 1, 100, and 1000 hens in an effort to encourage them to think more deliberately about the overall benefits of the change in farming systems.

Because we were interested in how people might value policies such as a nationwide ban on cages in layer production, we also asked people to bid to move all 284 million hens in the US from a cage to a free-range system. Of course, we could not actually control the lives of millions of hens, and thus these bids were hypothetical. Nevertheless, we asked participants to think carefully about their answer and submit a bid as if it we could actually hold the auction.

Half our participants bid in auction to improve the lives of laying hens, and the other half did the same for hogs. Similar to the egg auction, participants submitted bids to raise 1, 100, or 1000 sows and all their offspring in a shelter-pasture system instead of a confinement-crate system.[25] People were given information about the number of offspring associated with each sow, in addition to other pertinent information, such as expected life span. The auction rules were the same as the egg auction. After the bids were submitted, one person and one group of sows (1, 100, or 1000) were randomly chosen. If the randomly selected person's bid exceeded the secret price, the person paid the price and the given number of sows and her offspring were raised on a shelter-pasture system instead of the confinement-crate system. Otherwise, the person paid nothing and the hogs were raised in a crate system.

One final note: as the histograms in Figures 9.4 and 9.5 show, there are a few individuals who prefer the cage egg system over the free-range system, and a few who prefer the confinement-crate system over the shelter-pasture system. These individuals still participated in the auction, and if they won the animals would either be raised on the cage egg or confinement-crate system, depending on whether they attended an egg or pork session. So the auction was still for the individual to send the animals to the farm of their choice; the only difference is that the farm of their choice was different than most participants.

Public Good Auction Results

The value of the auction bid indicates the extent to which people believe the farm of their choice enhances animal happiness compared to the other farm, and the extent to which they will forgo money to help ensure that happiness. The results of the auctions are shown in Table 9.4, which display the amount of money people will pay to ensure hens live on a free-range farm and hogs live on a shelter-pasture farm.[26] The results show that the bids increased as the number of animals helped increased. The average bid when moving from 1, to 100, to 1000 hens was $0.98, $14.69, and $57.18. One interesting result, however, is that the value *per animal* is declining. The results indicate that average WTP *per hen* falls dramatically (from $0.98 to $0.15 to $0.06, respectively) as the number of hens increases from 1 to 100 to 1000. But why should the thousandth hen be any more or less valuable than the tenth hen?

Table 9.4 Public Good/Auction Results

Number of animals	Average bid	Average per animal	Median Bid	Percent of bids equal to zero
Number of laying hens moved from cage to free-range system				
1	$0.98	$1.08	$0.50	33%
100	$14.69	$0.19	$1.00	29%
1000	$57.18	$0.08	$2.00	29%
All in U.S. (hypothetical)	#$341.53	$0.0000014	$2.25	33%
Number of sows and offspring moved from confinement-crate to shelter-pasture system				
1	$2.85	$2.85	$0.10	40%
100	$7.72	$0.08	$0.30	38%
1000	$23.34	$0.02	$1.00	37%
all in U.S. (hypothetical)	$345.09	$0.000058	$1.75	36%

Notes: The egg auction involved 126 individuals, while the hog auction involved 134.

The result may be attributable to several factors. First, our desire to consume a good often falls as more of the good is consumed—something economists call diminishing marginal utility. For example, the first apple eaten in a day is much more tasty and satisfying than the fifteenth apple eaten in a day. Something deeper is going on though. As we described in Chapter 7, there is something emotionally appealing about helping *one* person or *one* animal. We can imagine what they look like and how they will feel; imagining the feelings of a thousand people or animals is a bit beyond our cognitive capacities. We sometimes feel a stronger moral obligation to help one animal than many animals.

It may also be that our brain is wired in a particular way; nature seems to have designed us to want to "help out." In fact, it may be that people don't really derive much satisfaction from improving animal well-being per se, but rather from "helping out"—from the act of supporting improved animal well-being. This phenomenon is referred to as *warm glow*.[27] We experience a *warm glow* when we receive satisfaction from the act of supporting improved animal welfare perhaps because it makes us feel like we are doing good. When people experience warm glow they care simply that something is done—not so much what or how much is done. Think about the cases where you have given money to a charity. In any of these cases do you have any idea of the tangible impact your contribution made? Probably not. One interpretation of the bid for the single animal is that it is an estimate of warm glow, rather than the direct satisfaction the person receives knowing the first animal has a better life, and the second animal has a better life, and so on.

If we are interested in the value of large-scale changes to animal welfare, the bids for improving the life of one animal is not going to tell us much. The bids for 1000 hens or sows and their offspring are a better indication of value from, say, new animal welfare regulations. The bids for 1000 animals are informative because they concern improvements in the lives of a large number of animals and because they refer to bids in a real auction where real money changes hands.

Consider the average bid for 1000 hens and 1000 sows and their offspring, and in particular, the large difference between the average and median bids. The average bid to move 1000 hens to a free-range system was $57 but the median was only $2. A similar result was found for hogs; the average bid to move 1000 sows and offspring was $23 but the median was only $1. The large difference between the mean and median can be explained by the presence of a few people who bid very high amounts. Consider an example. Suppose five people bid in an auction, biding amounts equal to 1, 2, 3, 4, and 100,000. The average bid is $(1 + 2 + 3 + 4 + 100,000) / 5 = 20,002$, but the median (the middle number) is only 3.

The results of the public good auction illustrate how the value of farm animal welfare is indeed influenced by a few extraordinarily concerned individuals. A few participants bid $1000 or more in the hen and hog auctions. However, the large majority of participants submitted bids of only a dollar or two to move 1000 animals and their offspring. Which measure is a better indicator of the public good value of farm animal welfare? This is a question with no definitive answer. The median tells us that most people are not willing to pay much—even if a 1000 animal lives are at stake. The average tells us that a few people *really* care. Perhaps, at this point, a skeptical reader wonders if the individuals who bid very large amounts really understood the auction. We took great care to make sure everyone understood the auction. There was ample time for questions. It was a real auction and participants knew it. We explicitly told the participants that if the winner bid hundreds or thousands of dollars and won the auction that we would collect the money. There is no doubt in our mind that the participants knew we meant it, and they knew the secret price of the auction (which is the price they would pay, if they won) could take any value.

In cost–benefit analysis (or in utilitarian calculation), the average bid is particularly meaningful. The total amount of money a group is willing to pay to change animal living conditions (i.e., the total benefit) is equal to the average bid times the number of people involved. For example, because 126 people bid in the hen auction, we can say that the total benefits this group would receive from moving 1000 hens from a cage to a free-range system is the average ($57) times the number of bidders (126) or $57 \times 126 = \$7182$. To the extent that our group of research participants represents Americans as a whole, the average bid ($57) can be used in conjunction with an estimate of the number of people (or households) in the US population to calculate the *total* public value to all citizens resulting from a change in animal welfare.

Asking Hypothetically...

The results just discussed are uniquely informative because they relate to questions where people were literally asked the amount of money they would spend in a real auction to improve the lives of 1000 real animals. The auction bids provide an accurate portrayal of Americans' concern for farm animals. The statistics are useful because they provide a *minimum* value people would place on improving the lives of *all* farm animals. If someone will pay $10 to make 1000 hens better off, then surely they will also pay at least $10 to make all 280 million hens better off.

We have argued that simply asking how much people will pay to improve the lives of all farm animals is not very appealing because such answers are prone to hypothetical bias, in which people say they will pay more than they truly will. Nevertheless, we thought it would be useful to ask one hypothetical WTP amount related to people's values for all animals. Because of the hypothetical nature of the question, though, we interpret people's answers as an upper bound for what they would really be willing to pay. By combining our answer of what people were really willing to pay in the auction of 1000 animals with what they said they were willing to pay for all animals, we have an upper and lower bound on what we believe to be people's true WTP to improve the lives of all hens or all hogs.

The last rows of results in Table 9.4 show how people answered the hypothetical question about all farm animals. The average bids are several times higher than the bids for 1000 animals, but not unreasonably large. These upper bound estimates imply that the people will pay an average of $341 to improve the lives of all laying hens and $345 to improve the lives of all hogs. The change in question is quite profound. For hogs, it would mean taking virtually every hog in the US, raised under the lowest standards of care, and moving them to an alternative farm that possesses the highest possible standards of care. The figures imply the potential for a very large estimate of the total benefits of a policy to improve all factory farms, but we must again be mindful of the median. Half of our participants were willing to pay less than $2.25 to move all hens to a free-range system, and half were willing to pay less than $1.75 to move all hogs to a pasture system.

The Value of Farm Animal Welfare

What are people willing to pay for cage-free rather than cage eggs? How much money would people be willing to forgo to allow all hogs in the US to venture outdoors? One cannot look to the prices of commodities at the New York Stock Exchange to answer these questions. There is no exchange institution in which buyers and sellers act to reveal their values for improved animal welfare. But there are policy-makers and business managers who want information on the potential demand for improved animal well-being; they want to know if the perceived benefits from the products outweigh the costs to provide them.

Businesses and marketing experts need to know consumer WTP to better understand consumer preferences, forecast new product success, and measure the effectiveness of promotional activities. McDonald's, for example, needs to know how their customers will react if they give in to the demands of the HSUS and PETA to source eggs from cage-free systems. Policy analysts need information on consumer WTP to forecast the outcome and consequences of initiatives like Prop 2 in California. For these reasons and more, this chapter was devoted to learning about consumer WTP.

Our attention has largely focused on our own experiments recording consumer expressions of farm animal welfare issues. This is because our study is a rarity in that it elicited real (non-hypothetical) WTP numbers for US consumers, and studied the private and public good portion of animal welfare separately. The nature of our experiments also allows us to calculate the value US consumers hold for producing eggs and pork across a large variety of settings. However, it should be noted that many other studies have also studied consumer demand for better animal care. While these studies have certain limitations that prohibit their inclusion in this chapter (many of them concern European consumers; for instance—each study has a strength in some other area that the reader may find fascinating). Thus, an Appendix is provided to this chapter detailing many such studies and their findings.

Consumer WTP for improvements in animal welfare was parsed into a *private* and *public* value. The private value relates to the additional money individuals will pay to ensure the food that they themselves eat comes from animals under improved living conditions. The public value relates to the money individuals will pay to improve the lives of farm animals irrespective of who eats the food products generated by the animals.

The results shown in this chapter provide evidence that many people have non-trivial private and public values. In fact, the private values elicited from our informed consumers suggest most were willing to pay the cost of providing improved living conditions. However, we should not take the results of these simple cost–benefit tests as sufficient justification for ushering in government regulation by being naive about the economics of political activity. Governments are rarely perfect and are easily manipulated. Quixotic crusaders for improved farm animal welfare may be surprised at the poor outcomes of government regulation. A key objective of this book is to warn the reader of naivety. The economics of farm animal welfare concern more than consumer values and farmer costs, but rather the interactions of consumers, farmers, activists, and politicians. Governments themselves are the subject of economic scrutiny, and that is where we now venture.

Appendix: Collection of WTP Studies Focusing
on Farm Animal Welfare

Study author	Year published	What did the authors do?
Bateman et al.	2008	Conducted in-person interviews with 200 people in Northern Ireland and asked about people's WTP (via a food tax) for animal welfare improvement programs for chicks, cows, hens, and pigs
Bennett/Bennett and Blaney	1997/2003	Conducted a mail survey among 591 British households, and asked people's WTP for a national policy to phase out the use of cages in egg production
Bennett and Blaney	2002	Conducted an in-person survey among 164 students at the University of Reading, England to study whether social consensus affects WTP for a policy to implement a more humane slaughtering technique for pigs
Bennett, Anderson, and Blaney	2002	Conducted an in-person survey among 119 students at the University of Reading, England to ask WTP for a policy to ban the export and import of live animals for slaughter and to ban the use of cages in egg production
Bernués et al.	2003	Conducted mail surveys among 2200 households in selected regions of England, Italy, France, Scotland, and Spain to determine the extrinsic attributes important in the purchase of beef and lamb
Brown et al.	2008	Conducted sensory evaluations of chicken breasts from standard, corn-fed, free-range, and organic production systems with about 300 taste testers
Carlsson et al.	2007a	Conducted a mail survey among 757 Swedish households, and asked people to choose between chicken fillet and ground beef packages that differed according to whether animals were given access to outdoors and/or were transported to slaughter

Carlsson et al.	2007*b*	Conducted a mail survey among 450 Swedish households, and asked people's WTP for eggs from free-range vs cage system
Chilton et al.	2006	Conducted in-person interviews with 200 people in Northern Ireland and asked about people's relative values for adopting animal welfare improvement programs for cows, pigs, hens, and chickens
Dickinson and Bailey	2002	Conducted non-hypothetical purchase experiments among 220 employees of Utah State University for beef and pork sandwiches produced under assurances for animal treatment (defined as "humane treatment procedures and no added growth hormones used in production of the meat")
Frank	2008	Compiled survey, consumption, and animal use data from a variety of countries and sources to determine whether animal welfare is related to a nation's economic standing
Frewer et al.	2005	Conducted a computer-based survey among 1000 Dutch households to study attitudes toward animal welfare in pig and fish production
Glass et al.	2005	Conducted a mail survey among 935 Irish households to determine consumers' WTP for several changes in pork production systems
Gracia et al.	2009	Conducted non-hypothetical purchase experiments with 202 consumers in Aragón (Spain) to determine the effects of different animal welfare labels on dry-cured ham
Hagelin et al.	2003	Reviewed over 56 previous surveys and opinion pools about the use of animals in experiments and research
Hall and Sandilands	2007	In-person interviews were held with 16 people in York, England to identify what attributes people thought affected chicken welfare
Lagerkvist et al.	2006	Conducted a mail survey among 285 Swedish households, and asked people to choose between pork chops that differed along 8+ attributes related to pig well-being

(continued)

Appendix (continued)

Study author	Year published	What did the authors do?
Liljenstope	2008	Conducted a mail survey among 1250 Swedish households, and asked people to choose between pork fillets that differed according to a 10+ of attributes related to pig well-being
Lind	2007	Conducted in-person qualitative interviews with 127 pork shoppers in Sweden to determine the attributes and values people associated with non-branded, branded, and organic pork
Lusk	2010	Utilized grocery store scanner data to estimate consumer demand for conventional, cage-free, and organic eggs in Texas and California over the time period surrounding Prop 2 in California
Lusk and Norwood	2009a	Conducted non-hypothetical purchase experiments among about 300 people in Dallas, Chicago, and Wilmington, where people bid to move pigs and hens from one production system to another, with no added growth hormones used in production of the meat
Lusk and Norwood	2008	Conducted a phone survey among 1019 US households and asked people a host of questions related to knowledge of and attitudes toward farm animal welfare, environmental, and/or no antibiotic certification
Lusk et al.	2007	Conducted a mail survey among 594 US households and asked people to choose between pork chops that differed according to whether they contained an animal welfare, environmental, and/or no antibiotic certification
Norwood and Lusk	Forthcoming	Conducted non-hypothetical purchase experiments among about 300 people in Dallas, Chicago, and Wilmington, where people bid to buy pork and eggs produced under different production systems with no added growth hormones used in production of the meat

McCarthy et al.	2003	Conducted a survey among 300 Irish households to determine correlations between beef purchase intentions and various beliefs and attitudes
Moran and McVittie	2008	Conducted an in-person survey among 318 English households to determine WTP for a policy to improve living standards for broilers
Napolitano et al.	2007	Conducted sensory evaluations of beef from a conventional production system and standard and a Podolian production system with higher levels of animal care, among other things, among 80 taste testers in Southern Italy; some people were informed of the difference while others were not
Ophuis	1994	Conducted sensory evaluations of pork from standard and free-range production systems with 172 taste testers in the Netherlands; some people were informed of the difference while others were not
Schnettler et al.	2009	Conducted in-person surveys among 770 meat consumers in Chile to determine the importance of information regarding animal treatment prior to slaughter when buying beef
Tonsor, Olynk, Wolf.	2009	Conducted mail survey among 205 Michigan households and asked people to choose between pork chop packages that differed according to whether sows were raised in gestation crate system or from a region that banned crates
Tonsor, Wolf, Olynk	2009	Conducted an on-line survey among about 1000 US residents and asked how they would vote on a referendum to ban use of gestation crates in hog production
Tonsor and Wolf	Forthcoming	Conducted an on-line survey among 2000 US residents and their WTP for mandatory labeling of pork from gestation crate systems
Videra	2006	Analyzed county-level voting data from the 2002 Florida ballot referendum to ban gestation crates

CHAPTER 10

Farm Animal Welfare and the State

*Translating Beliefs about Farm Animal Welfare
into Beliefs about the Role of Government
in Regulating Livestock Production*

"The statesman who should attempt to direct private people in what manner they ought to employ their capitals would not only load himself with most unnecessary attention but assume an authority which could safely be trusted to no council and senate whatever, and which would nowhere be so dangerous as in the hands of man who have folly and presumption enough to fancy himself fit to exercise it."

Adam Smith in *An Inquiry into the Nature and Causes of the Wealth of Nations* (1776)

"The curious task of economics is to demonstrate to men how little they really know about what they imagine they can design."

F. A. Hayek in *The Fatal Conceit* (1988)

Animal Welfare and the State

The relationship between individual food purchases and farm animal welfare has been confronted by asking consumers questions like, *"How would animal well-being change if you increased or decreased consumption of pork?"* or *"How much are you willing to pay for cage-free eggs?"* Questions about how to treat farm animals concern morality, so people often want to change not just their own behavior but other people's behavior as well. It is a human predilection to desire that the laws of a nation reflect the ethics of the individual. People with strong beliefs seek to force those beliefs onto others through public policy. Animal activists want more humanely produced food and they seek legislation and regulation to achieve the outcomes they desire. The debate about how farm animals should be raised is played out in the grocery store, the halls of Congress, voting booths, and the courtroom. No book on the economics of farm animal welfare would be complete without addressing public policy. In this chapter, we confront the issue of farm animal welfare and the *State* (all forms of government including Congress, courts, law enforcement, and the like). What policies might the citizenry want government to enact in the form of farm animal welfare? Under what conditions might additional regulations be beneficial and when might they be harmful?

For better or worse, the farm animal welfare debate is an important social and policy issue due to the power of animal advocacy organizations. The Humane Society of the United States (HSUS), People for the Ethical Treatment of Animals (PETA), Farm Sanctuary, and the like have achieved success in regulating farm production practices through ballot initiatives and by lobbying. Virtually all major animal advocacy groups, from the HSUS to PETA actively track and promote legislation on issues related to animal welfare. For example, at the time of this writing, Farm Sanctuary listed 30 pieces of federal legislation that it was tracking through the US Congress.[1] The University of Chicago law professor Cass Sunstein, who has written several articles and has co-edited a book on animal rights issues, was recently nominated by President Obama as regulatory czar, head of the Office of Information and Regulatory Affairs. It is unclear whether Sunstein will attempt to translate any of the animal rights ideas expressed in his writings into regulation. What is clear is that we are likely to witness new referenda, legislation, and lawsuits over animal welfare issues in the future, which suggests the need for closer examination of the relationship between animal well-being and the State.

This chapter addresses both positive and normative issues; such as, what are the effects of certain government regulations and what role should we advocate for government in regulating the well-being of farm animals? We caution the reader that these are not easy questions to answer, and how one chooses to answer the normative questions of what government *ought* to be doing will often stem not from data or logic, but will emanate from an ideological conflict between faith in the State and faith in the *individual* working out their preferences in the market. Nevertheless, our purpose in this chapter is much the same as it has been throughout this book: to provide the reader with the background to make an informed and reasoned decision for themselves.

Markets and Change

We have repeatedly argued that the choices people make in markets generally produce positive results, and that government attempts to interfere with the private decisions of individuals often (though not always) result in inferior and unintended outcomes. Policy-makers face an enormous informational challenge when trying to improve the well-being of citizens. Individuals are privy to information policy-makers are not. These include their personal preferences and idiosyncratic obstacles. No doubt, you prefer to eat different types of meals to your neighbor, and a chicken farmer in Arkansas must employ a different business model than a corn farmer in Iowa. Businesses and consumers acquire this information through trial and error and in response to their unique environments, and the wisest professor or the most competent government bureaucrat can only catch a glimpse of the information embodied in the many daily-life decisions. Much of this information is reflected in the prices that emerge in markets—note we say *emerge*, not set, because consumers play an active role in price setting, as does firm competition. The price of corn reflects an astounding amount of information, including the cost of fertilizer to the Iowa farmer and the intensity with which the Arkansas farmer needs corn to feed chickens. When people interact in an economy, both revealing and acting on market price information, the typical outcome is a level of wealth that no government plan, however wise or benevolent, can attain.

Alas, politicians are seldom wise or benevolent, but even if they were, they often face incentives that induce them to detract from public welfare to enhance their chances of re-election. Politicians are often made better off by bestowing benefits to special interest groups at the expense of those less politically organized. Because the prospects of getting re-elected depend on campaign contributions and because campaign contributions are given and influenced by special interest groups, a politician can, in a sense, become captured by a special interest. History reveals a tendency for businesses and other interest groups to shape regulation—an outcome called *regulatory capture*—in a manner that provides them with greater profits than if the regulation was never pursued in the first-place. Politicians who attempt to abolish or hinder such regulations are targeted by well-funded and highly organized groups who benefit from the regulations; regulatory capture ensues.

Market-based decisions are quite different to those made by governments. If a firm offers a product that nobody wants, it soon leaves business. If government enacts a bad policy, rarely is it repealed. Instead, lip-service is given to reform, and when the reform fails, more reform is pursued. Take, for example, farm support policies, which were enacted to provide financial assistance to struggling farmers during and after the Great Depression. Henry Wallace, the Secretary of Agriculture from 1933 to 1940, argued that the farm subsidies were, *"a temporary solution to deal with an emergency."*[2] But even though the average farm household income now exceeds that of the average US household, and even though there are now many fewer farmers, farm policies persist.

In stark contrast, consider how Wal-Mart makes decisions. Who decides whether to run a special or discount an item at Wal-Mart stores? You may be surprised that the answer is—*any employee.*[3] To be sure, much of Wal-Mart's success lies in smart management, but it is also attributable to management's humility and their willingness to adapt to bottom-up change. There is a surprising amount of variability in the offerings and prices across Wal-Mart markets. Our small town of Stillwater, Oklahoma (population of about 45,000) has two Wal-Marts on opposite ends of town. We were amazed when our wives informed us that the two Wal-Marts sell different items and post different prices. So different are the stores that we regularly shop at both Wal-Marts, even though one is very close and one is on the other side of town. Even such a small town as Stillwater has surprisingly high diversity in consumer preferences. Local managers and shelf-stockers have first-hand knowledge of the local conditions at their Wal-Mart.

It would be impossible for the central management at Wal-Mart to collect large amounts of information on shopper preferences in each store, much less process it in a prompt, meaningful manner. Thus, a surprising degree of decision-making is given to those closest to the consumer. The shelf-stockers do not make all the decisions, of course. Much of Wal-Mart's success seems to be attributable to a distribution system that is ruthlessly efficient and a stubborn insistence on paying low prices to its suppliers. These are the success of intelligent planning by Wal-Mart executives. The point, however, is that these executives delegate more authority to shelf-stockers than the reader surely imagines. Wal-Mart executives have apparently come to understand their inability to understand each local market fully and recognize the need for rapid change in response to market conditions. While Wal-Mart typically provides the goods we want at the low prices we relish, politicians rarely deliver the same kinds of regulations. Instead, we have farm policies in 2010 that were, in many ways, built to serve the needs of the world in 1940. Contrast this with Wal-Mart today, which bears little resemblance to the retailers that existed in 1940s.

If markets efficiently allocate goods to their most valued uses, and if politicians can suffer from regulatory capture and slow decision-making, what is the role of government? To say that the government should "do something" about animal well-being, we must identify situations in which one believes market outcomes are not ideal, or a situation when imperfect government is preferred to an imperfect market. It is not acceptable to claim that, because you desire an outcome (such as improved animal living conditions) that the market has failed to provide, the government should intervene. You would, no doubt, like to drive a nicer or sportier car, but would likely have a hard time providing a rational and systematic justification for the government providing you with one.

Economists have developed logical, compelling arguments that describe the settings in which government regulation can *potentially* improve upon market outcomes. There is disagreement among economists about the seriousness of these so-called *market failures*, and about the ability of government regulations to achieve the outcomes that our theoretical models predict. Although economists have similar training in how to evaluate public policies, they divide themselves on either side of the right–left political spectrum, just like non-economists. Economists are, after all, human.

Even in a supposedly "free" country like the US, the economy is replete with government regulation. Approximately one-third of all income is diverted to taxes, and as this book is being written Congress is fervently trying to secure greater control over the one-fifth of the economy called health care. Such extensive government involvement in daily life has come to fruition at the request of the citizens. The US, is, after all, a democracy. However much one dislikes government regulation, the citizenry apparently endorses an active government.

The question we ask in this chapter is: *what is the role of the government in regulating the well-being of farm animals?* In the sections that follow, we discuss some of the most compelling market-failure arguments for government intervention, and in some cases we calculate the benefits and costs of proposed regulations as they relate to the market-failure arguments. In each case, we also ask whether solutions to the market failures can be found outside the context of government regulation.

At the onset, however, we want to emphasize that we do not intend to calculate an all-encompassing measure of benefits and costs for policies such as a ban on cages in egg production or a ban on gestation crates in pork production. To derive such a measure would require making a series of untenable assumptions that render the measure meaningless. For reasons that will be discussed in what follows, it is difficult to precisely predict how consumers will respond to a ban. Thus, although consumer research can help us draw inferences about the effects of a ban, we are skeptical of our ability (or anyone else's for that matter) to articulate the consequences of a farm animal welfare regulation. In this chapter, we take a more modest approach. Rather than asking whether all the benefits of a ban exceed all the costs, we focus on particular market-failure arguments for government involvement. We ask whether our data support government intervention from the standpoint of several market-failure arguments.

Farm Animal Welfare as a Public Good

Unlike many of the items we purchase and consume, improved animal well-being possesses certain characteristics that suggest market outcomes may be less than desirable from the standpoint of society as a whole. First, farm animal welfare is a *non-rival* good, as the enjoyment you receive from improving the well-being of a farm animal does not prohibit others from enjoying the improvement as well. By contrast, an apple is a *rival* good; if you choose to eat and enjoy an apple, you have necessarily kept others from enjoying the same apple. Although you are in competition with other consumers for an apple, all consumers can share the satisfaction of an improved animal life. Second, animal welfare is *non-excludable,* meaning anyone can experience pleasure from better animal care, even if they did not consume the food or pay for the improvement. No one can stop you from benefiting from better animal care, if it occurs. Someone living in Australia might be happy that California banned cage egg production even though the Australian will not consume the food from that animal nor pay the cost of the policy. We have no way of excluding people from experiencing the pleasure of improved animal well-being. Conversely, the grocery store owns the apple and decides who will purchase it and at what price, making the apple an *excludable* good. Economists define *public goods* as goods that are both non-rival and non-excludable. Hence, farm animal welfare is a public good.

The public good aspect of farm animal well-being can also be stated in the negative. The sadness you experience from knowing a sow lives in a gestation crate does not end with you; others can experience the sadness equally as well. Moreover, we cannot keep you from experiencing sadness about an animal's living condition. Indeed, many of the people who care about animal well-being do not purchase eggs or meat, and as such, vegans have an interest in how meat, dairy, and eggs are produced.

Market suppliers can have a difficult time providing public goods because firms can have difficulty recouping the production costs. Specifically, the firms have difficulty charging everyone who benefits from the good. Suppose turkey producers decide to improve animal welfare, at significant cost. After testing these animal-friendly turkey products, they find consumers are not willing to pay the higher production costs. But wait: there are many consumers who never ate the turkey products but still took pleasure in the fact that the animals were happier. The fact that someone else ate the turkey did not impede their ability to benefit, nor could the producers exclude these other people from benefiting. If all these other people could be charged an amount just half of their benefit, it would be profitable to improve the birds' lives. Alas, these other people cannot be charged, and they do not voluntarily contribute money to the producers, so the animal-friendly turkey products are taken off the market. The normal interaction of buyers and sellers (i.e., the market) that normally does so much to improve humanity's lot fails to operate in the most ideal way in the case of public goods. This deficiency is a *market failure*.

The public good aspect of farm animal welfare can create situations in which people, if left to their own devices, will arrive at decisions that they all agree are inferior to a regulated outcome. This problem is best illustrated through example. Suppose *Jack* and *Jill* are deciding whether to buy crate-free (pork from confinement-pen system) or regular pork (pork from confinement-crate system), and suppose both people are exactly the same in so far as the benefits they derive from buying pork. Both are willing to pay a maximum of $6 for crate-free pork and a maximum of $4 for regular pork to satisfy their selfish private-good motives, motives that result because they believe crate-free pork might taste better, be safer, or because they personally enjoy buying pork from a more humane system. Suppose both shoppers enter a grocery store that has the following prices: crate-free pork sells at $5.00 and regular pork sells at $2.00.

At these prices, the private net-benefit, stated in dollar terms, both would get from buying regular pork is \$4.00 − \$2.00 = \$2.00 but is only \$6.00 − \$5.00 = \$1.00 for buying crate-free pork. Although they are willing to pay more for crate-free pork than regular pork, the price of crate-free pork is sufficiently high such that there is a higher net benefit (willingness-to-pay (WTP) minus price) from buying regular pork. This simple example is illustrated in Figure 10.1. Jack is better off buying regular pork regardless of what Jill does. Likewise, Jill is better off buying regular pork irrespective of Jack's decision.

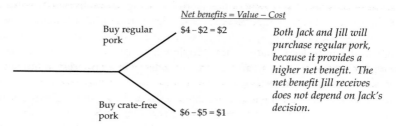

Figure 10.1 The pork purchasing decision considering only private motives

Now suppose that in addition to the private values shown in Figure 10.1, Jack and Jill also have public-good values. More humane animal care now has non-rival and non-exclusive properties. If Jack cares about the well-being of animals, he can benefit from Jill buying crate-free pork even if he does not. Assume that Jack experiences an increase in happiness, measured in dollars, of \$1.50 if Jill purchases crate-free pork. Jack and Jill are identical, so if Jack purchases crate-free pork, Jill is also happier by \$1.50.

Jill's \$1.50 public benefit derived when Jack purchases crate-free pork is slightly lower than her \$2.00 private WTP premium for crate-free pork because we presume Jill gets some extra satisfaction (\$0.50 worth in this case) from buying crate-free pork unrelated to animal well-being. Perhaps Jill thinks meat is tastier or safer when raised more humanely, or perhaps she takes particular comfort from knowing that the animal is made better off from *her purchases.*

The public-good problem is illustrated by Jack and Jill's purchasing dilemma, depicted in Figure 10.2. Jack's choice is to purchase either regular or crate-free pork and Jill's choice is the same. Thus, there are four possible outcomes: (1) both can buy regular; (2) both can buy crate-free; (3) Jill can buy regular while Jack buys crate-free; or (4) Jack can buy regular while Jill buys crate-free. Consider the benefits Jack and Jill would receive from these four outcomes:

(1) If both Jack and Jill purchase regular pork, each receives \$2.00 in net benefits. The net benefit is just the difference in the \$4.00 of *private* value and the \$2.00 purchase price.

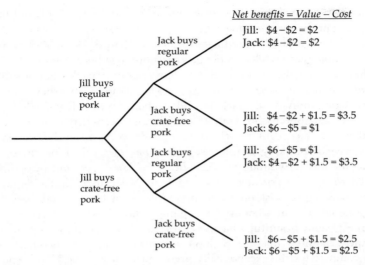

Net benefits = Value – Cost
Jill: $4 – $2 = $2
Jack: $4 – $2 = $2

Jack buys
regular
pork

Jill buys
regular
pork

Jack buys
crate-free
pork

Jill: $4 – $2 + $1.5 = $3.5
Jack: $6 – $5 = $1

Jack buys
regular
pork

Jill: $6 – $5 = $1
Jack: $4 – $2 + $1.5 = $3.5

Jill buys
crate-free
pork

Jack buys
crate-free
pork

Jill: $6 – $5 + $1.5 = $2.5
Jack: $6 – $5 + $1.5 = $2.5

Figure 10.2 Illustrating the public-good nature of farm animal welfare

(2) If Jill purchases crate-free pork but Jack purchases regular pork, Jack receives $2.00 in net benefits from regular pork as he did before, but now he also receives an extra $1.50 in extra benefit relating to his *public* value which results from Jill's purchase of crate-free pork. Jack's total net benefit is $3.50, while Jill's net benefit is only $1.00.

(3) Conversely, if Jack purchases crate-free pork but Jill does not, Jack receives a net benefit of $1.00 and Jill receives a large net benefit of $3.50.

(4) If both Jack and Jill purchase crate-free pork they each receive $2.50 in net benefits. This includes the $1.0 in net benefits from their *private* values plus $1.50 from their *public* values knowing the person's choice benefitted sows.

As we can see, the best choice for Jill now depends on what Jack does, and vice-versa. Jack's business is Jill's business, and Jill's business is Jack's business. What choice will Jill make? The answer to this question requires a bit of thinking.

Jill only has two choices: purchase regular or purchase crate-free pork. Jill knows that her well-being depends on what Jack does, so Jill has to think a bit about the choices Jack might make. If Jack purchases regular pork, what should Jill do? If Jack purchases regular pork, Jill is better off purchasing regular pork as well. If instead Jack purchases crate-free pork, Jill is still better off purchasing regular pork. *Thus, no matter what decision Jack makes, Jill is always better off purchasing regular pork.* Since Jack faces the same dilemma, Jack will also purchase regular pork. Both Jack and Jill purchase regular pork and find themselves receiving a rather low net benefit of $2.00. Both persons' decisions were entirely rational, but in a sense unfortunate, because they could both be made better off by working together.

What if Jack and Jill coordinated their purchases? What if they were a married couple that traveled to the grocery store together to buy pork? Figure 10.2 shows that if they both agreed to buy crate-free pork, they would both received a net benefit of $2.50, which is $0.50 higher than what they would have received acting individually. Jack and Jill can both see that they would be better cooperating and purchasing crate-free pork, and yet when they go shopping alone they both buy regular pork, and are made worse off in the process.

This type of thinking is referred to as *game theory* in economics. If you watched the award-winning movie *A Beautiful Mind*, you will recognize that Russell Crowe, who portrayed a young John Nash, used similar logic when deciding which girl to ask on a date. Nash (played by Crowe) and three friends were gathered at a bar when five young girls entered. All five were pretty, but one was especially beautiful. One of Nash's friends appealed to the principles of economics, claiming that competition was good, and that the male friends should compete for the exceptionally pretty girl. Then came Nash's epiphany. If all of the males compete for the pretty girl, *"we will block each other, not a single one of us is going to get her."* And if they then go for the other four girls, having failed to get the first girl, they would be shunned by these four too because, *"nobody likes being second choice."* Nash then proposes that they ignore the exceptional girl and ask out the four other girls. *"That's the only way we win,"* Nash argues. While the story is a product of the imagination of a Hollywood writer, the insight is real. The best result comes not just from everyone doing what is best for them, but what is best for themselves *and* everyone else.[4]

Just as Nash and his friends should coordinate their efforts to get a date, Jack and Jill should coordinate their purchasing decisions to increase the happiness they receive from buying pork. This is easy in theory but not in practice. Exactly how is Jill to harmonize her purchases with a man named Jack, whom she does not know? Is she expected to confront grocery store shoppers and attempt to strike an agreement? Consider also that the story of Jack and Jill is merely a metaphor for the millions of grocery shoppers. While the difficulties of two strangers striking a bargain are not insurmountable, imagine the difficulties of millions of people trying to take into account the interdependencies of their purchasing decisions? Actually, it may not be as hard as you think. A regulation banning the sale of regular pork solves the coordination problem.

There remains a paradox to the Jack and Jill story. We predicted that both Jack and Jill will purchase regular pork, but we also showed that they would both be better off if they purchased crate-free pork. Although they are worse off choosing the regular pork than crate-free, their decision is entirely rational. Accordingly, the grocery store will not sell crate-free pork because neither Jack nor Jill (or the millions of other people who Jack and Jill symbolize) will buy it. *The paradox is that Jack and Jill would be better off if the grocery store only sold crate-free pork, but the grocery store will only sell regular pork.* This is the public good problem.

The public good problem represents a market failure, which arises because of the inability of millions of shoppers to coordinate their decisions efficiently. The lack of coordination produces incentives for shoppers to *free-ride* off other peoples' purchases. We can see that the best possible outcome for Jill occurs when she purchases regular pork and Jack purchases crate-free pork. In this case, Jill enjoys the satisfaction of knowing that sows are being treated more humanely, but she also enjoys the satisfaction of not having to pay the cost of the crate-free pork. Thus, Jill has an incentive to *free-ride*, which simply means that she benefits at the expense of another. Free-riding is a potential problem when markets, not government, provide public goods.

When free-riding exists, people's individual actions in a market-based economy tend to generate too little of the public good as compared to some optimal amount that could be produced through coordinated action. The key to solving the public good problem is to eliminate the ability to free-ride. Jill should not be allowed to free-ride off the benevolence of Jack and his crate-free pork purchases. A simple and often effective solution is to have government provide the public good through taxation. Uniform taxation to the citizenry eliminates free-riders by forcing everyone to pay their "fair share." Another popular solution is for the government to outlaw certain production practices that are cheaper, yet less humane.

The government provision of public goods has a long history. National defense is a public good, and one that most would agree is best provided by the federal government. A private company might sell households in North Dakota a plan for protection from Canadian aggression, but Kansans could benefit from the protection without having to pay the cost. Public education is another public good, because education provides benefits not just to the student but to everyone else who encounters the student. As a result, most Americans are provided education directly from their local and state governments. National parks are another heralded public good, and almost everyone accepts government action to preserve National Parks. The logic behind each of these decisions is that the private actions of businesses and consumers are suboptimal and would lead to too little national defense, too little education, and too few parks as compared to what people really want and for which consumers are really willing to pay. If the government injects itself into the private affairs of people to ensure the adequate provision of all of these public goods, why not farm animal welfare as well?

The claim that government has the *potential* to benefit society does not imply that it necessarily will. While the public good nature of farm animal welfare indeed provides some motivation for government regulation, there are other economic reasons that make the case for government involvement dubious. Government has the potential to improve outcomes only when government is perfect or close to perfect. Not only must government possess a benevolent character towards the citizenry, and not only must it resist the lobbying of special interest groups, the government must have accurate information about preferences which it can shrewdly and efficiently process. For example, how will it know the values Jack and Jill possess? These are formidable obstacles to good government. It is true that markets must meet certain criteria before they contribute positively towards well-being, but the minimum standards for government, in our opinion, are even more formidable. As economists who openly endorse the benefits of capitalism, we must concede that market failures do occur. However, those who openly promote an active government should correspondingly concede that government failure is also possible.

The discussion of public goods requires far more than a concession that both government and markets may fail. A thorough discussion should also include the creative mechanisms by which society discovers non-governmental solutions to the public good problem. One prominent way in which people coordinate their actions to deal with what otherwise might be a market failure is through the creation of rules and social norms. For example, churches face the same potential for market failure as environmental protection and animal welfare. Church services are non-rival (unless the church attendance approaches capacity) and non-exclusive, as few churches would turn away anyone interested in attending. Yet the religious-social norm to tithe is strong enough to support countless churches.

Social networks, social norms, and other social institutions readily emerge to help individuals coordinate their food-purchasing decisions. Anyone over the age of 30 can attest to how quickly social norms have changed to encourage environmentally friendly actions. Few people are brave enough to litter or dump a used can of oil into a river in the plain view of others. Many people feel a moral imperative to recycle, and even take particular pride in the activity. Humans have an innate sense of morality, and most people understand the nature of the public good problem, if not the economic terminology we used to describe it. We desire to leave a prosperous world for future generations. We desire to leave wildlife and natural water systems intact. We desire to prevent the extinction of species. We take measures to do these things even if the market system provides us with inadequate profit incentives to do so. We respond to *all* incentives, even those that are not monetary, and social norms provide a powerful non-monetary incentive.

It is possible that social norms and customs will evolve to address the public good problems associated with farm animal welfare, and may ultimately be more effective (though perhaps slower) than government regulation. This is a view that is increasingly accepted. In 2009, the Nobel Prize in Economics went to Elinor Ostrum (along with Oliver E. Williamson) for her work documenting the myriad ways free societies overcome the public good problem.

In fact, current events suggest the evolution of social norms in regard to farm animal welfare. Animal advocacy groups like the HSUS and PETA have confronted food retailers in an attempt to force the use of more humane animal production practices. A great many retailers have announced the intention to purchase cage-free eggs, utilize better slaughtering techniques, and so on. Consider just one example among many. In 2008, the large supermarket chain Safeway emerged from talks with the HSUS agreeing to establish a preference for cage-free eggs, crate-free pork, and to show preference to food processors who utilize controlled-atmosphere stunning.[5] It is as if Jack and Jill, knowing the difficulties associated with harmonizing their food purchasing decisions, approached the grocery store manager and encouraged the manager to sell only crate-free pork. By placing pressure on food retailers to sell only a particular type of food, animal advocacy groups can act as surrogate for millions of consumers, helping them coordinate their decisions by negotiating with the food retailers directly.

As these words are being written, very few consumers buy cage-free eggs or demand more humane products. Our story of Jack and Jill, however, suggests consumers and HSUS may have similar views, even if they appear to make different choices. Jack and Jill sincerely care about the treatment of farm animals, and they were willing to pay more to obtain more humane meat. The market incentives, however, caused them to purchase the least humane meat. Yet if both Jack and Jill made donations to HSUS, giving HSUS the resources to shame their grocery store into selling crate-free pork only, both Jack and Jill would be better off (assuming their donations were not larger than the benefits of crate-free pork). If one did not understand the economics of the Jack and Jill story, it would appear as if HSUS forced Jack and Jill to purchase food they did not desire, but in reality, it took the bullying of HSUS to provide Jack and Jill with their desired pork. A little economics can make the world look like a very different place!

Specialty food retailers are also helping to overcome the animal welfare problem by selling only products raised under more humane standards. While regular pork may still be sold at conventional grocery stores, the transaction costs of shopping at two stores may be high enough to induce all consumers sympathizing with the animal welfare movement to shop at one store. Jack and Jill now shop at Whole Foods or Wild Oats, where the only pork products on sale are crate-free. Animal advocacy groups and specialty food stores represent the moral concerns of animal welfare sympathizers and attempt to translate that concern into action. The extent to which these concerns will change the average grocery store is difficult to predict, but the availability of cage-free eggs in most grocery stores attests to the fact that the food industry is changing in response to consumer concerns. Much of the increase in cage-free egg production has been realized without government help, indicating that the type of coordination Jack and Jill desire can be realized through mechanisms other than government regulation. In thirty years, cage egg production might be a relic of the past, not due to government bans, but to the activism of concerned consumers and the interest groups they support. Although such a future might seem a bit fanciful, the reader should bear in mind that less than a half-century ago, black Americans were unable to use the same bathrooms, water fountains, and restaurants as whites in many areas of the South. While government regulation forced the integration of schools, changes in social norms played major role in peeling back the official and unofficial Jim Crow laws that existed throughout the South. Social norms can change: drastically, and fast.

The story about Jack and Jill is useful because it allows us to put aside all the complexities of the "real world" and study a few important features of the world. These simple stories—what economists call economic models—can be dangerous if we do not acknowledge some of the simplifying assumptions of the story. The Jack and Jill story assumed that both people had identical preferences, whereas we know that no two people are perfectly alike. Consumers exhibit significant variations in their tastes and income, and while the flexibility of markets allow firms to respond to preference variation by selling different varieties of foods, governmental policies tend to be of the one-size-fits-all sort.

If the government (or the grocery store) allowed a vote on whether conventional pork should be banned in favor of only selling crate-free pork, both Jack and Jill would vote "yes." That is—there would be a 100 percent vote in favor of banning the very product they regularly purchase. Unlike the Jack and Jill story, there are some people who really do purchase crate-free pork, even if no one else does. There are some people that place no value on the public-good portion of animal welfare; for these people, your business is not their business.

There are private-good and public-good aspects of farm animal welfare. Free markets are adept at providing private goods, and possess the flexibility of accommodating differences in preferences. Sometimes the value of the public good is so large that government regulation is pursued, and the regulations separate individuals with disparate preferences into winners and losers. In these cases, the desirability of a regulation can be measured by the extent to which the benefits exceed the costs. Employing such a cost–benefit analysis is sometimes straightforward and accompanied by an unambiguous verdict; an example would be regulations in the 1980s targeting the restoration of the ozone layer, where the benefits exceeded the costs by such a large amount that little debate was needed.

In other cases, a cost–benefit analysis does little more than help the research clarify the issues involved. During 2002, one of the authors was involved in a large research project in North Carolina, the second-largest hog producing state in the US. The project entailed the use of cost–benefit analysis to identify better ways to handle hog manure. If a waste management technology yielded benefits higher than the costs, and this benefit-to-cost ratio was larger than any of the other technologies considered, hundreds of hog farms would be required, by law, to install the technology. The concept was simple, but in practice, an accurate cost-benefit required far more information than any group of researchers could obtain. Despite years of training preparing us for the task, and millions of dollars to cover expenses, it was absolutely impossible to determine whether a technology yielded benefits greater than the costs. In summary, there are instances when the case for public goods is so compelling that regulation may be desirable. There are three obstacles to effective regulation: a lack of information by the regulators, regulatory capture, and variability in preferences. In the next section we explore the public good component of farm animal welfare, and the extent to which regulation of livestock would suffer from the aforementioned obstacles.

Consequences of Farm Animal Welfare Legislation

The farm animal legislation considered in this chapter is a ban on a conventional method of production in favor for a more humane, and more costly, alternative in the egg and swine industry. There are other interesting regulations that could be considered, but production practice bans appear to be among the most commonly pursued policies.

To understand the consequences of production technology bans, one has to first acknowledge that the ban will be promoted by groups of citizens who are informed about livestock issues, but will primarily impact the uninformed citizenry. Thus, a policy has an *informed impact* and an *experienced impact*. A brief example should suffice: imagine cage egg production is banned, and that cage-free production will be substituted. Also, assume that the WTP of informed consumers for the ban exceeds the costs. The informed consumers can be thought of as a jury who scrutinizes all the evidence and renders a verdict. The *informed effect* comprises the benefits and costs of the ban as perceived by the informed consumers and includes private good benefits as well as public good benefits.

The verdict is heard outside the courtroom but the evidence leading to the verdict is not. Consumers see egg prices rise and notice all cartons bearing a "cage-free" label. Some consumers read about the ban in the newspaper and some heard about it through radio or television, but even these individuals do not understand the impetus or the consequence of the ban. Their purchases of eggs may fall due to the higher price, or it may rise in response to a perceived more humane product—there is no way to predict whether purchases will be higher, lower, or unchanged after the ban. The *experienced effect* pertains to the impact on uninformed consumers. Most likely, consumers will view eggs roughly the same after the ban as they did before. The higher production costs will increase costs, and consequently, prices. This price increase will then induce consumers to purchase fewer eggs. The only thing uninformed consumers really understand is that the price of eggs has risen, and they must do with less.

To help us resolve the conflict between the informed effect and the experienced effect, consider a hypothetical story. A city is considering the purchase of a large tract of forest to convert to a wetland. The wetland will help protect the city from infrequent but costly floods. A blue-ribbon commission consisting of scientific experts decides that the benefits of flood protection justify the large costs the city will incur to purchase the land. The money used to fund the wetland will be taken from the transportation budget, and will consequently reduce the rate at which the city repairs and repaves roads.

Although newspaper reporters were invited to all commission meetings and wrote stories in the local paper, very few citizens took the time to educate themselves about the issue. The citizens are unaware of the benefits created from the wetlands. After all, no one experiences a flood that did not occur. No one rejoices about the fact that a flood did not destroy their homes if no flood occurs. All the vast majority of the citizens experienced is a deterioration in the roads. The *informed effect* of the wetlands is unambiguously positive, meaning the benefits greatly exceed the costs. The *experienced effect* is negative; all citizens experienced deteriorated roads.

How should policy-makers reconcile the informed and experienced effects when they conflict? We believe that most citizens want policy-makers to make informed decisions. They understand that taxes fund projects that are desirable but that are off the average citizen's radar. We know our tax money funds espionage, which remain hidden from sight, and our tax money funds scientific exploration, which we will never understand. We expect our government to be thoughtful and sophisticated about the allocation of tax revenues. However, these experienced effects should not be ignored. If our hypothetical wetlands demand so much money that the city roads would change from deteriorated to dilapidated, citizen outcry would be substantial, removing whoever in office failed to maintain the roads.

The following sections seek to articulate the nature of the informed and experienced effects. We know of no method to scientifically reconcile informed and experienced effects and thus render unambiguous policy recommendations. All we can do is articulate the magnitude of the two effects, and reason through the consequences of different courses of action.

Banning Cage Eggs for Public-Good Reasons

A ban on cage egg production would presumably cause cages to be replaced by cage-free production methods. Suppose that a ban requires hens to be raised in a cage-free system that also provides outdoor access with shelter and predator system, which we will refer to as a *free-range* system. There are many different types of systems that could be deemed free-range, so we outline our definition here.

The cage egg ban will result in *experienced effects*, which refer to the increase in price resulting from the increase in production costs. With so few consumers being informed or educated about the ban, consumers will not view the desirability of eggs much different than before. With the higher price and no increase in demand, consumers will purchase fewer eggs. The experienced effect, which in this case economists might also refer to as *market* effects, is negative. Consumers purchase fewer eggs, forgoing the pleasure they formerly received from those eggs, and they pay a higher price for the eggs they do purchase. Knowing very little about the ban, they are unable to process how the ban affects eggs, and thus have received little benefit from happier animals.

Next we must consider the *informed effect*, which encompasses the change in the private and public good values of improved animal care, from consumers who are educated about the issue. Economists in this case might refer to the informed effects as *non-market* effects. If Jack and Jill only have private values for free-range eggs, then there is no coordination problem and there is no market failure. It is only when Jack and Jill have public values that a coordination problem exists. Consequently, the desirability of the cage egg ban is measured partly by whether the informed, public good component of improved animal welfare is larger than the cost of providing animal welfare. If it is, the results favor a ban on cage egg production.

Imagine a small community of 1001 citizens fed by the eggs of 1001 hens which are raised in a cage system. In effect, this means that each person's egg consumption is directly related to the existence of one hen.[6] Why does our fictitious community have 1001 people and hens rather than 1000? Because we want to focus on the value people place on the eggs from the 1000 hens that *other people than themselves* consume. The public good values relate not to animals providing the eggs *you* eat (the eggs from your one chicken), but from the well-being of layers that provide eggs for *others*. Every single person in this community is an *informed* consumer, receiving the same information as the participants in our experiments from the previous chapter.

Suppose the community mayor enacts a policy requiring all 1001 hens to be raised in a free-range system instead of a cage system. The policy will no doubt have market impacts, as the more expensive free-range system raises prices. It is also reasonable to assume that the demand for eggs may rise in response to a more humane egg industry (remember, consumers are informed). The impact of the higher price and a higher egg demand, we assume, causes egg consumption to be the same as it was before the switch to free-range.[7] Thus, we have the same number of hens before the ban as we did before: 1001.

The question is whether the magnitude of the public good values exceeds the costs of providing higher welfare. The mayor's proclamation means that 1001 hens are no longer allowed to be raised in a cage system. Research indicates that eggs produced in a free-range system are about $0.35 per dozen more costly to produce that those from a cage system.[8] A free-range system is the aforementioned cage-free system with a simple outdoor area. Our conversations with organic producers suggest that adding outdoor space is relatively inexpensive, and we estimate a $0.05 per dozen additional cost. The cost premium of a free-range system is thus $0.40 per dozen. Because a hen produces about 500 eggs over its 2-year life, the ban will increase costs by about $16.67 per hen or by about $16,683 for the entire community.

Assuming that the 1001 people in the community are similar in their preferences to the people we studied in last chapter's experiments, we would conclude that, on average, each citizen would be willing to pay $57 for the ban which improved the lives of the 1000 birds that produce food for the other 1000 people. When we multiply the $57 per person benefit times the 1001 people, the total public good value equals $57,057. Consequently, while the total cost of the measure is only $16,667, the total public good value is $57,057. Even if we underestimated the cost premium of the free-range system, it is doubtful the total cost would even approach one-half of the total benefit. The answer is clear: *the mayor's policy is a great success—the birds should be raised under in the free-range system*. Note that this conclusion is based solely on the public good benefits from informed consumers. If informed consumers also received a private good benefit, which we know they do, the benefits exceed the costs by an even larger amount.

Of course, large-scale changes such as those brought about by federal regulations would impact all 280+ million laying hens currently in existence in the US. The results from the small-scale community can readily be translated to make inferences about such large-scale changes. Extending the results of the 1001-member community to 280 million individuals is akin to replicating the small communities over and over. There is one twist, however, that indicates that the public good benefits might be even greater in larger society. Each time a new community is added and that community improves welfare standards, members *outside* the community will partake in the benefits as well. That is, a member of community A is willing to pay $57 on average to improve the living conditions of their community's 1000 hens, but presumably they would also be willing to pay some amount to improve the lives of hens in other communities too. The point is that if the public benefits exceed the costs for a 1001-member community, the same holds true for the community of all Americans. The thought experiment for the 1001 community was no trivial exercise; it constitutes the proof that the public good benefit for the entire US almost surely exceeds the cost—for informed consumers.

Another way to approach the policy is as follows. Suppose that the US bans the production of cage eggs, forcing the use of free-range methods. As stated previously, this would cost US consumers an amount equal to the 280 million hens times the $16.67 per hen cost, which equals $4,667,600,000 in total. We do not know the public good value of providing all 280 million hens a better life, but we know it must be at least as large as the value of providing 1000 hens with a better life. Given that there are 222 million Americans over the age of 18, with an average public good value of $57, the total public good value of moving all hens to a cage-free system with outdoor access is at least $12,654,000,000. *The costs of the cage egg ban are projected to be $4.67 billion, while the public good benefits are projected to be $12.65 billion.*

A closer look at the data reveals a peculiarity. Although the total public good benefits of banning cage eggs exceed the costs, very few Americans would approve of the measure. Recall that benefits are measured by the *total* amount of money *all* individuals are willing to forgo to improve the lives of 1000 hens. The total can be significantly influenced by the presence of a few people who are willing to contribute large sums of money.

The average value estimate of $57 was calculated from auction bids submitted by 126 individuals in three different cities. Thirty-nine people placed no value on improving the lives of the 1000 hens; their bid was precisely zero. Another 62 individuals submitted bids between $0 and $10. In fact, only eight individuals submitted bids larger than the average bid of $57 and there were two bids higher than $1000. Of these 126 participants, only 8 percent have a public good value greater than the per-person cost of $16.67. Thus, banning cage eggs would produce total public good benefits that are undoubtedly larger than the costs, but only 8 percent of the citizenry would actually derive public benefits from the ban.[9] Does this sound like something a politician should pursue?

Because total benefits exceed the total costs, it is theoretically possible for the 8 percent who place a high value on the policy to compensate the other 92 percent, such that everyone approves of the policy. For example, in our fictitious 1001-person community, the total amount of money individuals were willing to pay for the policy was $57,057. A compensation policy can be administered by having the mayor send a bill to each person exactly equal to their true willingness to pay for the ban. Because each person pays right up to their maximum WTP amount, we know they are indifferent as to whether the ban is a good idea. The mayor then pays the cost of the policy, which is only $16,667, leaving $40,390 on the city's balance sheet. Assuming the mayor is benevolent, he or she would redistribute these profits evenly to all citizens. Each household would receive a check for about $40. Each person was previously indifferent about the policy, but now they receive a check for $40. Every single person approves of the mayor's program! Even those who cared not one iota for the birds approve of the policy; these individuals paid nothing and received a $40 check.

While such an outcome is possible in theory, it is impossible to implement in practice. Charging each citizen a different tax would be a bureaucratic nightmare. Moreover, if citizens knew their answers to auctions and surveys would be used to set the price that their household would pay, they would not tell the truth about their desire for animal welfare improvements. Other such issues present formidable obstacles, such as the fact that citizens will feel injustice when they know that they pay a different amount than their neighbor. Whatever policy emerges to force the substitution of free-range for cage egg will probably force each person to pay a similar amount for the welfare improvement. If this occurs, most of the population would resent the policy. Politicians would suffer the inevitable consequence of unpopular policies, and the welfare improvement would be temporary.

Anything similar to the aforementioned policy is likely to convey large benefits to a few and impose losses on the majority of the citizenry. There is inequality in the distribution of animal welfare policy benefits, and in general, people tend to be averse to inequality. There is a general dislike of the rich being too much richer than the poor, even if the poor have a high standard of living. In fact, one of the factors motivating concern for farm animals probably stems from inequality aversion—from the feeling that humans derive too much benefit relative to the animals' suffering. Thus, it is sensible to ask whether even the 8 percent of individuals who presumably benefit from a ban on cage eggs might alter their support if they knew how unequal the benefits were shared by other citizens. Fortunately, economists have been hard at work in recent years trying to measure the extent of people's aversion to inequality. This work suggests that for many of the people whose public willingness to pay for the cage ban is greater than the costs—people with bids in the range of about $16.67 to about $50—that aversion to inequality would lead them to vote against such a policy even though their personal benefits would exceed the costs.[10]

In summary, the informed impacts of a ban on cage eggs in favor of a cage-free system with some outdoor access are large but unequally distributed across the citizenry. The question of whether the public good aspect of hen welfare justifies such policies then depends on how one views inequality. The total benefits, as measured by the maximum amount of money all citizens will pay for the policy, exceed the costs by a large amount. The benefits of the policy are distributed to a small portion of the population, and though it is theoretically possible for the winners to compensate the losers, it is naive to think such compensation programs are possible in practice. So long as each person is asked to pay roughly the same cost to implement the ban, only 8 percent are likely to be supportive. Supporting a cage egg ban because of the public good argument would mean disregarding the preferences of the vast majority to bestow a small minority with large benefits.

Banning Traditional Pork

The logic used to evaluate improvements in layer welfare can also be used to evaluate policies to improve the lives of hogs. Specifically, consider a proposal which bans the use of confinement-crate systems—what most people call factory farms—in favor of shelter-pasture pork (which could probably be sold under the Animal Welfare Approved label). The consumer experiments discussed in the previous chapter measured the public good value for raising 1000 sows and their offspring in the shelter-pasture system instead of the confinement-crate system. To employ the logic from the previous section, we need to calculate the number of people 1000 sows and their offspring can feed.

A typical sow will produce 44 offspring in two years, producing 5976 lbs of retail pork. Per capita consumption of pork is about 50 lbs per year in the US, so in two years the average person will consume 100 lbs of pork. Thus, 1001 sows and their offspring will supply pork to a community of 59,820 people.

As before, assume that the policy replacing confinement-crate production with shelter-pasture production does not change the amount of pork consumed. Our research indicates that, on average, people will pay $23 to ensure 1000 sows and their offspring are raised on the shelter-pasture pork (instead of confinement-crate pork). When this average is multiplied by the 59,760 people, we find that the community is willing to pay a total of $1,375,860 to fund the policy. Research suggests that it costs about $0.11 per lb more to produce retail pork in the shelter-pasture system,[11] implying that to produce the 5,981,976 lbs of retail pork in a shelter-pasture system, the policy will entail costs of $658,017. *The verdict: the benefit of the policy is $1,375,860 while the costs are $658,017.*

These estimates can easily be extrapolated to the US population at large. We can continue to add communities of 59,760 people until we reach a total population that is the same size as the US. As with the egg layers, our public benefit estimates will represent a lower-bound because individuals in Community A enjoy knowing about the welfare improvements made in Community B, and vice-versa. The US produces about 15 billion lbs of retail pork every year. Assuming production levels stay roughly the same after the ban implies that the cost of producing pork will increase by $1.65 billion. We also know that, on average, each person's public good value for the change is at least $23. Given that there are 222 million Americans over the age of 18, the total benefits are $5.1 billion. Thus, the minimum amount that adult Americans are willing to pay to ensure a better life for hogs producing meat that they do not eat ($5.1 billion) exceeds the costs ($1.65 billion). Thus, considering only public good values, the policy provides benefits much larger than the associated cost.

Again, however, the aggregate results are heavily influenced by the bids of a few people who bid very high values in our experiments. Recall that although the average bid was $23, the median was only $1. If the costs of the ban are uniformly distributed across the adult population, each person would pay about $11 for the ban over two years. Yet, data from our auction suggest that only 8 percent of the adult population receives a public good benefit more than $11.[12] As with the egg policy, the total public benefits of the policy to improve hog welfare are greater than the costs, but the policy benefits only a small percentage of the population. The theoretical compensation scheme described in the egg policy could be applied here, allowing all individuals to benefit from the policy, but the practical issues involved with administering the redistribution of benefits create insurmountable obstacles.

Thus, we find ourselves in the same situation as in the layer case. The *total* benefits definitely exceed the cost. If a compensation scheme could be developed to redistribute benefits from those few who feel most strongly about animal welfare to those who are less enthusiastic, everyone could benefit. Obstacles to this redistribution of benefits are, however, probably insurmountable.

Difficult Decisions

Imagine ten friends out for a night on the town. Everyone would like to visit the nightclub Studio 54, but the cover charge is $100. The price is too high for nine members of the group, but there is one holdout, Charles, who still wants to go to Studio 54. In fact, Charles really wants to visit the club tonight. Charles' friends value the club less than $100, but suppose Charles is willing to pay up to $5000 to go. Should the group of friends go to Studio 54? Only Charles would benefit, but the value he extracts from entering the club is so large that it overwhelms the sadness of his other friends. The total benefits of entering Studio 54 exceed the cost, but only one person actually benefits.

The solution to the dilemma is simple. Charles should pay the cover charge for all of his friends. He still extracts large benefits, as he values the club at $5000 but only pay $1000 in cover charges. His friends benefit too; they paid no cover charge but still were able to enjoy a night of dancing at Studio 54. The ability to redistribute benefits from winners to losers can make the all the lounge-lizards better off, and the same is true for animal welfare. If those who care most about farm animal welfare could compensate those who do not, a perfect government pursuing higher animal welfare standards could generate benefits for every American. Without such a compensation scheme, the best course of action is less clear.

Suppose Charles is prohibited from paying his friends' cover charges. It is up to you to decide whether the group should enter Studio 54. Is it better to bring tremendous joy to one man and slight discomfort to nine, or is it better that the group visit another club where all ten enjoy small benefits? If you were in charge of farm animal welfare policy, you would face a similar dilemma as it relates to the public good benefits.

Considering only the public good benefits, most Americans are not willing to pay the cost of improving farm animal welfare, at least not for the food that other people eat. We began this section by asking whether government involvement in regulating animal welfare is justifiable based on the argument that there is a public good. Now is an appropriate time to pause, and revisit why only public good values are being considered. The *private* value is the benefit you receive from purchasing more humane meat that is independent from the purchasing habits of other people. For this reason, there is no need for government to play a role in providing you with the private value. If the private value you receive is greater than the price you pay, you will purchase the meat; otherwise, you will not. There is no need for government to help you.

The *public* good value, on the other hand, is the value you receive from improved animal care that depends on the actions of other people. Remember that Jack received a public good value when Jill purchased cage-free pork. Because these public good values for each person depend on the actions of everyone else, there are benefits from coordinating purchasing habits. Two people can mutually agree to purchase more humane meat. Interest groups may shame grocery stores into purchasing more humane meat. New grocery stores may emerge that sell only humane meat. Or, government can help coordinate people's purchases by banning inhumane practices.

Because there must be a public good value to motivate the need for government, we have considered only public good values when evaluating the potential for a governmental role in the farm animal welfare debate. Our results indicate that improvements in animal well-being deliver large benefits to a small minority of people, at the expense of the majority. Enacting such a policy would be consistent with the utilitarian philosophy, but would not be consistent with our ideas of democracy.

What You Don't Know Can Hurt You

The Information Problem

If there is one salient fact we have learned talking with thousands of people about farm animal welfare, it is this: *people do not know much about the way farm animals are raised*. Even one of our own children recently held up a drumstick asking where it came from, and was astounded to discover it was the leg of a chicken! Their reactions are a reminder for those of us with an agricultural background. Modern agriculture has become so efficient and productive that most of us carry on with our daily lives without much thought as to where our food comes from. This fact is often lamented by farmers, who grow frustrated when they are criticized by people who know little about raising animals. However, it is generally a beneficial feature of life that we do not need to know how food is made. Do you know how your iPod, shoes, or stapler were made?

Our complex modern economy allows us to enjoy the fruits of all these goods without having to become a specialist in each. The extreme alternative is a Robinson Crusoe economy, where each person must personally produce everything they want to consume. In market-based economy, each person concentrates on work to produce a few select goods, and then trades the product of their labor for everything else they want in life. The consequence of this specialization is greater wealth for everyone. Think about it: would you be wealthier as Robinson

Crusoe on a deserted island or as a part of the modern economy? However, despite the enormous material benefits that come from engaging in a capitalistic economy, there are times when our lack of knowledge about how goods are produced is problematic.

The lack of knowledge problem can lead to *information asymmetries* which, in turn, can result in a market failure. Information asymmetries arise when the buyer or seller of a product has better information than the other party. In the case of animal welfare, producers know exactly how meat, eggs, and dairy foods are produced, but consumers do not. Because it is cheaper to produce animal-based foods under low levels of animal care, and because consumers do not know how the animals are raised, producers have an incentive to only provide food produced under low levels of animal welfare. Consumers may expect producers to provide low standards of animal care even if labels suggest high welfare standards.

How could a lack of knowledge lead to a market failure? Consider an extreme example where all consumers actually know how most farm animals are raised and dislike purchasing products from factory farms. In such a world, there might cease to be a market for animal foods at all unless there were a credible means for producers to signal the quality of their products to consumers. Consumers would refrain from purchasing animal foods, knowing that the only products delivered to the market are the low-quality products. George Akerlof, the Nobel Prize winner in economics in 2001, identified this as the *market for lemons* problem. Some used cars are *lemons* and Akerlof pointed out that owners of used cars are more likely to try to get rid of lemons than well-functioning used cars, and buyers, recognizing this fact, will choose to refrain from buying all used cars if they cannot identify the lemons from the non-lemons. The consequence of this information asymmetry is the absence of a market for used cars.

Economists typically confront the asymmetric (or lack of) information problem in the following manner. They may seek to determine the *value* of the information to consumers, or the *cost of ignorance*. The idea is that information changes what consumers choose to buy. For example, it may be that the information we provided to our experimental participants, if given to consumers in general, would cause them to alter purchasing habits. Perhaps they may purchase fewer eggs or only cage-free eggs. By calculating how much money the consumer would be willing to pay to have obtained the information earlier in their life, and hence made different purchases, we know the extent to which a lack of information poses a problem. Often economists use this sort of reasoning as a motivation for public policy. That is, they encourage policies that would lessen the cost of ignorance. In the case of animal welfare, however, we believe this sort of thinking is not very useful because consumers will generally remain uninformed regardless of the farm animal welfare policies pursued by the State.

Regardless of the extent to which food is labeled, the average consumer will always know little about how an animal is raised. If consumers cannot discern the level of animal care, the farmer has the incentive to set low welfare standards. There is little incentive to incur the expense of raising happy animals if the only information the consumer will have is a pretty picture and an appealing brand name—which all products have, by the way. Even if farmer Bill wants to set high standards for personal reasons, so long as other farmers are willing to raise animals under worse conditions the market will force Bill to mimic the others. In turn, consumers know they are uninformed, and the natural cynic within them will cause them to surmise that farm animals are raised in unappealing conditions. Farmers then expect consumer cynicism, providing them even less incentive to employ high welfare standards—and so on. Because the product label does little to convey information, the market provides little incentive to employ high welfare standards.

Government could establish animal welfare standards, and enforce the standards. Standards are often effective. For example, it seems that certification standards set and enforced by the government have contributed to the growing market share of organic foods. Yet government is not necessary for, nor does it ensure, that food production will occur as advertised. The fact that a market, albeit a small market, exists for animal-friendly foods suggests that the asymmetric information problem can be overcome. Over time the market for cage and crate-free products may grow and come to dominate grocery shelves. After all, you know almost nothing about how the majority of the goods you purchase are made. How do you know the gasoline you pump into your car is really gasoline, and how do you know the computer you order over the internet will really work?

Markets work not because you understand how the goods are made, but because firms must establish and maintain a good reputation to thrive. Every market is plagued with asymmetric information, and reputation effects often help to eliminate the textbook market for lemons problem. Asymmetric information can be a bigger problem in food production because there are large numbers of farms, few of which sell through their own label. Government can help facilitate the coordination between food retailers and individual farms by establishing and enforcing standards that are mutually agreeable to both parties. The private sector can play this role as well. There are a number of private food labels, such as Niman Pork Ranch and Beeler's Pork, that place great pride in their reputation. Their business-model centers on their reputation for excellent animal care among food retailers and consumers. If firms attempt to maintain a good reputation in the face of pressure from activist groups, the motivation for government action to solve information asymmetries becomes less convincing.

A salient issue is how to reconcile the preferences of the informed consumers in our study and the uninformed consumers that comprise most of the general populous. First, consider the reasons why we say that most consumers are "uninformed." In the consumer valuation experiments discussed in the last chapter, we provided people with extensive information on modern egg and hog production systems, including unbiased information about why certain practices were used along with pictures of differing systems. We found that after this presentation, over 70 percent indicated they were *more* concerned about the well-being of farm animals than they were before. Only about 1 percent indicated that the information made them less concerned, and the remaining 29 percent said their opinions were unchanged. These data suggest that if people were more informed as to how animals are actually raised, the farm animal welfare debate would be more prominent in the news.

Our data also reveal that consumers have inaccurate perceptions about the livestock industry. Table 10.1 shows, for example, that only 44 percent of people believe that the eggs they buy from their grocery store are from a cage system, and, on average, consumers believe only 37 percent of all eggs produced in the US come from a cage system. However, actual retail data indicate that over 90 percent of eggs produced and consumed in the US are produced in cage systems.[13] Similarly, although about 30 percent (17% + 12%) of people believe the pork they purchase comes from pasture or organic systems, the actual percentage is lower than 5 percent. Grocery stores are not selling what shoppers think they are buying.

The presence of consumers with incorrect beliefs poses particular challenges for thinking about the benefits and costs of public policies. The experiments we described in the last chapter showed that *informed* consumers, on average, prefer

Table 10.1 Consumer Beliefs about How Laying Hens and Hogs are Raised

Production system	Belief about which system they normally purchase eggs & pork	Belief about the percentage of egg & hog production taking place in each system
Egg (N = 127)		
Cage	44%	37%
Cage-free (barn)	17%	22%
Cage-free (aviary)	7%	16%
Cage-free (aviary w/outside)	21%	14%
Organic	11%	12%
Pork (N = 133)		
Confinement-crate	40%	45%
Crate-free (confinement-pen)	18%	20%
Enhanced-confinement	13%	14%
Shelter-pasture	17%	11%
Organic	12%	10%

that eggs and pork be produced in systems that provide higher animal welfare, even after accounting for the higher cost. Suppose that government bans on cage eggs and confinement pork were administered based on the results from our consumer experiments. The ban would not take effect in a setting of *informed* consumers, but rather in a world of *un*informed consumers. Informed consumers familiar with modern livestock agriculture might be pleased with the ban, but the majority of people will not understand exactly why the ban has taken place, the production methods that will be substituted for the banned methods, and what the ban implies for the level of animal care. Moreover, most consumers will not even be aware of the ban, only the higher prices that follow. Thus, the only probable impact of the ban is a frustration with higher food prices.

What we have is a dissonance between *in*formed consumers and *un*informed consumers. It is necessary to inform research participants about livestock production, otherwise they will be unable to use the value expression tool described in the previous chapter. People often change their preferences when they become informed about a topic, and this fact complicates our understanding of the potential effects of regulation. For example, when calculating the benefits and costs of a policy to label or ban eggs from cage systems, should we use the WTP values from informed or uninformed consumers? Should we take into account the fact that most people will experience only economic harm from farm animal welfare regulations, despite the fact that that harm turns into benefit if those people are better informed?

To illustrate the complexities, recall our finding from the last chapter that consumption of cage-free eggs provides a net benefit of $0.20 per dozen to the informed consumers (i.e., people were willing to pay, on average, $0.55 per dozen more for cage-free eggs than cage eggs, and the extra cost of cage-free eggs is $0.35 per dozen, providing a net benefit of $0.55−$0.35 = $0.20 per dozen). Suppose a shopper named *Janet* purchases one dozen eggs per week every week. Initially, *uninformed Janet* knows very little about how eggs are produced. Then suppose Janet attends one of our research sessions or watches an internet video, and becomes *informed Janet*. Like the average participant in our experiments, suppose informed Janet receives a net benefit from purchasing cage-free eggs of $0.20 per dozen relative to what she would receive purchasing cage eggs. How does informed Janet's purchasing pattern differ from uninformed Janet's? One would suspect that that informed Janet will begin purchasing cage-free eggs and that she might also begin purchasing more than a dozen eggs per week. If Janet had previously purchased a dozen eggs per week and now learns that eggs are now more valuable to her, it is logical to conclude that Janet will now purchase more than one dozen eggs each week. If the quality of orange juice increased and the price remained the same, would you not drink more?

In this scenario, Janet is better off because she derives more value from her purchases. Furthermore, producers are better off because they sell more eggs. Although this seems to be a win–win, there might be a flaw in the logic. In particular, the assumption that informing people about farm animal welfare will cause them to not only substitute towards more humane foods, but also increase their total food consumption may be dubious.

Janet's story is missing one important element. When Janet becomes educated about how eggs are produced, it is true that she places a greater value on cage-free eggs than cage eggs. However, the information will likely cause Janet's willingness to pay for cage eggs to fall. Uninformed Janet always thought that hens had spacious room and laid eggs in nests. Informed Janet now understands that hens in cage systems live in close quarters and have no nests. Thus, the provision of information might increase the *premium* placed on cage-free eggs, but might decrease the overall willingness to pay for cage eggs. After *uninformed Janet* becomes *informed Janet*, she places a greater premium on cage-free eggs relative to cage eggs, but her overall level of egg consumption might fall, stay the same, or rise.

This is not just some theoretical possibility. For example, most consumers know little about milk production, and when asked how much they would pay for a quart of milk in real money experiments similar to the ones we conducted for eggs and pork, researchers at Cornell University found that people's willingness to pay for conventional milk depended heavily on whether they knew organic or rBST-free milk was available for sale. The mere presence of rBST-free and organic milk stigmatized conventional milk. The amount people were willing to pay for conventional milk fell $0.35–$0.50 per quart when they learned that rBST-free and organic milk were available.[14] These results suggest that as people become informed, their overall consumption of milk might very well fall.

The story goes to show that just because informed consumers express a net benefit for cage-free eggs does not mean that total egg consumption would rise as cage eggs are replaced with cage-free eggs. Because of the difficulties involved with projecting whether total egg consumption will rise or fall after, say, a ban on cage eggs, the effects on informed consumers are often best understood on a per egg or per pound consumed basis, just as we stated them in the previous chapter. Although it is difficult to know how total industry profits and total consumer benefits would change from a cage egg ban, we can calculate how the average consumer would be affected for each egg they consume. Thus, although we are dubious about the ability to calculate the overall welfare changes for consumers and producers from a ban on certain production practices, the per-egg or per-pound *net benefit* calculations can provide a piece of information useful for judging the merits of policy.

Calculating the Experienced Effect

We now delve more deeply into the question of the empirical magnitude of the effects caused by lack of information. Calculating the *informed effect* of a change in egg and pork production methods is useful because it tells us what would happen if everyone understood the farm animal welfare issue. However, even if *informed* consumers approve of a large-scale change that improves animal welfare, *uninformed* consumers will only be cognizant of a negative *experienced effect* in the form of higher food prices. A consumer who sees egg prices rise 30 percent but knows nothing about the improved care for layers will only experience harm. Consequently, policies that improve animal welfare have two effects (1) the *informed effect*, constituting the value consumers place on improving animal welfare if they were informed, and (2) the *experienced effect*, which is the *actual* impact consumers directly feel.

If a ban on the use of cages in egg production passed, it is very likely that most citizens would be unaware and uninformed of the change. Although some people would surely hear something about it in the news, the actual effect of this media discussion is likely to be negligible. On the other hand, all shoppers would witness a rise in egg prices. Higher prices mean reduced egg consumption, and most people would *experience* a loss from the policy. It is true that shoppers would replace the now higher-priced eggs with other foods; however, the extra beef or soybeans now consumed are, by definition, valued less than the eggs shoppers gave up as a result of higher prices. If this were not true, shoppers would have given up eggs for beef and soybeans before the price change.

To illustrate the *experienced effect*, consider the following story. Harry currently consumes 100 lbs of pork each year. A ban on gestation crates is passed, which increases pork production costs. Harry understands very little about hog production or the consequences of the ban. Even though Harry might be pleased with the ban and might be willing to pay more for crate-free pork, his lack of knowledge implies that his demand for pork is unchanged after the ban. When we say *demand* we are not referring to the quantity of pork Harry consumes, but rather the relationship between the quantity of pork Harry would purchase and pork price. By saying that Harry's demand is unchanged, we are saying that at the same price, he purchases the same amount, and that he responds to a price increases exactly as he would have responded before the change in animal welfare.

After the ban, hog farmers experience higher costs, and some of these costs are passed on to Harry in the form of higher prices. Harry responds by consuming less pork. Before Harry ate 100 lbs each year, now he eats 75 lbs. The *experienced effect* for Harry is negative; not only does he consume less pork, but pays a higher price for the pork he does consume. The losses from the *experienced effect* can be stated in dollars. We first multiply the increase in price Harry pays by his new consumption level. If price rises $1 per lb, then we know that Harry is worse off by $1 × 75 = $75 per year from the pork he still consumes. However, Harry is also worse off because he consumes 25 fewer pounds of pork. Harry loses the enjoyment he previously experienced from the 25 additional pounds of pork consumption. The calculation of the economic value of the 25 lbs loss to Harry is more complex, requiring an elaborate discussion of economic models, as well as geometry. (This discussion along with futher technical details can be found in this chapter's Appendix.)

Using existing estimates of consumer demand for pork and eggs in the US, we calculate the magnitude of the *experienced effect* for three policies: (1) a ban on cage eggs in favor of cage-free eggs; (2) a ban on confinement-crate pork in favor of crate-free pork; and (3) a ban on crate pork in favor of shelter-pasture pork. The results are shown in Tables 10.2 and 10.3. To be clear, the calculations assume *all* consumers know absolutely nothing about how the lives of farm animals change, and they do not even necessarily understand why prices rise after the ban.

The calculations are negative and rather large: consumers are worse off when they have to pay higher prices without being made aware of any offsetting quality enhancements. The *experienced effect* from the cage-egg ban is relatively large at $1.78 billion. The reason is that converting from cage egg to cage-free egg production is costly, and consumers are not very responsive to changes in the price of eggs. Because there are very few substitutes for eggs, especially in bakery products, as the price of eggs rises, consumers have fewer substitutes and must simply pay higher prices. This explains the predicted large price increase of 21.2 percent for eggs that would result from a ban on cage eggs. Despite this 20 percent increase, egg production is only curtailed by 4.24 percent. The consumer *experienced effect* for pork is less costly because the impacts on

Table 10.2 The *Experienced Effect* of Three Policies as Experienced by Uninformed Consumers

Livestock industry change	Consumer experienced effect ($ per year for U.S. population)	Consumer experienced effect ($ per year, per person)	Percentage increase in price	Percentage decrease in consumption
Converting all cage eggs to cage-free	−$1,872,704,705	−$6.16	21.18%	4.24%
Converting all confinement-crate pork to crate-free* pork	−$738,660,007	−$2.43	1.72%	1.20%
Converting all confinement-crate pork to shelter-pasture pork	−$1,244,799,974	−$4.10	2.91%	2.04%

Note: *Crate-free pork refers to pork produced using the confinement-pen system.

Table 10.3 The *Experienced Effect* of Three Policies as Experienced by Meat and Egg Producers

Livestock industry change	Producer experienced effect (million $ per year for U.S. producers)
Converting all cage eggs to cage-free	−$187,270,470
Converting all confinement-crate pork to crate-free* pork	−$258,531,002
Converting all confinement-crate pork to shelter-pasture pork	−$435,679,991

Note: *Crate-free pork refers to pork produced using the confinement-pen system.

pork production costs are smaller.[15] Moreover, consumers have better substitutes for pork (e.g., beef and chicken) and thus have more alternatives to compensate for the price rise. The percentage change for the pork price and quantities are smaller, and the total *experienced effect* is smaller. In one sense, the total numbers are large, in the billions for eggs and one-half billion for shelter-pasture pork. Yet, when expressed on a per-person basis they are rather small. The economic harm imposed on uninformed consumers equals $5.86 per person per year for eggs and $0.74 to $1.50 per person per year for crate-free and shelter-pasture pork, respectively.

In addition to the losses uninformed consumers would experience from the price increase, producers are worse off. Forced to adopt more costly practices, producers will only partly be able to pass along higher costs to consumers. Producers face higher costs and lower sales, both of which translate into lower profits. Table 10.3 shows that egg producers would lose $179 million in aggregate profitability if a cage egg ban went into effect and consumer demand for eggs remain unchanged. Likewise, if consumers do not change their demand for pork after a ban on confinement-crate, pork producers would lose $78 million per year after adopting the confinement-pen system and would lose $158 million per year if forced to adopt the shelter-pasture system. These losses may be smaller than the consumers' losses, but they affect a smaller number of people: there are more meat consumers than there are meat producers.[16]

Our analysis suggests that citizens who are informed about the farm animal welfare debate may desire government action, but the policy will raise prices and deliver benefits only to the few who become educated about the new livestock production regulations. The stark contrast between the informed and experienced outcomes suggests the need to provide education to consumers. If all consumers could receive the same information as our subjects, they would be likely to make different food purchasing decisions. Does the value of this information justify the expense of information services?

We do not know. What we do know is that it is Pollyanna-like to believe the government could provide the objective and full information the participants in our experiments receive. One reason relates to the resources it would consume. Our subjects spent approximately half an hour learning about the production practices of one industry (eggs or pork) and asking questions. An hour's time is worth about $20 to the average US worker.[17] Thus, to get people to give up the time to listen to 30 minutes' worth of education cost, on average, $10 to the person in terms of lost time they could have spent doing other things. Recall, we had to pay the participants about $75 to attend our research session just to get a reasonable response rate. Another reason is that government is not adept at broadly communicating information. Government agencies regularly publish excellent, unbiased information on a variety of matters relating to agriculture, such as the nutritional content of an ear of corn, but they do not spend large amounts of resources advertising the information.

How should the difference between informed and uninformed consumers factor into the policy debate? If legislators acted as benevolent public servants (which they frequently do not), they might pass regulations despite knowing that the citizenry would experience economic harm as a result, but with the understanding that if the citizenry fully understood the issues they would applaud the legislation. That is, politicians might act out of a sense of paternalism. But, do politicians have the incentives to act paternalistically or will they be voted out of office by an uninformed public which only witnesses, in this case, higher food prices? There is no clear answer. Consumers may assume politicians are acting in their best interest even if they do not understand why certain actions are taken, but ultimately, it seems politicians themselves have to defend the actions taken in a way that is convincing to the public.

Countless laws are passed each year, and the citizenry often cannot parse and identify specific pieces of legislation. What laws caused hamburgers to be safer than fifty years ago? Which legislators are responsible for converting the two-lane highway to a four-lane highway? Who is more responsible for the recent financial crises: Barney Frank, Ben Bernanke, Allen Greenspan, George W. Bush, or James Cayne (as former CEO of *Bear Sterns*)? You do not know, and we do not know. After all, economists are still debating what caused and what cured the Great Depression!

Supporting farm animal welfare legislation may harm or benefit legislators who support the legislation. The asymmetric information problem does not destine a politician to pursue better livestock care at her own peril. The informed and experienced effects have differing influences in the policy formation process. First, the *informed effect* (assuming it supports better animal care) provides legislators with scientific evidence to support the pursuit of livestock production regulations. When they run for re-election and opponents attack them for rising food prices, they can assert that, *"Researchers found that consumers want this legislation!"* Second, the *experienced effect* allows politicians to carefully survey the consequences before supporting animal welfare policies. Based on the results in Table 10.2, a politician may support a ban on confinement-crate production in pork but not a ban on cage-egg production. The reason for this nuanced stance on livestock welfare is that the pork ban will have negligible effects on pork prices, while the egg policy would increase egg prices by 20 percent. The experienced effect is considerably larger for the egg policy than the pork policies as well. It is understandable that a Congress member would want to avoid being blamed for a 20 percent rise in egg prices—nor would he or she want to be responsible for the billions of lost profits to the egg industry and the effect this would have on his or her re-election campaign.

One final point on consumers' knowledge is important. That informed consumers may make different choices than the uninformed does *not* necessarily mean that consumers are irrational, nor is this observation necessarily proof that public policy is needed. Acquiring information is costly and we live our lives in ignorance of much of the world around us. The choice to acquire information is a choice like any other. We implicitly or explicitly choose to be uninformed because it is simply too costly to be otherwise. Indeed, one can be "rationally ignorant" and even "rationally irrational" about certain beliefs if the costs to the individual of making an error are very small or if the costs of information are high.[18] Acquiring information is costly and the dollars and time to obtain information could have been used elsewhere. Apparently most people have decided *not* to use their own money and time to acquire information about animal welfare, and as such it is dubious to think that a person would be made better off by the government spending the person's money (via taxation), as opposed to the person spending their time and money where they wish. Although it is admittedly tempting to use our superior information about farm animal welfare to override the choices of the citizenry, economics demands humility. The fact that information is a result of peoples' actions requires us to acknowledge that ignorance about modern farm systems may derive from a lack of concern for farm animals. Ignorance is not just a description of the individual's understanding, but an indicator of the person's demand for information and interest in the topic.

Despite our concern with uninformed consumers, people are becoming more informed. The advent of the internet has drastically reduced the cost of disseminating information, and groups like HSUS and PETA have effectively capitalized on these trends, as is evidenced by examples like the YouTube.com video *Meet your Meat*. If the costs of acquiring information about animal production have fallen in recent years, we would certainly expect such trends to affect consumer behavior. Indeed, data indicate that although organic, cage-free, and free-range eggs represent a very small part of the overall egg market, their market share in the US has doubled over the past five years, and no doubt much of this trend is a result of increased consumer knowledge. Better, cheaper information can also work in favor of the livestock industry. We encourage you to visit YouTube.com and search the terms "dairy animal welfare." There you will find numerous videos that take you on a tour of dairy farms and give you the opportunity to hear the farmer's side of the animal welfare debate. While it is true that consumers are more uninformed about food production than ever, it is also true that an interested individual can easily access a wealth of free, largely accurate, and entertaining resources on how the modern farm operates. We hope this book is also a useful resource as well!

Externalities

If I sell you my used car, presumably we both would benefit. I value the car less than the price at which it is being sold (otherwise I would not sell it) and you receive a car that you value more than the price you have paid (otherwise you would not have bought it). The transaction would only occur if you valued the car more than I did, and the price is negotiated between my minimum willingness-to-sell and your maximum willingness-to-pay. Both you and I gain from the trade. In fact, by definition, the buyer and seller should always be made better off from a transaction. This is why we tend to think of the market as a beneficial institution in society.

However, there are situations where a market transaction makes the buyer and seller better off, but harms a third party. This external harm to a third party who is neither or buyer a seller is an *externality*.[19] A conventional example of an externality involves a firm generating pollution in the process of manufacturing a good in a factory. The firm makes money from the factory production and consumers benefit from their purchases, but the air or water pollution imposes costs on those who breathe the air or drink the water—neither the seller nor the buyer of the good accounts for the cost of the air/water pollution that is imposed on other consumers or firms.

Without externalities there is no need to interfere with the buying and selling of goods, as the buyers and sellers are made better off through their voluntary transactions. When externalities exist the benefits to the buyers and sellers may be less than the cost of the externality, and society as a whole can be made better off if it works with the buyers and sellers to address the externality.

The farm animal welfare debate is in some ways a debate about externalities. When someone eats a food that is made from animals that suffer, everyone who cares about animal suffering is negatively affected. A significant number of people today do not like the conditions in which animals are raised. They seek to redress the externality by forcing food producers to employ more costly, humane production practices.

There is another deeper and potentially more important externality involved in livestock production. Trade between livestock producers and meat consumers imposes a potential cost on a third party—*the animal*. When someone goes to the grocery store and buys a pound of chicken breast, the store benefits, as does the consumer, but there is a third party—the chicken. The chicken obviously incurs harm as a result of the transaction—it is dead after all! Although that is not quite fair, as the chicken owes its existence to the fact that it will one day be harvested. Yet if the chicken is raised in poor conditions, and the chicken could have been made happier if raised in different conditions, the suffering the chicken experiences so that a person may have cheaper food is an externality borne by the animal.

When externalities are imposed on humans they take actions to address the harm. When surface waters deteriorate due to waste treatment plants, consumers lobby for stricter environmental laws, sue the companies causing pollution, and attempt to shame the companies into mitigating the externalities. Externalities are attended to because those affected stand up for themselves. This is something animals are unable to do.

How can we require consumers and producers of food to account for the externalities they generate in the form of farm animal welfare? One strategy is to tax food to offset the externality. These are called *Pigouvian taxes*. The tax would be applied to all foods whose production inflicts animal suffering, and the tax would be proportional to the suffering entailed. If hogs are thought to suffer more than layers, the per-hog tax would be higher than the per-layer tax.[20] The tax would lead to higher prices for food that causes animal suffering, inducing consumers to switch to alternative foods where less suffering occurs. The tax money could even be used to help farmers pay for better production facilities. Of course, the logic of the Pigouvian tax dictates that food produced from happy animals should be subsidized. Beef production, it could be argued, affords the animals a relatively natural, pleasant life. This contribution to animal happiness is what economists call a *positive* externality. In this case, animals are a third party benefiting from the food production. Consumers do not account for the fact that their beef purchases are responsible for the existence of more happy cows, and thus the subsidy lowers the price of their food, encouraging more cow happiness. Animal advocates who like the logic of a tax on animal suffering but detest the subsidy for animal happiness may want to revisit the logic underpinning their activism.

The externality derived from animal suffering occurs partly because there is no market for animal well-being. If people could assert their desire for better animal care by, in a sense, purchasing animal happiness in a market, it would then become profitable to treat animals compassionately, enhancing the level of livestock well-being. Indeed, this is what happens every day when consumers pay higher prices for cage-free eggs and prefer to eat at the restaurants that use only Animal Welfare Approved pork.

Other solutions to the externality include bans on certain production practices, like in California's Proposition 2 in 2008. Lobbying government to force reductions in the animal welfare externality is an obvious solution, but government involvement is not always necessary. As argued by the Nobel Laureate Ronald Coase, however, when externalities exist there are incentives for the third party that is harmed to negotiate and contract with the producer creating the harm.[21] This observation might seem silly, given that animals cannot bargain, but as we have seen there are many organizations that will bargain on behalf of animals. These organizations *have* influenced the way certain restaurants procure meat/eggs and livestock producers operate. Externalities can be mitigated by private party negotiations without resort to costly policy initiatives.

The question of whether the externality created by humans, but borne by animals, justifies government involvement is a complex issue. In some ways, the question is really one about philosophy rather than economics. The *economic* insight is that government involvement might improve market outcomes if one is willing to take the philosophical position that animal suffering does matter and should be reflected in governmental policy. Additionally, the philosophical position that consumers should consider their impact on livestock may allow the creation of a market for animal welfare, most probably through the sale of animal-friendly foods. If such externalities deserve our attention, it is prudent to revisit the issues in Chapter 7 related to the double-counting of human and animal benefits.

What Do People Want from Their Government?

This chapter has described reasons why government involvement in regulating farm animal welfare could be desirable. Rather than asking economists what government should do, perhaps we should ask the citizens. Some indication of preferences for government is revealed in voting behavior. A majority of voters in Arizona, California, and Florida have voted in favor of ballot initiates that effectively ban the use of gestation crates in pork production and/or cages in egg production. County-level analysis of the voting data in Florida indicates that counties that were more likely to vote in favor of the ban on gestation crates tended to be more urban (less rural), have a higher share of females, have a higher proportion of Clinton/Gore voters in the 1996 election, have more adherence to the Catholic faith, and fewer Evangelical Christians.[22]

The existence of only a few ballot initiatives limits our ability to describe people's preferences for government. Consequently, we turn directly to citizens to ask what they think. To wrap up this chapter, we report the results from two sets of surveys that have been previously discussed in the book. One was a telephone phone survey administered to a random sample of over 1000 US households in the summer of 2007, and the other was a survey given after our consumer valuation experiments in Dallas, TX, Chicago, IL, and Wilmington, NC, in the summer of 2008.

Consider some of the questions we asked after our consumer valuation experiments. This subject pool is diverse and is a good—though imperfect—representation of US consumers. To be sure, these people were no experts in animal welfare, but recall that these were not people naively answering questions on the phone; they received extensive information on animal production systems, including the pros and cons of various production methods. Their answers provide insights from informed consumers on what they desire in farm policy.

Subjects were asked a series of questions regarding their views on the role of government in regulating animal welfare (see Figure 10.3). The results indicate that people generally oppose banning production practices. Why, then, do citizens in other states vote for cage bans so enthusiastically? The difference between our findings and the results of actual ballot initiatives in many states (where a majority of people vote for bans) can be explained, in part, by the fact that people self-select into voting booths (i.e., a random sample of voters is not the same thing as a random sample of citizens) and the fact that our respondents received objective information on farm animal welfare. More important, however, is the fact that our subjects indicated a belief that government should pass and enforce anti-cruelty legislation. If voters in real referenda thought the practices they were banning are indeed cruel, then our results

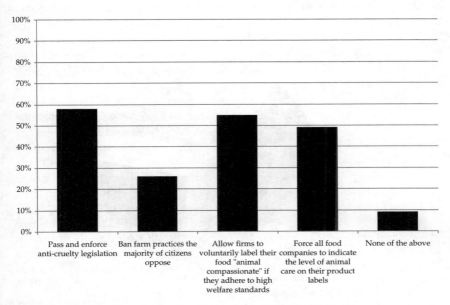

Figure 10.3 Percentage of consumers (N = 288) who believe government should ... (*respondent could choose multiple statements*)

and the referendum results are not in conflict. If the subjects in our study resemble the Californians who voted on Proposition 2, then they may have approved of the proposition because they felt the animals were suffering on the conventional livestock farm.

A slight majority of our respondents approved of firms voluntarily labeling food raised under higher welfare standards, and almost as many thought government should force companies to indicate the level of animal welfare on their products (how they would do this is not clear). What these results show is that our subjects generally favored allowing consumers the right to determine the level of animal welfare on products they purchase, but they favor banning cruel practices. How consumers believe the farm animal welfare debate should be resolved likely depends on whether they think farm animals suffer (69 percent indicated that they believe farm animals should not suffer, 29 percent thought animals' feelings are not important, and only 1 percent thought that farm animals should be guaranteed a happy and content life). Differentiating food products according to animal welfare allows consumers to decide which practices cause suffering, which is the central consumer concern.

Respondents were also asked whether they favored bans on animal practices they do not approve of, even if they could easily find food using practices they do approve. We asked this question because it provides a measure of the extent to which people perceive there to be a public good problem (i.e., whether people think your business is their business). As Figure 10.4 shows, almost one-third

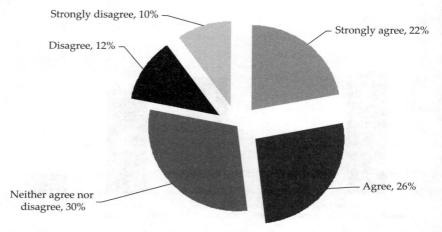

Figure 10.4 Percentage of consumers (N = 288) who agree with statement: *Even if I could easily find egg or pork products that meet my standards of animal care, I would favor governmental bans on eggs produced under lower standards of care.*

issued no opinion, stating they neither agreed nor disagree. Of those who did issue an opinion, 69 percent favored such a ban. Thus, a strong majority of the respondents who had an opinion expressed one indicating that a public good or externality problem exists. Even if the food they prefer is available, they wish to control the food you eat.

We now turn to the US nationwide survey we conducted in 2007. This survey was much larger in terms of the type and breath of questions asked, and the sample of respondents was larger and more representative than that for the sample just discussed. The responses given in this survey represent the statements of uninformed people who may have had very little knowledge about animal production or animal welfare. In this sense, the sample is likely quite representative of the US population.

Recall that the motivation for government involvement in animal welfare is the idea that there exists a market failure. The market failure stems from an externality, or public good properties of animal welfare. We asked respondents a series of questions to determine the extent to which they believed markets could address animal welfare concerns. As shown in Table 10.4, 52 percent of surveyed consumers believe their personal food choices have a large impact on the well-being of farm animals and only 36 percent thought their choices did not have much of an impact. When asked whether they agreed with the statement,

Table 10.4 Percentage of Consumers (N = 1019) Who Agree with Statements Regarding the Market Outcomes

Statement	Strongly agree	Agree	Neither agree nor disagree or don't know	Disagree	Strongly disagree
My personal food choices have a large impact on the well-being of farm animals	25%	27%	12%	19%	17%
Food companies would voluntarily improve animal welfare, and would advertise as such, if people really wanted it	32%	36%	9%	12%	11%
Farmers and food companies put their own profits ahead of treating farm animals humanely	36%	28%	15%	12%	9%

Source: Norwood, Lusk, and Prickett, 2007.

"food companies would voluntarily improve animal welfare, and would advertise as such, if people really wanted it," 32 percent strongly agreed and 36 percent
agreed. Only 23 percent strongly disagreed or disagreed with the statement.
Thus, consumers have faith that their food choices matter and that food
companies will provide the products they want. Taken alone, this suggests little
need for government involvement. However, there is an indication that
people do not necessarily believe unregulated outcomes are best for animals,
as 64 percent of people believe that farmers and food companies put their own
profits ahead of treating farm animals humanely. Taken as a whole, these results
show that while consumers express some faith in market outcomes, they are
somewhat skeptical, and they also indicate strong support for government
involvement to prevent cruelty.

As shown in Table 10.5, 68 percent, of respondents believe government
should take an active role in promoting farm animal welfare, and 75 percent

Table 10.5 Percentage of Consumers (*N* = 1019) Who Agree with Statements
Regarding Government Involvement

Statement	Strongly agree	Agree	Neither agree nor disagree or don't know	Disagree	Strongly disagree
The government should take an active role in promoting farm animal welfare	42%	26%	8%	10%	14%
I would vote for a law in my state that would require farmers to treat their animals more humanely	55%	20%	9%	7%	9%
Farmers should be compensated if forced to comply with higher farm animal welfare standards	37%	33%	8%	12%	10%
Housing pregnant sows in crates is humane	10%	8%	18%	19%	45%
Housing pregnant sows in creates for their protection from other hogs is humane	21%	24%	23%	16%	16%
Housing chickens in cages is humane	13%	18%	14%	18%	37%

Source: Norwood, Lusk, and Prickett, 2007.

say they would vote for a law in their state requiring farmers to treat their animals better. If consumers believe markets can address animal welfare concerns, why would they also request government involvement? There are at least two possible explanations. First, the term "farm animal welfare" can have a number of interpretations. People may interpret it to include basic animal cruelty, such as health neglect and lack of feed, which rarely occurs on for-profit farms. State or local government agencies enforce laws against animal cruelty, so respondents may consider this a natural function of government. Second, the survey suggests that compassionate animal treatment is a basic value, as 95 percent of respondents agreed with the statement, "It is important to me that animals on farms are well cared for." The laws a government establish reflect the values of its constituency, so respondents may feel it important to codify this value into law, even if it is ultimately unnecessary.

While on the topic of government regulation, another important result emerged from the survey. Almost 70 percent of respondents agreed that farmers should be compensated if forced to increase their production costs to comply with more stringent welfare standards. Although it is clear from these results that people desire government involvement, what type of involvement consumers want is less clear. The survey results also indicate that consumers expect improvements in animal welfare to increase food prices, but to also produce safer, better tasting food.[23] Other questions administered on the survey indicate that consumers believe human welfare issues to be much more important than animal welfare issues, but consumers desire progress on the animal welfare front even if human problems cannot be immediately solved.

Because one of the primary paths animal activists groups have pursued in improving animal welfare is the use of ballot initiatives to ban cages, we asked people whether they believed such practices were humane. Not surprisingly, housing chickens in cages and pregnant sows in crates is deemed inhumane by a majority of individuals (64 percent for hogs and 55 percent for hens). However, when the question was slightly re-worded to ask whether housing sows in crates *for their protection from other hogs* was justified, the percentage of people who believed the crates were inhumane dropped from 64 percent to 32 percent.[24] Thus, although cages and crates have a negative impression by the general public, their perceptions are easily swayed by providing some justification for the practices.

Last, we asked a few questions about who should make farm animal welfare decisions. These are interesting questions because not only are people debating the role that science should play in the animal welfare debate, but there is substantive disagreement on whether and how issues related to farm animal welfare should be regulated. Much of the controversy in the animal welfare debate stems from questions over who should have the authority to decide how farm animals are raised. Our survey participants were asked to indicate the extent to which they agreed with the following two statements "decisions about animal welfare should be left to experts, and should not be based on public opinion" and "scientific measures of animal well-being should used to determine how farm animals are treated, not moral or ethical considerations." Based on responses to these two questions, people were placed into one of four categories: (i) *scientific elitist*; (ii) *moral elitist*; (iii) *scientific populist*; or (iv) *moral populist*—as shown in Table 10.6.

The names of these categories may be self-explanatory, but in case they are not the definitions are as follows. The terms *elitist* and *populists* are mutually exclusive terms used to describe whether someone thinks experts should be responsible for making decisions on how farm animals are raised (elitists), or whether the decisions should be determined by attitudes of the society at large (populists). When deciding whether gestation crates or group pens should be used to house sows, the elitist would consult a panel of animal scientists, whereas a populists would consult the survey results of a representative sample of the US.

The *scientific* and *moral* terms within these labels indicate whether the person believes that animal welfare decisions should be determined by scientific measurements (scientific) or moral considerations (moral). People aligning themselves with the *scientific* label are likely to want to use scientific measurements such as productivity and stress hormone levels to determine animal living conditions. People assigned to the *moral* label may also want scientific input, but might want to use other criteria as well. For example, is it ethical to own animals? This is a question science cannot answer. The *moral* crowd is willing to go beyond science. Claiming it is unethical to confine animals to crates so small

Table 10.6 Consumer Opinions about Who Should Make Decisions about Farm Animal Welfare and on What Basis. Percentage of Consumers (N = 1019)

Consumer type	Should decisions about animal welfare be left to the experts rather than the public or based on the views of the public?	Should decisions be made on the basis of scientific measures of animal well-being or moral and ethical considerations?	Percentage in each category
Scientific elitist (SE)	expert	scientific measures	36%
Moral elitist (ME)	expert	moral considerations	21%
Scientific populist (SP)	public	scientific measures	18%
Moral populist (MP)	public	moral considerations	25%

Source: Lusk and Norwood, 2008.

they cannot turn around entails a mixture of perceptions and personal values. This claim reflects not only the perceived well-being of animals in cages but a statement about the responsibilities of humans towards animals. Of course, values may play a role in determining what the scientific crowd believes scientific measurements imply about what we "should do." All Americans probably identify somewhat with each label. The discrete categories they choose to describe themselves reveal to which categories they are most sympathetic.

Roughly half of respondents (57 percent), believed that decisions about animal welfare should be left to the experts, as opposed to public opinion. A slight majority of respondents (54 percent) also believed that decisions about animal welfare should be based on scientific measures of animal well-being rather than moral and ethical considerations. Table 10.6 reports the percentage of people that can be classified into each of the four categories of views towards the governance of animal welfare: 36 percent were scientific elitists, 21 percent were moral elitists, 18 percent were scientific populists, and 25 percent were moral populists. Another, similar study regarding technology in general calculated that 54 percent of US citizens consider themselves scientific elitists.[25] Though this percentage is higher than those reported in Table 10.6, they share the same result that most Americans are scientific elitists. This implies that, compared to technology in general, people are more interested in public opinion and moral concerns when it comes to the issue of farm animal welfare. We found that scientific populists are the smallest segment of the population, comprising 18 percent of the current sample related to animal welfare. Where people fall on the continuation of how they believe farm animal welfare decisions should be made is also related to their views about farm animal welfare. For example, those individuals classified as scientific elitists were the least concerned about farm animal welfare issues, whereas individuals classified as moral populists were most concerned. Information about a potential benefit of gestation crates was most effective at changing the opinions of scientific elitists.[26]

A Tentative Verdict on Eggs and Pork

Now is the time to bring all the information together in one place for a *tentative* verdict on whether government action should be used to improve living conditions in the egg and pork industries. The *private* value people place on cage-free eggs and crate-free pork suggest that most *informed* people are willing to pay the cost of higher welfare standards. If all consumers were informed, then there would be no need for government to ensure a better life for layers and hogs. The market would (perhaps slowly) evolve such that traditional eggs and pork would be replaced with animal-friendly products.

The case for government involvement in food markets becomes stronger when *public good values* exist, which are values that consumers place on improving the well-being of animals distinct from the animals used to produce the meat, dairy, and eggs that the person consumes. The results of our experiments indicate that people do exhibit public good values that, in aggregate, far exceed the cost of producing more animal-friendly products. It was noted, however, that the aggregate public good benefits were largely driven by the values of a small minority of people. If the costs of the ban were distributed on a per-person basis, the per-person costs would exceed the public good benefits for the vast majority of citizens.

The citizenry possesses varying views on how policy should be formed. The issues of disagreement concern whether policy should be motivated by scientific measurement or morals, and whether expert opinion should trump popular opinion. Scientific experts tend to suggest that cage-free egg production produces higher levels of animal well-being than cage egg production, and that shelter-pasture hog production produces higher levels of animal well-being than confinement production methods. Moreover, confinement hog production using group-pens is preferred to gestation crates.

This is a generalization of our reading of the literature. Any one scientist may claim that we are wrong. For example, some scientists argue that cage-free and cage egg production systems each have their advantages and disadvantages, and one method is not unambiguously better. However, other scientists claim that cage-free is more conducive to higher well-being than cage systems. We interpret the synthesis of these assertions to be in favor of cage-free methods, for the same reason that the average of the numbers "0" and "2" is a positive number, 1. These scientific opinions are informed by both scientific measurements, and ethical considerations.[27]

Surveys of informed and uninformed consumers indicate that consumers generally disapprove of farms utilizing small cages, and informed consumers are willing to pay the additional cost of more humane livestock systems. Regardless of whether one is a scientific elitist, moral populist, or a combination of those two, the answer to the question of what we should do about farm animal welfare is to improve it by pursuing cage-free egg systems, group-pen gestation systems for sows, and shelter-pasture pork systems. This is not an opinion, but an impression made from consumers in telephone surveys and economic experiments.

President Truman once quipped he needed a "one-armed economist," because he grew tired of hearing economists say, "but on the other hand...". It can be frustrating when economists are reluctant to commit to a stance, but a good economist must be forthright about the complexities of social issues. Although we have outlined a large body of knowledge suggesting that the American consumer desires better care of livestock, and governmental policy has the potential to improve human welfare by improving animal welfare, we are not adamant that government *should* regulate food production. Just because Americans approve of *some* changes in livestock production does not imply that they approve of *any* change. Perhaps the reader has experienced disappointment at the failure of politicians they once championed. The same can occur with animal welfare. Consumers may support policies but disapprove of the change that occurs. This has serious implications for how animal welfare should be pursued. Bad market outcomes are rapidly corrected, as unprofitable ventures are disbanded by entrepreneurs, but bad policy can persist for decades.

Our research results stem from informed consumers, but most people are not informed and never will be. If, based on the above findings, the government banned cages in egg production and crates in pork production, most consumers will face higher prices without the information to perceive the benefits of animal welfare regulations. Uninformed consumers will be hurt economically in the form of higher prices and foregone consumption. The estimated size of this *experienced effect* is in the billions for eggs and millions for pork. When expressed as an annual per-person cost, however it is small—in the neighborhood of $6.00. Industry profits would fall too, forcing some farmers out of the market and others to earn smaller profits.

Many of the public goods provided by the government have a similar experienced effect. We are only vaguely aware of the foreign aid that our government provides to developing countries, but we feel the pain of paying taxes. We barely notice many of the regulations that protect the environment, but again, we feel the pain of paying taxes. Yet many people approve of such measures when informed. To deny regulation the ability to provide public goods because their benefits are not immediately obvious to the majority of Americans would imply the removal of many programs of which people, once they learn about them, approve. A sophisticated policy regime would actively seek policies that address problems not immediately obvious to the average person. It is simply the type of government that we believe the average person wants and expects.

One of the reasons for calculating the experienced effect is to try to honestly anticipate the market consequence of animal welfare regulations. Egg and pork prices would likely rise, and our consumption fall. We would feel the pain of these changes just as we bear the burden of paying taxes for the many other public goods whose benefits are outside our immediate attention. The benefits produced by such regulations would be intimately experienced by farm animals though, and the citizenry, when and if it learns about them.

If we have left you with some uncertainty about whether government regulation is needed to improve the lives of farm animals, our job has been accomplished. In principle, a ban on certain production practices can provide people with improvements in the eggs and pork they buy at a cost they are willing to pay—assuming people are informed and assuming our estimates of costs are correct. In addition, a government-enforced ban could eliminate the public good problem, reduce externalities imposed on animals, and make the products people buy correspond more accurately with what people actually think they are buying. In practice, however, government regulations are slow to respond to changing consumer demands and political outcomes are prone to manipulation by vested and powerful interest groups. Moreover, regulated outcomes impose a single outcome for all consumers with little sensitivity for the diversity in individuals' tastes and beliefs.

When possible, we believe the best approach forward is to seek solutions that preserve the benefits of the market (allowing information revealed in prices to direct resources, the incentives for trial and error experimentation by producers and consumers, and decentralization of power) while avoiding the market failures that government action seeks to rectify.

Appendix: Calculations and Methods Supporting Results

This section utilizes the elasticities given in Table 8.A.1, the production costs detailed in Chapter 9, an equilibrium displacement model,[28] and linear supply and demand curves to calculate the changes in price, quantity, producer surplus, and consumer surplus as a livestock industry replaces its conventional production technology with a system thought to improve animal well-being. The three scenarios considered are (1) replacing cage egg with cage-free egg production; (2) replacing confinement-crate with confinement-pen production; and (3) replacing confinement-crate with shelter-pasture production. The old and new system will possess different supply-and-demand elasticities. Because the producer and consumer surplus numbers are approximations, we simply employ the elasticities associated with the conventional system.

The methodology is standard in economics but will be unfamiliar to the non-economist. Readers interested in learning more about the equilibrium displacement models are encouraged to consult chapters 2 and 3 of our textbook: *Agricultural Marketing and Price Analysis*.

Cage to Cage-Free Eggs

Switching from cage to cage-free eggs is expected to increase the cost of retail egg production by $0.35 per dozen. The price of retail eggs is about $1.50 per dozen, so this entails a (0.35 / 1.5 = 0.2333) 23.33 percent increase in retail egg production costs. If there is no demand change associated with this cost increase, the price of eggs is projected to rise by [(2) (0.233) / [2 + 0.2] = 0.2118] 21.18%. This is a $1.50 × 0.2118 = $0.3177 increase per dozen, making retail egg prices ($1.50 × 1.2118 =) $1.82 per dozen. The quantity of eggs produced and sold is expected to change by [(−0.20) (2) (0.233) / [2 + 0.2] = −0.0424] − 4.24%. Egg prices rise 21 percent and egg consumption falls 4 percent. This assumes the supply and demand elasticities detailed in Chapter 8, and the equilibrium displacement model method described in pages 78–83 in Norwood and Lusk (2008).

The US per capita egg consumption is 250 eggs[29] and the US population is 303,824,640.[30] Approximately 95 percent of all egg production takes place in cage egg facilities,[31] which implies that (0.95) (250 × 303,824,640)/12 = 6,013,196,000 dozen eggs are consumed each year. If cage eggs were replaced with cage-free eggs, the quantity of eggs consumed would fall by 4.24% or 5,696,712,000 × 0.0424 = 241,540,589 dozen eggs. Assuming consumers were only aware of the price increase and not cognizant of how farm animals were affected, the change in producer and consumer surplus can be calculated as:

Change in producer surplus = (change in price − change in costs) (original quantity) (1 + 0.5 × (percent change in quantity))

Change in consumer surplus = − (original quantity) (original price) (percent change in price) (1 + 0.5 (percent change in quantity)).

For the cage egg ban, the calculations are:

Change in producer surplus = (0.3177 − 0.35) (6,013,196,000) (1 + 0.5 × (−0.0424)) = −\$187,270,470.

Change in consumer surplus = − (6,013,196,000) (1.5) (0.2118) (1+0.5(−0.0424)) = −\$1,872,704,705.

Readers who calculate these equations by hand may compute a slightly different number. Our numbers were computed in a spreadsheet, which allows for many more decimal places than shown here.

Confinement-crate to Confinement-pen Pork

Now turn to hogs, and the conversion of confinement-crate pork to confinement-pen pork. Switching production methods is expected to increase the cost of retail pork production by \$0.065 per lb. Sources suggest the price of retail pork is about \$2.80 per lb,[32] so this entails a [0.065 / 2.8 = 0.0232] 2.32% increase in retail pork production costs. If there is no demand change associated with this cost increase, the price of pork will rise [{(2) (0.0232) / [2 + 0.70]} = 0.0172] 1.72% and the quantity of pork produced and sold is expected to change by [(−0.70) {(2) (0.0232) / [2 + 0.70]} = −0.0120] −1.20%. Multiplying 0.0172 by the \$2.8 retail price yields a retail pork price change of \$0.0481 per retail lb. This assumes the supply-and-demand elasticities detailed in Chapter 8, and the equilibrium displacement model method described in pages 78–83 in Norwood and Lusk (2008).[33] The per capita consumption of pork is 50.8 lbs,[33] and the US population is 303,824,640,[34] which multiplied by one another equal 15,434,291,712 total lbs of pork consumed by Americans. We have already seen that the cost increase is \$0.065 per retail lb and the higher cost will reduce pork consumption by 1.20 percent or 15,434,291,712 × 0.0120 = 185,211,501 lbs. The welfare effects of the cost increase are calculated using the formulas above.

Change in producer surplus = (change in price − change in costs) (original quantity) (1 + 0.5 × (percent change in quantity)).

Change in consumer surplus = − (original quantity) (original price) (percent change in price) (1 + 0.5 (percent change in quantity)).

Plugging in the appropriate numbers, we calculate,

Change in producer surplus = (0.0481 − 0.0650) (15,434,291,712) (1 + 0.5 (−0.0120)) = −$258,531,002.

Change in consumer surplus = − (15,434,291,712) (2.8) (0.0172) (1 + 0.5 (−0.012)) = −$738,660,007.

Readers who calculate these equations by hand may compute a slightly different number. Our numbers were computed in a spreadsheet, which allows for many more decimal places than shown here.

Confinement-crate to Shelter-pasture Pork

Instead of converting from crate to crate-free pork, now consider a conversion from confinement-crate to shelter-pasture pork. This pork costs $0.11 more per retail lb to produce, which divided by the $2.8 per lb price constitutes a (0.11 / 2.8 = 0.0393) 3.93% rise in costs. Using the equilibrium displacement model in Norwood and Lusk (2008) and the supply-and-demand elasticities in a previous chapter, if no change in pork demand occurs, the price change is [{(2) (0.039) / [2.7]} = 0.0291] 2.91%, making the new price $2.8 × 1.0291 = $2.8815, and the quantity of pork will change by [(−0.70) {(2) (0.039) / [2.7]} = −0.0204] −2.04%. Switching all pork from confinement-crate to shelter-pasture pork increases costs by $0.11, increases price by $0.0815, and will lower pork production by 2.04% or 15,434,291,712 × 0.0204 = 314,859,551 lbs.

As before, the welfare changes are:

Change in producer surplus = (change in price − change in costs) (original quantity) (1 + 0.5 × (percent change in quantity)).

Change in consumer surplus = − (original quantity) (original price) (percent change in price) (1 + 0.5(percent change in quantity)).

Plugging in the appropriate numbers, we calculate,

Change in producer surplus = (0.0815 − 0.1100) (15,434,291,712) (1 + 0.5 (− 0.0204)) = −$435,679,991.

Change in consumer surplus = − (15,434,291,712) (2.8) (0.0291) (1 + 0.5 (−0.0204)) = −$1,244,799,974.

Readers who calculate these equations by hand may compute a slightly different number. Our numbers were computed in a spreadsheet, which allows for many more decimal places than shown here.

CHAPTER II

The Question Before Us

"There are known knowns. These are things we know that we know. There are known unknowns. That is to say, there are things that we know we don't know. But there are also unknown unknowns. There are things we don't know we don't know."

Donald Rumsfeld at a 2002 press conference

Possibilities

This book has journeyed back to the earliest days of animal domestication and then forward into modern industrial farms. Delving into questions of ethics and animal sentience, we then compared the costs of improving animal care to the benefits. Our goal was to shed some new light on the animal welfare debate through the lens of economics. We do not claim to know with certainty what is truly ethical in terms of the treatment of farm animals. Nor do we have strong views on whether our current system of livestock production is humane or whether farmers should be encouraged (or forced) to adopt more humane production methods. We are passionate about only one thing: for the reader to acquire information helpful for making reasoned and informed decisions about the farm animal welfare debate.

The animal welfare debate is a complex topic involving issues of biology, history, farm husbandry, animal sentience, philosophy, and *economics*. The complexity of the topic creates uncertainty about the effect of public and private initiatives to improve farm animal well-being. Uncertainty does not imply a lack of knowledge. After spending years caring for farm animals and studying agricultural economics, we have hopefully learned something. Yet, we remain agnostic about the appropriateness of many details of farm animal production and welfare policy. Speaking honestly and objectively about farm animal welfare requires a nuanced connection of the knowns and unknowns, and anyone who claims to have complete certainty is probably a fraud or an ideologue.

What We Know

Life is a journey that is certain to have a beginning and an end, but what happens in the middle is fraught with uncertainty. Understanding and resolving life's uncertainties makes life more enjoyable, providing a sense of control and engagement. One of life's uncertainties is the extent to which our actions—our food and voting choices—might cause animal suffering. So what do we know with confidence about farm animal well-being?

First, we are quite confident that the lives of egg layers are improved by moving from a conventional battery cage system to a cage-free system—not just any cage-free system though; the stocking density most be reasonably low and the facility must be reasonably enriched to enhance animal health and allow for natural behaviors. The cage-free standards designed by the United Egg Producers are good standards, but a few additional modifications might be desirable. Likewise, the evidence is clear that sow well-being would be improved by replacing the confinement-crate system with a better alternative. Simply replacing the gestation crate with a group-pen is a positive step for the sows, but it is a marginal improvement. It is very likely that sows in either the confinement-crate or confinement-pen systems are perpetually frustrated. The confinement-enhanced or shelter-pasture system represents an enormous improvement. In fact, the two systems are probably capable of generating among the highest levels of sow well-being possible.

Other livestock industries could be improved as well, by: selecting broilers with better leg structures; giving veal calves space to turn around; providing all dairy cows with dry flooring; and constructing shelters in beef feedlots. These are all changes that are certain to improve animal well-being.

Although animal well-being can be enhanced in most typical animal production systems, we are quite certain that the overall level of animal welfare is higher in the broiler, dairy, and beef industries than the egg, pork, and veal industries. Beef cattle in particular experience high levels of well-being. A movement to improve the lives of egg-laying hens or sows would substantially reduce animal *suffering*, whereas an improvement in the beef industry would only make *happy* animals happier—both are praiseworthy, of course.

The reader may find these assertions obvious. We mention them because many people employed in the animal production industries so often make arguments to the contrary. To be clear, we assert that the lives of most hens and sows can be improved by moving to alternative production systems. However, we are not asserting that these alternatives necessarily *should* be adopted. People do not live life with the goal of maximizing farm animal well-being. Improving farm animal well-being would mean giving up other things that we enjoy, and the extent to which people are willing to make such trade-offs is far less certain. However, this should not blind us to the fact that farm animal welfare *can* be improved relative to current farm practices.

Another certainty is the fact that ultimately it is *consumers*, not farmers, who decide how farm animals are raised. Farmers have no more choice over what production system to employ than McDonalds had over whether it would sell the McLean Burger—if it does not sell, it is out of the question. Any firm that does not satisfy consumers' desires is a firm destined for bankruptcy. Small businesses and large businesses, farmers and Microsoft, are all servants to the consumer.

Improving animal welfare will certainly increase production costs at the farm. People who argue otherwise are necessarily asserting that farmers are either too ignorant or too malevolent to improve animal welfare at no cost to themselves. (Believe it or not, there are people who assert that improving animal welfare will lower costs; these beliefs are without merit.[1]) Another fact of which we are certain is that increases in farm production costs will cause food prices to increase. It is doubtful that farmers can pass the entire cost increase onto consumers in the form of higher prices, but it is equally doubtful that farmers will fully absorb the higher costs that go along with higher animal welfare standards. Regulations demanding improved animal care will impose some economic burden on the farmer *and* the consumer. Even food processors, wholesalers, and retailers will be adversely affected.

Economic analysis applied to the topic of animal welfare provides new insights that heretofore have been largely unappreciated in current writings on farm animal welfare. For example, by focusing on outcomes and the quantitative links between consumers and farm animals, Chapter 8 shows that even if cage-free production is more humane than cage egg production, this does not imply that it is more ethical to consume cage-free eggs. When efficiency differences between cage and cage-free production are ignored, we lose sight of the fact that cage-free production requires more hens. If cage-free hens have higher welfare than their caged counterparts, but suffer nonetheless, consuming cage-free eggs may translate into greater overall suffering. The reason is simple: we may prefer to have fewer hens suffering a great amount than to have many hens suffering a small amount.

A more rigorous analysis of the relationship between food choices and animal well-being also raises ethical conundrums. Vegans are apt to boast that their diet causes fewer animals to suffer. To the extent that the non-vegan diet is derived from animals that suffer, this is true. However, some animals arguably experience more positive than negative emotions. The vegan who fails to eat food derived from happy animals thereby prevents happy animals from existing. So who is more ethical: a vegan whose diet prevents some animal suffering but also prevents happy animals from existing, or the omnivore whose diet brings some happy animals into existence but causes other animals to suffer? The logic asserting it is good to prevent the existence of suffering animals by altering food choices also dictates that it is good to make food choices that bring happy animals into existence. Arguments to the contrary are illogical.

Likewise, consider the effect of higher animal welfare standards on farm animals that arguably live a good life. Suppose a law is passed mandating the construction of shelters in cattle feedlots. The increased cost of beef production raises the price of beef, inducing consumers to consume less. As a result, fewer beef cattle will exist. But remember, there is good reason to believe beef cattle live overall pleasurable lives. Improving welfare standards may increase the welfare of animals that are born, but the same action will cause many animals that would have otherwise experienced a "happy life" to be unborn. If we conclude that increasing the number of animals who live in misery is a bad thing, logic requires that decreasing the number of animals who live in happiness is also a bad thing.

Consider another certainty discussed in Chapter 7. Some animals receive poor treatment because they are unable to defend themselves and fight for their own rights. Chapter 7 showed that when humans take up the animals' cause by arguing that the welfare of animals should be included alongside the welfare of humans in the cost–benefit analyses of government policies, we should logically ignore human altruism toward animals. That is, if a policy benefits animals by giving them a more pleasant life, and if the policy benefits Jim because he is happy knowing that animals' lives are improved, the desirability of the policy should not depend on the benefit Jim receives. When we choose to give animal misery and merriment similar consideration to that of humans, logic dictates that we should ignored the happiness we get from knowing animals are better treated when conducting cost–benefit analysis.

Our consumer research uncovered a number of results that have helped resolve some of the uncertainties in the farm animal welfare debate. In telephone surveys, US consumers rate the importance of farm animal well-being well below other societal concerns; such as human poverty, food safety, the environment, and the financial well-being of US farmers. However, people still express an interest in improving farm animal well-being. People do not believe farm animals should suffer and are largely supportive of anti-cruelty legislation; however, they are less supportive of outright bans on livestock production practices. It seems that a distinguishing characteristic people use in judging whether public policy is warranted is the question: *does the animal suffer?* Most people do not think farm animals should be treated like the family pet, but they do believe that farm animals should not suffer.

We also invited hundreds of people to participate in real money, real food, and real animal experiments to determine what they were willing to pay for improvements in farm animal well-being. When consumers are informed about farm animal welfare and current production practices, they are willing to pay a premium for animal-friendly foods that is, on average, larger than the cost.

Informed consumers will voluntarily pay the increased cost to obtain cage-free or free-range[2] eggs over cage eggs. Informed consumers will voluntarily pay the increased cost to obtain crate-free pork and pork from a system that provides ample shelter and access to pasture.

People are even willing to pay large sums of money to improve the lives of farm animals, even if they will never consume food produced by the animal. These are referred to as *public values*, as they refer to food consumed by other members of the public. The total magnitudes of the public values alone are large enough to justify the cost of improving the lives of hogs and layers. However, the average public value is profoundly influenced by a vocal but small portion of the population. For example, only 8 percent of our respondents would be willing to pay the per-person cost to move 1000 hens from a cage to the cage-free system, even though the aggregate public good benefits exceed the costs.

These insights alone do not justify more invasive regulation of livestock farms. However, an individual seeking legislation that will improve the lives of farm animals will find much evidence supporting their initiative in this book.

What Remains to be Learned

Conducting research can be a humbling experience. Human behavior is anything but simple. Research often uncovers more questions than answers. Although we have learned many things about the farm animal welfare debate, there remains much to learn. Acknowledging this fact is a good first start.

One intriguing finding is the dramatic divergence between the retail price premiums charged for cage-free eggs and the many estimates of the production costs differences. Data collected from thousands of grocery stores across the US reveals that cage-free eggs are priced, on average, 120 percent higher than cage eggs. This contrasts sharply with several budgetary studies that place the differences in farm-level costs between cage and cage-free systems at only 20 percent. In a competitive market, one would expect price differences between two otherwise similar products to roughly reflect costs differences. Maybe the cage-free market is simply not competitive yet, but will be in time.

An alternative explanation is that the large cage-free premiums result from the differences in preferences of grocery store shoppers—not just differences in farm-level costs. Grocery stores compete for customers based on their ability to offer low prices and a variety of items. Most consumers purchase eggs and desire the least expensive eggs in stock. Hence the lower price premiums for cage eggs. Other consumers seek premium eggs, and, just as inexpensive wine signals low wine quality, if stores charged a low premium for cage-free eggs some consumers would be skeptical of the eggs' quality. Food retailers do not apply the same per-dozen markup for all eggs; data made available to us reveal that food retailers make a higher profit for each dozen of brown, cage-free eggs sold than white eggs produced in a cage system. Grocery stores pay a higher price for cage-free eggs, but the price premiums they charge far exceed the price premiums they pay wholesalers. The increased cost at the farm level may be of lesser importance compared to the ability and desire of certain consumers to pay high prices.

Should cage eggs be banned, cage-free eggs would then become the staple egg item. Retailers would then feel obliged to lower their cage-free egg prices. When cage-free eggs become the standard egg, retailers will cease marketing the eggs to their more affluent customers and will attempt to provide low prices to the ordinary consumer. When this happens, the premiums consumers pay at the store will better reflect the differences in production costs of cage-free eggs. At least, that is our hypothesis. Because cage eggs have not been banned, our hypothesis remains untested, and is thus categorized as a known unknown.

Most consumers are ignorant of farm production practices, and regardless of what happens to the labels placed on the eggs they will never take the time to learn about production practices. Although we know what *informed* people are willing to pay to have, say, cage-free rather than cage eggs, it is less clear how both informed and uninformed people will alter the *quantity* of pork or eggs consumed should animal welfare improve.

We do not know the consequences of the many ballot initiatives to ban cages and gestation crates which are appearing across the US. Policies often have unintended consequences that are difficult to foresee at the time of the vote. Banning production practices in certain states and not others is likely to cause a re-organization of where animal products are produced, but may have very little effect on how animals are actually raised. Ballot initiatives might serve to heighten awareness of current farm animal production practices, and thereby lead to increased consumer demand for cage-free and crate-free products. While we can be reasonably certain that such policies will cause the price of animal food products to rise, we are less certain about how human (and even animal) well-being will change.

We primarily devoted our attention on the egg and pork industries, those which many believe to cause the most animal suffering. As a result, our experiments neglected the question of animal welfare improvements in other animal production sectors. Little is known about the cost of breeding broilers with fewer leg problems. Almost no attention has been given to the cost of improving the lives of beef and dairy cows. Because so little veal is produced and consumed, economists have not and probably will not seek to measure the benefits and costs of improving veal calf welfare.

Finally, changes in how livestock are raised may cause changes in food safety. Both sides of the animal welfare debate argue that food will become less safe if the other side wins. Both positions can claim to be right because there is little research on the topic. Take the typical hog farm. Placing so many pigs in such a small space might encourage the spread of disease, but such farms feed anti-biotics which discourage the spread of disease. Such farms often use high-tech systems to remove manure, reducing the ability of disease to spread from hog feces to hogs, which might be cost-prohibitive in systems that require larger space (or outdoor) space requirements. Moreover, the cost of adopting food safety systems often declines, on a per-pound basis, as a farm gets larger, which seems to run against many people's beliefs that "small" farms produce safer meat than "large" farms. Which livestock production system provides the safest food? We do not know. But, changes in animal welfare standards could affect food safety—for the better or for the worse.

The Future

The modern world provides greater opportunities for human happiness than any preceding era. At least for now, the struggle against nature has largely been won. The stubborn will and ingenuity of the human mind has created an environment vastly different from the natural world from which we emerged. Our lives vary little with the seasons. For most people living in the developed world, the fear of predators and the anxiety of certain, future hunger is found only in our imagination.

It is only in our imagination that we can relate to the humans that first domesticated animals. Those animals did not merely accompany our glorious journey. Livestock not only accompanied our rise to affluence, but played an integral role getting us here. At this junction, we must ask ourselves: did we leave animals behind? Is it fair to revel in our bounty while our livestock are confined to small, barren cages? Or, have cattle, pigs, and chickens been mere tools all along? *This is the question before us.*

What can be expected in the future? Concern over the treatment of farm animals is unlikely to be a mere fad. As consumer incomes rise and as fewer consumers have a connection with agriculture, we can expect more calls to improve the lives of farm animals. Expect new referenda, protests of food retailers, new lawsuits, and appeals to the public evoking sympathy for animals. Expect animal agriculture to fight back, for example, by introducing legislation prohibiting the use of referendums for farm animal issues. Expect a patchy, incoherent set of laws and court rulings to emerge. Livestock industries are currently protected from anti-cruelty laws because they are "customary" prac-tices. Should a court rule that cruel practices cannot be justified based on their popularity, livestock producers could face stiff animal cruelty charges. This would lead to a series of precedents, and also motivate new legislation that would alter the farm animal welfare conversation.

People involved in the animal welfare debate often become, understandably, frustrated. Animal welfarists striving to regulate farms within the US may feel deflated when they see new, unregulated farms rise across the border. Farmers grow tired of people who have never stepped foot on a farm telling them how animals should be raised. The average American may never understand why some animal welfare improvements that cost pennies per pound are not adopted. Anyone who joins the farm animal welfare debate will soon discover the difficulty of moving the debate even a small step towards their side. This is a good sign. The inability to alter society's behavior without the citizens' consent is a trademark of democracy. One person should not be able to dictate how farm animals are raised; one farmer should not be able to control the internet. One animal rightist should not impose their preferred diet on their neighbor—and the unwillingness of markets to make marginal improvements should not be confused with the *inability* of markets to improve the lives of farm animals.

The farm animal welfare debate will not converge to a simple struggle for market share between cage and cage-free eggs. That battle for market share will take place, but it is a skirmish compared to the more substantial debates. The key debates will take place in the voting booth, the courtroom, the public arena, perhaps in newly designed markets, and, most importantly, in the heart and mind of the consumer. When the debate is at hand, it is our hope that sharp reasoning accompanied by a keen understanding of economic consequences, will prevail.

ENDNOTES

Chapter 1

1. Gardner, 2002; NASS, 2002.
2. Sunstein, 2004.
3. Posner (2004, p. 63).
4. The estimate of the number of laws introduced each year can be found in Rollin (2004). For discussion on regulation in Europe, see Moynagh (2000) or the topical list on the European Union's web page: <http://ec.europa.eu/food/animal/welfare/index_en.htm>.
5. Ginsberg and Ostrowski, 2007. <http://www.raw-food-health.net/NumberOfVegetarians.html>.
6. Wolfson and Sullivan, 2004.
7. Carbone, 2004.

Chapter 2

1. Holldobler and Wilson, 1994; Mozoyer and Roudart, 2006.
2. Hominids are the family of great apes, including chimpanzees, gorillas, humans, and orangutans.
3. Flandrin, 1996; Tannahill, 1988.
4. Khaitovich et al., 2008.
5. Flandrin, 1996; Tannahill, 1988.
6. Davis et al., 1928 p. 225.
7. Tannahill, 1988; Kritzman, 1996; Diamond, 1997; Mozoyer and Roudart, 2006.
8. Baker, 1913.
9. Mozoyer and Roudart, 2006.
10. Clutton-Brock, 1989.
11. Ruminants are animals that have hooves, have a stomach divided into four compartments, and chew cud. Ruminants include most of the herding animals, such as sheep, cattle, deer, and antelopes.
12. Diamond, 1997.
13. Ibid.; Montanari, 1996; Anderson, 2004.
14. Diamond, 1997.
15. Tannahill, 1988.
16. Clutton-Brock, 1989.

17. *The Economist*, 2007.
18. Clutton-Brock, 1989.
19. Ibid.; Mozoyer and Roudart, 2006.
20. Ibid.
21. Horan, Shogren, and Bulte, 2008.
22. Loosely taken from Powell, Shennan, and Thomas, 2009.
23. Tannahill, 1988; Diamond, 1997.
24. Soler, 1996, p. 52.
25. Ibid.
26. Tannahill, 1988. It may seem unusual to associate pigs with meat eaters, but that is only because they are rarely fed meat in our modern society.
27. Montanari, 1996.
28. Frassetto, 2008, p. 4.
29. Stark, 2004.
30. Spencer, 2000, p. 160.
31. Manchester, 1993; Spencer, 2000; Frassetto, 2008.
32. Exodus 23:11; Leviticus 25:7; Deuteronomy 22:4, 6–7, 25:4.
33. Hills, 2005.
34. Spencer, 2000.
35. Spencer, 2000; Tristram, 2006.
36. Anderson, 2004.
37. Hills, 2005.
38. Anderson, 2004.
39. Ibid., p. 158.
40. Tristram, 2006.
41. Anderson, 2004.
42. Spencer, 2000, p. 76.
43. Gandhi, 1957, pp. 426–7.
44. Tannahill, 1988; Spencer, 2000.
45. Wirth, 2007.
46. Spencer, 2000, p. 279.
47. *British Medical Journal*, 2009.
48. Tannahill, 1988; Manchester, 1993.
49. Anderson, 2004.
50. Tannahill, 1988, p. 247.
51. Ibid.; Spencer, 2000; Stuart, 2006.
52. *Time Magazine*, 1926.
53. Gardner, 2002.

Chapter 3

1. Beers, 2006; Wirth, 2007; *Animal Rights History*, 2009; Dixon, 2009; Massachusetts Body of Liberties, 1641.
2. Beers, 2006.

3. Phelps, 2007.
4. Ibid.
5. Gough, 2008.
6. Baker, 1913.
7. Gardner. 2002.
8. USDA data available at: <http://www.agmanager.info/livestock/marketing/database/default.asp>.
9. Capper, Cady, and Bauman, 2009 *b*.
10. Economic Research Service, 2003.
11. Economic Research Service, 2007.
12. Davis et al., 1928, p. 225.
13. Data were obtained from the Livestock Marketing Center. These data indicate that the nominal retail prices of beef, pork, and poultry have fallen an average of $0.021, $0.017, and $0.018 per pound per year, respectively from 1960 to 2008. This downward price trend implies a per-pound price decline from 1960 to 2008 of $1.012, $0.797, and $0.876 for beef, pork, and chicken, respectively, which further implies a 43.1%, 45.6%, and 101.7% decline in prices for beef, pork, and chicken relative to the mean prices over this time period. Given consumption shares for beef, pork, and poultry of 32%, 25%, and 43% in 2008, these figures imply a weighted-average price decline of about 69% for all meat.
14. The 69% figure only represents *price* changes, but there is not a 1-to-1 correspondence between price and cost changes. For most situations, the price change will be something less than the cost change. Moreover, changes in consumer demand can cause prices to change as well—and in ways that would tend to mask the benefits of technological change. For example, from 1960 to 2008, the US population increased over 68% according to the US Census Bureau. Thus, there are now more than 123 million more people in the US than there were in 1960, which means that there are many more people willing to pay a particular price for a pound of meat. All else equal, this would cause meat prices to rise to meet the demand increases. That meat prices have actually fallen means that the changes in technology have more than offset the growth in demand. That is, increases in supply due to technological change have been larger than the outward shifts in demand. In summary, the cost reductions resulting from technological change must be greater than 69% to for retail meat prices to have fallen this much over the time period.
15. We arrived at this estimate using the food demand model reported in Raper et al. (1996). They used household expenditure data to estimate consumer demand for 10 separate food categories, including meat. Another category in their demand system is food away from home. Their data indicate that 28% of at-home consumption is meat. Thus, in addition to assuming meat prices change by 69%, we also assume prices of food away from home rise by 0.28*69% = 19% (assuming the composition of people's diets is the same at home and away from home). We used their estimated linear-expenditure demand system to recover the consumers' indirect utility function, which was used to calculate the compensating variation resulting from the 69% meat price and the 19% away-from-home food price change. Their data indicate that the compensating variation resulting from these price changes is $14.07 per household per week in 1992 dollars, which is

$21.59/household/week in 2008 dollars. This amounts to approximately $1122 per household per year.

16. Although it may not be initially obvious, in a simple one-good supply–demand setting, the answer to this question amounts to asking who is more sensitive to changes in price. If consumers' purchase decisions are less influenced by price changes than producers' production decisions, then consumers will derive a larger share of the benefits.

17. We use the three-sector model outlined in Lusk and Anderson (2004) with the only alteration being the use of long-run supply elasticities of 2, 2, and 10 for beef, pork, and chicken, respectively. These long-run supply elasticities are similar to those used by Brester et al. (2004). Moreover, some have argued that in the long run, these industries exhibit constant returns to scale, which would imply a perfectly elastic long-run supply curve. We use the Lusk and Anderson (2004) model to identify the farm supply shocks necessary to produce the 43.1%, 45.6%, and 101.7% decline in retail beef, pork, and poultry prices calculated as described in note 1 and then calculate the producer surplus changes resulting from these shocks.

18. There is evidence to suggest that the adoption of enhanced food safety practices is much less expensive for larger than smaller firms (Unnevehr, 1996). As a result, there are greater incentives for larger firms to adopt food safety technologies than smaller firms. In fact, research shows that larger cattle slaughter plants score 44% higher on an index of food safety ratings than smaller plants (Economic Research Service, 2004).

19. Meat and livestock marketing has dramatically changed in recent years and contracting, quality-based "grid" pricing, and vertical integration are now widely used in beef, pork, and poultry production. From 2002 to 2005, the use of such "alternative" marketing arrangements accounted for 38% of the fed beef cattle volume, 89% of the finish hog volume and 44% of the fed lamb volume sold to packers (GIPSA in Research Triangle Institute, 2007). Virtually all poultry and egg production occurs using production contracts or in a vertically integrated supply chain. Larger livestock producers and meat packers are much more likely to use these alternative marketing arrangements (perhaps due to economies of scale), and such alternative marketing arrangements have resulted in higher quality meat products. For example, as just noted, the vast majority of hogs produced are sold under some kind of "alternative" marketing arrangement, and those that are not are primarily sold by small producers. Furthermore, research shows that hogs sold under these quality-grid pricing mechanisms are "consistently associated with higher quality hogs than negotiated (spot [cash] market) purchases" (GIPSA in Research Triangle Institute, 2007, p. ES-11).

20. Norwood and Lusk, 2008.

21. Steinbeck, John, 1962, p. 83.

22. Schardt, 2003; Hoffmann et al., 2007. It should be noted, however, that contamination of vegetables typically occurs because the vegetables come into contact with livestock manure or meat.

23. Interestingly, one egg farmer we visited who produced both cage and cage-free eggs told us his employees will not eat the eggs from cage-free systems.

24. MacDonald and Korb, 2008.

25. Hoppe et al., 2007.
26. The Wendell Murphy story is derived from Warrick and Stith (1995) and conversations with N.C. hog producers.
27. Davis et al., 1928.
28. Ibrahim, 2007.
29. Unti, 2004; Beers, 2006; Phelps, 2007.
30. Unti, 2004, p. 54.
31. Arnold, 2009.
32. Unti, 2004.
33. Economic Research Service, 2006.
34. <http://www.peta.org/factsheet/files/FactsheetDisplay.asp?ID=107>.
35. Specter, 2003; Bailey, 2008.
37. Driver, No Date; Phelps, 2007; Montgomery, 2007; Norwood, 2009.
38. See, for example, the list at <http://www.hsus.org/farm/camp/victories.html>.
39. From various statistical highlights at the National Agricultural Statistics Service (NASS) website: <http://www.nass.usda.gov/>.
40. *Sacremento Bee*, 2008.
41. *The Beef Site*, 2010; Elliott, 2010.
42. For a more detailed list of companies and Universities see: <http://www.hsus.org/farm/camp/victories.html>.
43. Norwood and Lusk, 2009.
44. Smith, 2009*b*.
45. *Animal People News*, 1993.
46. Kilian, 2008.
47. *Feedstuffs Foodlink*, 2008; *Progressive Newswire*, 2008.
48. HFA, No Date (two refs); Miller, 2009.
49. For more information on these European laws, see Matheny and Leahy, 2007.
50. Hills, 2005.
51. Osbourne, 2009; Rutten, 2009.
52. Blankstein, 2008.
53. O'Shaughnessy, 1997.
54. Phelps, 2007.
55. PETA. No Date. "Animal Rights Uncompromised: PETA on 'Pets.' "
56. In all fairness, it should be noted that the routine euthanasia performed by PETA is not driven by their opposition to pet ownership, but due to their perception that the animal would suffer so much misery in its life that it is better off dead.
57. We recognize that the use of the term *extinction* is, however, an exaggeration because some people would no doubt raise livestock as pets.

Chapter 4

1. Koch, 2008/2009.
2. Gazzaniga, 2008.
3. Gladwell, 2005.

4. Pinker, 1997, p. 143.
5. Pinker, 1997.
6. Ibid.
7. Dawkins, 1993.
8. Smith, 1817, part 1.
9. Rising, 2009.
10. Dawkins, 1993.
11. Spencer, 2000, p. 2.
12. Kagel, Battalio, and Green, 1995.
13. Koch, 2008/2009.
14. Wiepkema, 1997.
15. Broom and Johnson, 1993.
16. Hatkoff, 2009.
17. Zimmermann et al., 2009.
18. Numerous sources indicate that the founding prophet of Scientology, L. Ron Hubbard, believed it possible that plants could feel pain.
19. Silverman 1978.
20. Dawkins, 1993; Hills, 2005.
21. Hatkoff, 2009.
22. Appleby, Mench, and Hughes, 2004.
23. Enrichment refers to the provision of amenities that allow animals to perform natural behaviors other than eating and producing meat or eggs. Examples are sand for dust-bathing, areas birds can scratch in the dirt and look for food, and perches.
24. Appleby, Mench, and Hughes, 2004.
25. Rogers and Kaplan, 2004.
26. Rollin, 1995.
27. HSUS, 2009; Hatkoff, 2009.
28. Spears, 2007; Agrillo et al., 2009.
29. Dawkins, 1993, p. 161; Appleby, Mench, and Hughes, 2004; Hatkoff, 2009.
30. Bertenshaw and Rowlinson, 2009.
31. Hatkoff, 2009.
32. *The Economist*, April 4, 2009a.
33. Hatkoff, 2009.
34. Watanabe, 2009.
35. Dawkins, 1993, p. 161.
36. Seligman, 1995.
37. Broom and Johnson, 1993.
38. Hatkoff, 2009.
39. Ibid., 2009. HSUS, No Date, "About Chickens."
40. Hatkoff, 2009; HSUS, No Date, "About Pigs."
41. Hatkoff, 2009; HSUS, No Date, "About Cattle."
42. Blackmore, 2004.
43. Dawkins, 1999.
44. Dawkins, 1993, p. 177.
45. Pinker, 1997.

Chapter 5

1. Mench, 1998.
2. Brambell Report, 1965, p. 13.
3. Farm Animal Welfare Council Press Statement, 1979.
4. Arnold, 2009.
5. Lyubomirsky, 2008.
6. Curtis, 2007, p. 574.
7. These numbers are obtained from a number of published papers and conversations with egg producers.
8. Gregory, 2009.
9. Bell et al., 2004.
10. United Egg Producers Certified. 2008, p. 11.
11. Singer and Mason (2006) document this precise scenario on a hog farm.
12. Houpt, 2005.
13. Meager, 2009.
14. Lynch et al., 2000.
15. McFarlane, Boe, and Curtis, 1988.
16. Appleby, Mench, and Hughes, 2004.
17. Boyle, et al., 2002.
18. Norwood and Lusk, 2009.
19. LayWel, 2004.
20. Rhodes et al., 2005.
21. Surowieck, 2004.
22. Bracke et al., 2002a.
23. Bracke et al., 2002b.
24. Hall and Sandilands, 2007.
25. Ibid.
26. De Mol et al., 2006.
27. Bracke et al., 2002a.
28. Bracke et al., 2002b.
29. Baker, 1913; Davis et al., 1928; Deyoe, Ross, and Peters, 1954.
30. Wright, 1936.
31. These numbers taken from Bell (2005) and conversations with numerous egg producers.
32. An example is the Rhode Island Red bird, which is brown in color. Some birds that lay brownish eggs are not brown in color; for example, the Plymouth Rock hen is black and white but lays brown eggs. However, "brown bird" is a popular term for any bird that lays brown-tinted eggs, and thus the term is used in this book.
33. *Nature* 461: 267–71.
34. Appleby, Mench, and Hughes, 2004; Laywel, 2004.
35. Appleby, Mench, and Hughes, 2004; HSUS, 2006a.
36. It is a bit unclear exactly why consumers are willing to pay a premium for brown eggs. For example, the nutritionist Marion Nestle (2006) argued, "From a nutritional standpoint, eggs are eggs. Turning eggs into a "designer" food is a great way

to get you to pay more for them but there are less expensive and easier ways to get vitamin E, selenium, lutein, and omega-3s from foods. If you do not give a hoot about how the eggs are produced, buy the cheapest ones you can find. The shell color makes no nutritional difference" (Nestle, 2006).

37. Chang, Lusk, and Norwood, 2009*b*.
38. Bell, 2005; personal conversations with US egg producers who utilize cage and cage-free systems.
39. Rollin, 1995.
40. European Food Safety Authority, 2007. <http://www.efsa.europa.eu/en/efsajournal/pup/97rhtm>.
41. Messens, et al., 2007.
42. Laywel, 2004. On page 30 the report states, "With the exception of conventional cages, we conclude that all systems have the potential to provide satisfactory welfare for laying hens."
43. Lundeen, 2009.
44. Aerni et al., 2005.
45. Lymberry, 2002.
46. EEC, 1991.
47. Watson, 2008; Smith, 2007.
48. United Egg Producers Certified, 2008; Smith, 2008*a*, p. 3.
49. LayWel, 2004, p. 30.
50. De Mol et al., 2006.
51. Rodenburg et al., 2008*a* and *b*.
52. Baker, 1913; Davis et al., 1928.
53. Baker, 1913, p. 1168.
54. YouTube.com has several excellent videos describing the 1948 contest, e.g., <http://www.youtube.com/watch?v=SG7TIQk1UiM&feature=related>.
55. Parkhurst and Mountney, 1988.
56. Savory and Maros, 1993; HSUS, 2008*a*; Parkhurst and Mountney, 1988.
57. Katanbaf, Dunnington, and Siegel, 1988.
58. Parkhurst and Mountney, 1988.
59. United Egg Producers Certified, 2008.
60. HSUS, 2008*a*.
61. Dawkins, Connelly, and Jones. 2004. "Chicken welfare is influenced more by housing conditions than by stocking density." *Nature* 427: 342–4.
62. National Chicken Council, 2005; Delaware Poultry Industry, Inc., No Date; Dawkins et al., 2004.
63. Hall and Sandilands, 2007.
64. Kestin et al., 1992.
65. McGeown et al., 1999.
66. Danbury et al., 2000.
67. Knowles et al., 2008.
68. Kestin et al., 1992; Sanotra et al., 2001.
69. Davis et al., 1928, p. 362.
70. Parkhurst and Mountney, 1988.
71. National Chicken Council, 2005; HSUS, 2008*a*.
72. Turner, Garces, and Smith, 2005.

73. Ray, 1985.
74. *The Times*, 1909.
75. Tannahill, 1988; Clutton-Brock, 1989; Kritzman, 1996.
76. *The Economist*, May 9, 2009*b*.
77. Quoted in Cronon, 1991.
78. Davis et al., 1928, p. 225.
79. Baker, 1913; Davis et al., 1928; Deyoe, Ross, and Peters, 1954.
80. Hueth, Ibarburu, and Kliebenstein, 2007.
81. Baker, 1913, p. 974.
82. Deyoe, Ross, and Peters, 1954, p. 176.
83. Brambell Report, 1965.
84. Baker, 1913; Deyoe, Ross, and Peters, 1954.
85. Young, 2009.
86. Dhuyvetter et al., 2007.
87. Taken from authors' measurements.
88. Kilian, 2008.
89. Hotzel et al., 2009.
90. Boggess, 2007, and personal communication with farmers.
91. McGlone et al., 2004; Task Force Report, 2005; Curtis et al., 2009.
92. Bracke et al., 2002*a* and *b*.
93. Taken from authors' measurements.
94. Lay, 2002; Stassen, 2003; Kliebenstein, Stender, Huber and Mabry, 2006.
95. Economic Research Service, 2005.
96. Taken from authors' measurements.
97. Prosch, Christensen, and Johnson, 2001.
98. Antibiotics are supposed to be given to sick animals in organic systems too, but an animal receiving antibiotics must be sold in the non-organic market and at a steep discount. For this reason, organic farmers will be reluctant to prescribe antibiotics to their animals.
99. This quote is displayed prominently on the institution's website.
100. See Seibert and Norwood, forthcoming.
101. Rollin, 1995, introduction to chapter 6.
102. Data from Livestock Marketing Information Center.
103. The Red Meat Yearbook published by the Economic Research Service indicates that, between 1970 and 2004, an average of 35 million head of cattle and 2.592 million calves were slaughtered each year. A veal calve will weigh about 400 lbs (*Weekly Veal Market Summary*, 2009) and an adult cow will weigh about 1,250 lbs when slaughtered. Thus, the cattle industry produces about 1250 × 35,000,000 = 43,720,000,000 and 400 × 2,592,000 = 1,036,000,000 lbs of live cattle/calves each year. Assuming one pound of live-weight produces 0.417 lbs of retail beef (ERS, 2009) for both cattle types, this translates into 18,243,750,000 and 432,345,600 lbs of retail meat from beef and veal, respectively. This implies that veal meat is 432,345,600/18,243,750,000 = 2.37% of all beef produced from cattle and calves. However, another source (InfoPlease, 2010) indicates that per capita consumption of beef and veal is 65.2 and 0.5 lbs per year, respectively, making veal consumption 0.5 / 65.2 = 0.0077 or 0.77% of beef production.

104. In a typical dairy herd of 100 lactating cows, there will be 100 cows being milked and 18 cows experiencing a dry period. Each of these cows will give birth to 1 calf each year, yielding 118 calves. Not all calves are born alive and some die soon after birth, so assume only 112 calves live to potentially become veal calves. Half of the calves will be males and will be females. The large majority of female calves are retained for the milking herd, and about 35% of the male calves will be harvested for veal. Thus, a herd of 100 milk cows will produce approximately 20 veal calves. Alternatively, the number of veal calves per 100 milk cows can be estimated using national agricultural statistics. Using these statistics, if you divide the number of calves slaughtered annually by the number of milk cows for 1970–2004, the average equals 0.24. This implies that for every 100 milk cows there will be 24 veal calves. However, some of these veal calves would have been born in the previous year. Assuming the birthing of the calves are uniform across the year, and given that they are slaughtered at 17 weeks of age, approximately (17 / 52 = 0.33) 33% of calves slaughtered in one year were born in a previous year. Multiplying 0.66 times 20 veal calves yields 13 calves born in any one year for a herd of 100 milk cows. Taking the two estimates of 13 and 20, we assume a dairy herd of 100 lactating cows will produce 16 veal calves each year (see Burdine, Maynard, and Meyer, 2004; *Weekly Veal Market Summary*, 2009; ERS, 2006).
105. American Veal Association, Not Dated, <http://www.americanveal.com/VEAL_WHITE_PAPER_R1_0706.pdf>.
106. HSUS, 2008*b*.
107. Smith, 2009*c*.
108. MacDonald et al., 2007.
109. APHIS, 2007.
110. Ibid.
111. Espejo et al., 2006; APHIS 2007; Leach et al., 2009.
112. APHIS, 2007.
113. Kanter, Messer, and Kaiser, 2009.
114. University of Missouri-Columbia, 2009.
115. APHIS, 2007.
116. Tucker and Weary, 2001–2; Cassandra et al., 2001; Eicher et al., 2006.
117. Rollin, 1995; Bertenshaw and Rowlinson, 2009.
118. Ursinus et al., 2009.
119. Beers, 2006.
120. Various statistics from the National Agricultural Statistics Service (NASS) website <http://www.nass.usda.gov/>. Accessed in November 13, 2009.
121. The statistic is based on the following thought experiment. Let us assume that the length and width of a cow, as a percentage of the animal body weight, is the same as a pig. A pig that is close to harvest will weigh 250 lbs and its body will occupy 5 square feet of space. This, the floor area occupied by the pig's body is 5/250, when expressed as a percentage of its weight. Cattle are approximately 1200 lbs when ready for harvest, and are thus projected to occupy (5 / 250) (1200) = 24 square feet of ground space. When comparing the 250 square feet per animal space allotment to the 24 square foot ground area occupied by cattle, it is evident that cattle in feedlots have much, much more room that pigs in factory farms. Thus, pigs have about 5 / 250 = 0.02 square feet of space for each pound, whereas cattle have about 250 / 1200 = 0.208 square feet of space for each pound, meaning cattle have about 0.208 / 0.02 = 10.4 more space than hogs per pound.

122. Davis, Stanton, and Haren, 2002.
123. The argument was made by Schlosser on the television show, *The Colbert Report* which aired June 3, 2009. The argument was also a central premise in the movie *Food, Inc.*
124. We have toured numerous meat harvesting facilities and find the technology used to minimize risk of food poisoning is quite amazing.
125. Hancock and Besser, 2006.
126. Cheng et al., 1998.
127. Federal Register, 2007.
128. Finishing cattle on a grain diet is often thought to be a relatively new practice. However, there are many references from 100 or more years ago demonstrating the prevalence of the practice (Williams, 2010).
129. *Federal Register*, 2007.
130. Umberger et al., 2002.
131. Rossi, 2006; Loos, 2009.
132. An interesting and plausible explanation for the earlier maturation of girls was given in a speech by Trent Loos in 2008. Cows can be brought into heat earlier by placing them in close proximity with a mature bull. Young girls in the modern world, perhaps, mature earlier from their incessant exposure to more mature males on television.
133. Wierup, 2001; Griener, 2009.
134. Hayes and Jensen, 2003.
135. Dangour et al., 2009.
136. There is no study that definitely proves or disproves this claim. Both conventional and organic fertilizer essentially applies the same basic elements to the soil: nitrogen, phosphorus, and potassium. Obtaining higher yields requires more of these elements in both systems, and the extent to which either system results in greater nutrient runoff depends on the manner in which it is applied. We can envision scenarios where either system can produce more nutrient runoff. Soil erosion may be more prevalent in organic farming, as it must rely more on tillage than pesticides to reduce runoff; however, the use of manure as fertilizer acts to "create" soil to a greater extent than chemical fertilizers. Pesticide runoff only occurs with conventional fertilizers, but this may be negated by reduced soil erosion or reduced nutrient runoff. The extent to which each system produces greenhouse gases is absolutely indeterminate, and varies across crops and settings. Sustainability is a topic so ill-defined, and (perhaps counter-intuitively) is not determined by which production system can be repeated for longer periods of time. The ability of the world to consistently provide adequate nutrition depends far more on the political environment and the freedom bestowed upon the private sector. Sustainability is not determined by whether an agricultural system uses non-renewable resources, but the degree to which the agricultural industry is given the incentives to respond to changes and to invent and employ new technologies. The only true facts regarding conventional and organic food are that their nutritional contents are similar and that organic food costs more. Whether organic food is more environmentally friendly or more sustainable is indeterminate.
137. Organic Foods Production Act of 1990.
138. <http://www.goveg.com/feat/paulmveg/>.

139. Morales, 2009.
140. Lusk and Norwood, 2009*b*.
141. Capper et al., 2009*a*; Muirhead, 2009; Peter et al., 2010.
142. Weber and Matthews, 2008.
143. Morrison, 2010. This reference also contains a number of scientific journal references backing this claim.
144. Capper, Cady, and Bauman, 2009*b*.
145. Capper, Cady, and Bauman, 2009*a*.
146. Allison, 2007.
147. Lichtenberg, 2004.
148. European Food Safety Authority, 2007; Messens et al., 2007.
149. Wasserman, 2006.
150. You can watch Mike Rowe tell this story at <http://www.youtube.com/watch?v=r-udsIV4Hmc>.

Chapter 6

1. This story is taken from Gigerenzer (2007), but can be found in many other writings as well.
2. Gigerenzer, 2007; Haidt, 2001.
3. Posner, 2004, p. 65, emphases added.
4. Ibid., p. 67.
5. Khaitovich et al., 2008.
6. This is not to say that humans could not have evolved to our current state without meat. It is possible, just much less likely.
7. Posner, 2004, p. 68.
8. Menand, 2001.
9. Gigerenzer, 2007.
10. People could also answer with responses of "neither agree nor disagree" or "don't know."
11. <http://www.reason.com/blog/show/123776.html>.
12. American Pet Products Manufacturers Association. 2005–6, National Pet Owners Survey.
13. *Animal People News*, 1993.
14. Singer, 1975.
15. Norwood, Lusk, and Prickett, 2007; Prickett, Norwood, and Lusk, 2010.
16. Bentham,1823.
17. During our intellectual experiments with utilitarianism, it often struck us that utilitarianism seemed to amount to little more than saying good is better than bad. People generally define what is "good" and what is "bad," and dictate the long-run and unintended consequences of the "good" and "bad," such that the philosophy supports what they generally like and not what they dislike. Put differently, utilitarianism is often used to support prior beliefs of the user and not to help the individual discover their beliefs. When this occurs, utilitarianism becomes a tautology, and it would be more transparent to simply say what one likes and dislikes, without the masquerade.

18. Matheny, 2006.
19. In reality if Frank eats one more chicken, total chicken consumption will increase by slightly less than one chicken. As Frank demands more chicken the price will slightly rise, inducing others to eat less. These complexities are ignored here to focus attention on the mechanics of utilitarianism.
20. *Business Wire*, 2005.
21. Robin, 2005.
22. PETA, No date.
23. Singer, 2004, p. 8.
24. Lusk and Norwood, 2009*b*.
25. These criticisms of utilitarianism are ones for which Singer and other philosophers no doubt have a ready answer. The criticisms are listed here not because they are robust to counter-argument, but to describe the arguments that people make against utilitarianism, regardless whether the response is valid or invalid. If there is one thing we have learned from reading the works of ethical philosophers; it is that no one ever, ever wins the debate.
26. Ware, 2006.
27. Regan, 2004, p. 203.
28. From Wise, 2004, p.27.
29. For a nice discussion on this issue, see Fehr and Gächter, 2000.
30. Dawrst, 2009. In case the reader wonders, neither of the authors is called Dawrst.
31. Stark, 2004.
32. Singer, 2004.
33. Posner, 2004, p. 64.
34. These results are also discussed in Norwood and Lusk, 2008.
35. In practice, we hired marketing research companies in each location to randomly recruit members of the general population of the respective areas to attend 90-minute sessions in groups of 25. When recruited, the respondents did not know the subject matter to be discussed was farm animal welfare. Thus, there is no reason to believe that one specific type of person chose to participate and another type did not. The results presented here are based only on the 263 people who provided complete answers to all the survey questions.

Chapter 7

1. Robbins, 1932, p. 16.
2. Ibid., p. 24.
3. Bastiat, 1848.
4. Copenhagen Consensus Center, 2008.
5. Sumner et al., 2008.
6. Ibid.
7. Norwood, Lusk, and Prickett, 2007; Prickett, Norwood, and Lusk, 2010.
8. See the data and discussion in Chapter 9.
9. Read, 1958. Also, a PowerPoint presentation of the *I, Pencil* essay can be downloaded at <http://cafehayek.com/2008/08/i-i-pencil.html>.

10. Smith, 2009*d*.

11. Singer, 2004, p. 88–9.

12. Regan, 2004, p. 335.

13. Data from Raper et al., 1996. The figure assumes the allocation of expenditures on food categories is identical at home and away from home.

14. Inferred from data and model in Raper et al., 1996.

15. Norwood and Lusk, 2008.

16. This is not the place to argue the merits of capitalism versus socialism, but this sort of argument must be recognized for what it is and proponents of such an argument should be willing to describe exactly the kind of world they envision that has equal incomes.

17. Brown, Cranfield, and Henson, 2005.

18. Simonson and Drolet, 2004; Schwartz, 2007; Anthes, 2009.

19. Morwitz and Fitzsimons, 2004.

20. Carter-Long, 2002; Hsee and Rottenstreich, 2004; Bailey, 2009.

21. Ovaskainen and Kniivila, 2005.

22. List and Gallet, 2001; Little and Berrens, 2004.

23. Stark, 1996, p. 161.

24. Lusk and Shogren, 2007.

25. List, 2002.

26. See for example: List, 2003; Plott and Zeiler, 2005.

27. The median of a series of numbers is determined by first ordering the numbers from the smallest to largest values, and then choosing the middle number. If there is an odd number of numbers the median will be a unique value, whereas if the number of numbers is even, one should take the average of the two middle numbers. For example, the median of [1 2 3 4 5] is 3, while the median of [1 2 3 4 5 6] is 3.5.

28. We provide a more formal and detailed treatment of this issue in one of our working papers: "Speciesism, Altruism, and the Economics of Farm Animal Welfare."

29. We do not view these arguments as controversial. See, for example: Appleby and Hughes, 1997.

30. Kagel, Battalio, and Green, 1995.

31. Matthews and Ladewig, 1994.

32. Animal scientists typically measure the relative value of one amenity over another, such as the value of food versus social contact, by the ratio of the demand elasticities. For some utility functions, these ratios are meaningful in that they relate to the ratio of marginal utilities. For example, the ratio of demand elasticities in a Cobb-Douglass utility function is proportional to the ratio of marginal utilities (only economists would know the Cobb-Douglass function). Ratios of marginal utilities are important because they define marginal rates of substitution between goods, and thus yield willingness-to-pay values.

Chapter 8

1. Economic Research Service, 2009.

2. Foreign Agricultural Service, 2006.

3. HSUS, 2006*b*.
4. Davis, 2003; Matheny, 2003.
5. One exception to this statement is the writing of Dawrst (2009), *The Importance of Wild-Animal Suffering*, which concludes that wild animals do suffer, and greatly. We encourage the readers to peruse this writing. It is one of the most interesting and well-researched narratives that is not officially published by any organization.
6. Corn yields approximately 163 bushels per acre, and each bushel weighs 56 lbs, so one acre of corn yields 9128 lbs of corn. Considering that each lb of corn provides 81 calories, each acre of corn provides 163 * 56 * 81 = 739,368 calories. If each person consumes 2000 calories, one acre of corn can feed 370 people for one day.
7. Laying hens consume about 15 lbs of feed in total before they start laying eggs at 17 weeks of age. When they lay eggs they will consume about 4 lbs of feed for each dozen eggs, and will lay 509 eggs throughout their life. Feed consists mostly of corn, so this implies it requires (15 + (4) (115−7) (7)) /509 = 5.42 lbs of corn for each egg produced. If an acre of corn is devoted to egg production, it will provide 9128 /5.42 = 1684 eggs. A scrambled egg contains 167 calories, so an acre of corn used for egg production will yield (1684) (167) = 281,250 calories. This implies that to feed 370 people for one day using only corn requires 739,368 /281,250 = 2.63 acres.
8. One pound of corn produces 0.35 lbs of broiler meat, and one pound of retail broiler meat provides 47 calories. Thus, one pound of corn utilized for broiler production yields 0.35 * 47 = 16.45 calories. One acre of corn, which produces 9128 lbs of corn, can thus produce (9128) (0.35) = 3195 lbs of broiler meat, or (3195 * 47 =) 150,156 calories. This implies that to feed 370 people for one day using only corn requires 739,368 /150,156 = 4.92 acres.
9. Some animals live longer than others and thus produce more output in their lifetime. The longer the animal life, the smaller the number of animals affected from the production of a pound of meat/milk/egg. Almost everyone would agree, however, that living for a longer period of time in a suffering state is bad. For example, breeder-hogs (sows) and laying hens lead similar lives (both are housed in very small cages), but a sow will be harvested at three years of age while the hen is harvested a few months earlier. If one believes sows suffer more because they exist in such conditions for a longer period of time, then this should be taken into consideration when one establishes the welfare score they give for each animal (e.g., breeder hogs should be given a lower score than laying hens).
10. The number 0.71 in the equation for crate-free pork replaced the number 0.74 for conventional pork because we assume the demand elasticities for the two products differ (see Table 8.A.1 in the Appendix).
11. The authors are aware that humans are actually omnivores.
12. Cowen, 1996, p. 754.
13. We have also prepared a more user-friendly version of the worksheet, which is available online at: <http://asp.okstate.edu/baileynorwood/Survey4/Default.aspx? name=eatingguide>.
14. See Prickett, Norwood, and Lusk (2010) for more detail on these results as well as information about how the results were derived.
15. WTP is expressed in the units of dollars per pound. The numerator multiplies the EEAT score by the utility per EEAT score unit, making the units of the numerator

utility. The denominator is the disutility of \$1, which is in the units' utility per dollar. Thus, dividing the numerator by the denominator cancels out the units of utility and leaves us with the units of dollars. Moreover, the units are dollars per pound, as the action was to purchase one pound of shelter-pasture pork instead of regular pork.

16. See <http://www.peacefulprairie.org/freerange1.html>.
17. Beers, 2006; Prickett, Norwood, and Lusk, 2010.
18. Bell, 2005; personal telephone conversations with egg producers.
19. Bell, 2005; personal phone conversations with egg producers.
20. McGlone et al., 2004; Task Force Report, 2005.
21. Kliebenstein et al., 2004; Dhuyvetter et al., 2007.
22. EPA, 2007.
23. Taylor, 1992; Wilson, Greaser, and Harper, 1994; McKenna, 2001; Food Safety and Inspection Service, 2006; HSUS, 2008*b*.
24. The pounds of retail meat derived from a pound of live weight is assumed to be the same for veal calves and cattle.

Chapter 9

1. List and Gallet, 2001; Little and Berrens, 2004.
2. Chang, Lusk, and Norwood, 2009*a*; Lusk and Norwood, 2009*c*.
3. The scanner data was purchased from Information Resources Inc., and was obtained from their InfoScan data base comprised of approximately 15,400 Grocery, 15,800 Drug, and 3000 Mass Merchandiser stores across the United States.
4. *Almost* all eggs. The market share is calculated as the eggs of a particular product sold divided by the total egg sales of regular eggs plus cage-free eggs plus organic eggs. A few unique varieties were excluded, such as "natural" eggs and eggs with additional Omega-3 fatty acids.
5. Sumner et al., 2008.
6. Businesses often define the profit margin as the difference between price and cost, divided by the price.
7. There are certain fixed costs to managing each variety of food within a grocery store. When these fixed costs are spread over fewer units of a good, the per-unit costs are higher.
8. Lusk, 2009.
9. You can see the exact presentations we gave at the book website: <http://asp.okstate.edu/baileynorwood/Survey4/Default.aspx>.
10. <http://asp.okstate.edu/baileynorwood/Survey4/Default.aspx>.
11. You do not have to rely on our descriptions of the tool. You can see it and answer the questions for yourself for eggs at <http://asp.okstate.edu/baileynorwood/Survey1/Intro.aspx> or for pork at <http://asp.okstate.edu/baileynorwood/Survey2/Intro.aspx>.
12. The feed attribute was included to measure consumer preferences for organic food. Including the attribute allowed the consumer to express their preferences for feed-related attributes, which helps ensure they accurately represent their true

preferences for animal care. Some consumers may have a dislike for conventional foods, and might wish to express this dislike during the research session. These consumers may dislike conventional foods because of hormone use in pork production, and care nothing for farm animal welfare. If we do not give these consumers the option of choosing foods from animals raised without artificial hormone use, they may express their dislike for conventional foods by falsely claiming to desire higher animal welfare standards. By giving them an outlet to express their displeasure towards hormone use, we ensure they express more truthful preferences for farm animal welfare.

13. Consider a simplistic mathematical demonstration. Suppose there were two attributes being considered for eggs: barn space per hen and the flock size. In Step 2.A a person indicates that the importance of barn space per hen is 8, while flock size receives an importance rating of 4. When transitioning from Step 2.A to Step 2.B, the initial importance score given for barn space per hen is calculated as $8/(8 + 4) = 0.66$, and the initial importance score for flock size is $4/(8 + 4) = 0.33$. With this method, attributes rated higher receive a higher score, and the weights for all attributes sum to 100. The user may then adjust these importance scores however they like.

14. Participants were told that if they bought one of the items we would give them a small Styrofoam cooler and ice to keep it cool if they were not immediately headed home. The coolers and ice were worth very little, and thus would not interfere significantly with the level of bids. Even if it did affect the level of bids, it is only the differences between the bids that are studied.

15. It is important to note that the data reported in Table 9.2 are not based on the raw bids for the systems described in Table 9.1 obtained directly from the auction, but are rather bids for four systems very similar to those studied in our research sessions. Recall that because we have a complete mapping between people's preferences for attributes of production systems and economic values, we can calculate (or project) the values for any potential system described by the underlying attributes. Here, we show projected values for systems we feel are most representative of the options currently available in the US.

16. Sumner et al., 2008.

17. Ibid.

18. Reach and Discover, 2004. The remark was made using Canadian dollars, and he actually stated 5 cents per dozen. In 2004 the exchange rate was about US$1.35 for CA$1. This implies that CA$1 is equivalent to US$1.35, or that 5 Canadian cents are equivalent to $5 * 1.35 = 6.75$ US cents.

19. Tonsor, Olynk, and Wolf, 2009.

20. Plain, 2006; Aho, 2008.

21. Seibert, Lacey, and Norwood 2011.

22. Ibid. We use Farm A from Seibert and Norwood (forthcoming), because more information is available for Farm A and it represents a more aggressive system for improving animal welfare.

23. Seibert and Norwood, forthcoming.

24. There were a few people whose answers in the value expression tool indicated that they believed hens to have higher well-being in the cage system as compared to the

free-range system. All participants bid to move animals from their least to their most preferred system. For these people with "unusual preferences," they bid to move hens from a free-range to a cage system. Thus, when we analyze the data on public willingness-to-pay to go from cage to free-range, these particular people's bids are entered as negative values.

25. As with the hen auctions, there were a few people who preferred the crate system to the shelter-pasture system. These people were asked to bid to move hogs from the shelter-pasture system to the hen system, and their bids were entered as negative numbers.

26. The bids for individuals who preferred that the animals live on a cage-egg or confinement-crate farm are multiplied by negative ones when statistics such as the average bid are calculated. If someone is willing to pay $5 to ensure that hens are raised on a cage farm, then the value of those hens living on the cage farm is $5, and the value of the hens living on a free-range farm is −$5. The idea is that a person will accept hens being raised on a cage farm so long as they are compensated $5, and thus the "value" enters −$5.

27. Andrioni, 1990.

Chapter 10

1. Farm Sanctuary Website. 2009.
2. *Reason Foundation*, 2009.
3. Roberts, 2009.
4. *A Beautiful Mind*, 2002.
5. This list can be viewed at <http://www.hsu3.org/farm/camp/victories.html>.
6. This assumption is actually quite close to being true. There are about 280 million laying hens in the US and a similar number of adults over the age of 18. The average American will consume about 500 eggs every two years, and the average hen lays eggs for two years, providing 500 eggs during this period.
7. In our experiments, informed consumers expressed private values for cage-free eggs that were slightly above cage-free egg prices. The free-range egg prices were much higher than what we believe the costs are. Consequently, by assuming that consumers purchase the same number of eggs in a free-range and cage system, we are ignoring a net benefit from private good consumption, and the experiment here thus underestimates the value of a cage-egg ban.
8. Sumner et al., 2008.
9. In reality, 8% is probably a lower-bound estimate because it corresponds only to the improvement in 1000 hens' lives, not *all* hens' lives. Recall that we also asked hypothetical question regarding how much money the participants would pay to move all hens to a cage-free system. Even these bids suggest that only about 18% of public values are higher than the per-person cost. Because the question is hypothetical, we interpret the figure as an upper-bound estimate. All together then, our estimates imply that if every person is asked to pay the per-person same cost of improving the welfare, the percentage of people who have public good values exceeding the costs is between 8% and 18%.

10. This statement is based on the model of inequality aversion proposed by Fehr and Schmidt (1999) and on the average values for the distribution of advantageous and disadvantageous inequality aversion parameters reported in their paper.
11. Sows are less productive in a shelter-pasture system, and there will need to be more than 1001 sows to provide each American with 50 lbs of meat each year. These additional sows are not counted in the benefits. It is assumed that the additional sows experience high welfare in the shelter-pasture system, and including them would only increase the value of the policy. The cost numbers are from Seibert and Norwood (2009).
12. Using the hypothetical bids for placing all hogs under the shelter-pasture system, only 25% submit bids greater than $11.
13. Proprietary data obtained from Information Resources, Inc. See Chapter 9.
14. Kanter, Messer, and Kaiser, 2009.
15. The increase in production costs as a percent of the retail price is smaller for pork than eggs.
16. The number of meat producers may be larger than an initial impression would suggest. The term "meat producers" includes not just farmers, but the companies who harvest live-animals and process them into meat, the wholesalers and distributing company that coordinate the movement of processed meat to grocery stores, and of course, the grocery stores themselves. Because it is impossible to count the number of people involved in the production of meat, or the extent to which each person is affected by the policy, we refrain from attempting to calculate a per-person costs for meat producers.
17. <ftp://ftp.bls.gov/pub/special.requests/lf/aat39.txt>. The average weekly income is $722, which when divided by an assumed 40 hours worked per week, equals $18.
18. Caplan, 2001.
19. Positive externalities, where a third party benefits from the transaction, can also exist. When the US took actions to restore the depleted ozone layer in the 1980s, all other countries benefitted from the actions of Americans. This third-party benefit is a positive externality. The astute reader may also note that this positive externality is similar to free-riding, discussed earlier.
20. While the taxes would likely be administered on a per-pound basis, the tax should be calibrated to reflect the number of brains associated with each pound sold and the intensity by which each animal suffers. It is the brain that suffers, not the pound, and the intensity of a single brain suffering indicates the degree to which a negative externality is produced.
21. Coase, 1960.
22. Videra, 2006.
23. For a fuller discussion of these survey results, see <http://asp.okstate.edu/baileynorwood/AW2/InitialReporttoAFB.pdf>.
24. One-half the sample received the question with the added phrase "for their protection from other hogs," whereas the other half of the sample did not. Respondents received one question or the other but not both.
25. Gaskell et al., 2005.
26. Lusk and Norwood, 2008.
27. Remember, "ethical considerations" here also include peoples' perceptions about what science would say, if science were perfect.

28. Detailed in Lusk and Norwood, 2008.
29. Infoplease, 2010.
30. US Census Bureau, 2009*a*.
31. Gregory, 2009; Lusk and Norwood, 2009.
32. LMIC, 2008.
33. Infoplease, 2010.
34. US Census Bureau, 2009.

Chapter 11

1. Periodically, a new technology may develop that simultaneously improves welfare and lowers costs. When this happens, firms will improve welfare voluntarily and will boast about their commitment to animal welfare. The impetus for adopting the technology is higher profits, not animal welfare. When we discredit claims asserting that enhanced welfare and lower costs go hand-in-hand, we assume that the change leading to better animal care was motivated by animal welfare, not by a new technology that lowers costs and just happens to benefit animals.
2. Free-range in this context refers to a cage-free facility with an outdoor area that contains both shelter and predator protection. Some consumers envision the word "free-range" describing a system where chickens are simply let loose into a pasture. That is not the free-range definition used here.

REFERENCES

A Beautiful Mind. 2002. DreamWorks Entertainment.

Aerni, V. M., W. G. Brinkhof, B. Wechsler, H. Oester, and E. Frohlich. 2005 "Productivity and Mortality of Laying Hens in Aviaries: A Systematic Review." *World's Poultry Science Association*. 61: 130–42.

Agriculture Research Service (ARS). USDA National Nutrient Database. United States Department of Agriculture, Washington, DC.

Agrillo, Christian, Marco Dadda, Giovanna Serena, and Angelo Bisazza. 2009. "Use of Numbers by Fish." *PLoS ONE*. 4 (3, March) : e4786.

Aho, Paul. 2008. "Substitution Effect to the Rescue—Eventually." *Arbor Acres*. 16 (2, April).

Ahn, Byeong-Il and Daniel A. Sumner. 2006. *Assessment of the Political Market Power of Milk Producers in U.S. Milk Pricing Regulations*. California: Agricultural Marketing Resource Center. Agricultural Issues Center, University of California.

Allison, R. 2007. "Organic Chicken Production Criticized for Leaving a Larger Carbon Footprint." *Poultry World*. 161(3): 8.

American Veal Association. Not Dated. *The Science of Veal Calf Welfare and Nutrition*. Washington, DC: American Veal Association.

Anderson, Virginia De John. 2004. *Creatures of Empire*. New York: Oxford University Press.

Andrioni, J. 1990. "Impure Altruism and Donations to Public Goods: A Theory of Warm-Glow Giving." *Economic Journal*. 100: 464–77.

Animal Rights History. 2009. Online library of animal rights literature and legislation. Available at <www.animalrightshistory.org>. Accessed February 25, 2009.

Animal People News. May 1993. <www.ActivistCash.com>.

American Pet Products Manufacturers Association. 2005–6. No longer available on their website.

Anti-Defamation League. March 20, 2009. "Animal Rights Extremists Target the University of California." <http://www.adl.org/main_Extremism/university_of_california_animal_rights_extremism.htm>. Accessed July 14, 2009.

Anthes, Emily. 2009. "Building Around the Mind." *Scientific American Mind* April/May: 52–9.

APHIS (Animal and Plant Health Inspection Service). 2007. Animal Health Monitoring and Surveillance. *Dairy 2007. Parts I–IV*. Washington, DC: United States Department of Agriculture.

Appleby, M. C. and B. O. Hughes. 1997. *Animal Welfare*. Cambridge, UK: CABI Publishing.

—— Joy A. Mench, and Barry O. Hughes. 2004. *Poultry Behavior and Welfare*. Cambridge, MA: CABI Publishing.

Arnold, Kyle. 2009. "Farm Groups Support Bill." *Tulsa World*. February 4: E2.

Bailey, Ronald. 2008. "PETA Asks Ben & Jerry's to Use Human Milk to Make Ice Cream." Reason Online September 24.

——2009. "The Death of One Man is a Tragedy, the Death of Millions is a Statistic." Reasononline. January 7. <http://reason.com/blog/2009/01/07/the-death-of-one-man-is-a-trag>.

Baker, A. H. 1913. *Livestock: A Cyclopedia*. Kansas City: Thompson Publishing, p. 1168.

Barberg, A. E., M. I. Endres, J. A. Salfer, and J. K. Reneau. 2007. "Performance and Welfare of Dairy Cows in an Alternative Housing System in Minnesota." *Journal of Dairy Science* 90: 1575–83.

Bastiat, Frédéric. 1848. "What is Seen. What is Not Seen and" *Selected Essays on Political Economy*. New York: Irvington-on-Hudson, ch 1. <http://www.econlib.org>.

Bateman, I. J., D. Burgess, W. G. Hutchinson, and D. I. Matthews. 2008. "Learning Design Contingent Valuation (LDCV): NOAA Guidelines, Preference Learning and Coherent Arbitrariness." *Journal of Environmental Economics and Management* 55: 127–41.

Beers, Diane L. 2006. *For the Prevention of Cruelty*. Athens, OH: Swallow Press/Ohio University Press.

The Beef Site. 2010. "Newspaper Investigates Welfare of Imported Meat." <http://www.thebeefsite.com/news/29945/newspaper-investigates-welfare-of-imported-meat>. Accessed February 17, 2009.

Bell, Don. 2005. "A Review of Recent Publications on Animal Welfare Issues for Table Egg Laying Hens." Paper presented at the United Egg Producers Annual Meeting, October.

——Bill Chase, Adele Douglass, Patricia Hester, Joy Mench, Rugh Newberry, Margaret Shea-Moore, Larry Stanker, Janice Swanson, and Jeff Armstrong. 2004. "UEP Uses Scientific Approach to Establish Welfare Guidelines." *Feedstuffs* 76 (11, March 15): 8.

Bennett, R. M. 1997. "Farm Animal Welfare and Food Policy." *Food Policy* 22: 281–8.

——and R. J. P. Blaney. 2002. "Social Consensus, Moral Intensity, and Willingness to Pay for Address a Farm Animal Welfare Issue." *Journal of Economic Psychology* 23: 501–20.

————2003. "Estimating the Benefits of Farm Animal Welfare Legislation Using the Contingent Valuation Method." *Agricultural Economics* 29: 85–98.

——J. Anderson, and R. J. P. Blaney. 2002. "Moral Intensity and Willingness to Pay Concerning Farm Animal Welfare Issues and the Implications for Agricultural Policy." *Journal of Agricultural and Environmental Ethics* 15: 187–202.

Bentham, Jeremy. 1823. *Introduction to the Principles of Morals and Legislation*, 2nd edn, chapter 17. London: Oxford Publishing.

Bernués, A., A. Olaizola, and K. Corcoran. 2003. "Extrinsic Attributes of Red Meat as Indicators of Quality in Europe: An Application for Market Segmentation." *Food Quality and Preference* 14 (4, June): 265–76.

Bertenshaw, Catherine and Peter Rowlinson. 2009. "Exploring Stock Managers' Perceptions of the Human–Animal Relationship on Dairy Farms and an Association with Milk Production." *Anthrozoos* 11: 59–69.

Blackmore, Susan. 2004. *Consciousness: An Introduction*. New York: Oxford University Press.

Blankstein, Andrew. 2008. "Animal Rights Extremists Target UCLA Research in Arson Attack." *Los Angeles Times* November 29. <http://articles.latimes.com/2008/nov/29/local/me-animal-arson29>.

Boggess, Mark. 2007. "NPB Sow Housing Calculator." Sow Housing Forum. Des Moines, Iowa, June 6.

Boyle, L. A., F. C. Leonard, P. B. Lynch, and P. Brophy. 2002. "Effect of Gestation Housing on Behaviour and Skin Lesions of Sows in Farrowing Crates." *Applied Animal Behaviour Science* 76(2): 119–34.

Bracke, M. B. M., B. M. Spruijt, J. H. M. Metz, and W. G. P. Schouten. 2002a. "Decision Support System for Overall Welfare Assessment in Pregnant Sows A: Model Structure and Weighting Procedure." *Journal of Animal Science* 80: 1819–34.

————————2002b. "Decision Support System for Overall Welfare Assessment in Pregnant Sows B: Validation by Expert Opinion." *Journal of Animal Science* 80: 1819–34.

Braga, Jacinto and Chris Starmer. 2005. "Preference Anomalies, Preference Elicitation, and the Discovered Preference Hypothesis." *Environmental & Resource Economics* 32: 55–89.

Brambell Report. 1965. "Report of the Technical Committee to Enquire into the Welfare of Animals Kept Under Intensive Livestock Husbandry Systems." Presented to Parliament by the Secretary of State for Scotland and the Minister of Agriculture, Fisheries and Food by Command of her Majesty, December.

Brester, G. W., J. M. Marsh, and J. A. Atwood. 2004. "Distributional Impact of Country-of-Origin Labeling in the US Meat Industry." *Journal of Agricultural and Resource Economics*, 29(2): 206–27.

Broom, D. M. and K. C. Johnson. 1993. *Stress and Animal Welfare*. Boston, MA: Kluwer Academic Publishers.

Brown, Jennifer, John Cranfield, and Spencer Henson. 2005. "Relating Consumer Willingness-To-Pay for Food Safety to Risk Tolerance: An Experimental Approach." *Canadian Journal of Agricultural Economics* 53(2–3): 249–63.

Brown, S. N., G. R. Nute, A. Baker, S. I. Hughes, and P. D. Warriss. 2008. "Aspects of Meat and Eating Quality of Broiler Chickens Reared Under Standard, Maize-Fed, Free Range or Organic Systems." *British Poultry Science* 49: 118–24.

Buck, Pearl. [1931] 2005. *The Good Earth*. Enriched Classic Edition. New York: Simon and Schuster Paperbacks, p. 75.

Burdine, Kenneth, Leigh J. Maynard, and A. Lee Meyer. 2004. "Understanding the Market for Holstein Steers." American Agricultural Economics Association Annual Meeting Selected Paper, August.

Business Wire. 2005. "Monitoring the Vegetarian Market in the US Where Meat-Free Hot Dogs Look the Same as Their Meat-Based Counterparts." August 30. <http://findarticles.com/p/articles/mi_moEIN/is_2005_August_30/ai_n14937123>. Accessed July 19, 2008.

Caplan, B. 2001. "Rational Ignorance versus Rational Irrationality." *Kyklos* 54: 3–26.

Capper, J. L., R. A. Cady, and D. E. Bauman. 2009a. "Demystifying the Environmental Sustainability of Food Production." *Cornell Nutrition Conference.*

<http://wsu.academia.edu/documents/0046/7264/2009_Cornell_Nutrition_Conference_
Capper_et_al.pdf>. Accessed February 23, 2010.

Capper, J. L., R. A. Cady, and D. E. Bauman. 2009*b*. "The Environmental Impact of Dairy
Production: 1944 Compared With 2007." *Journal of Animal Science* 87: 2160–7.

Carbone, L. 2004. *What Animals Want: Expertise and Advocacy in Laboratory Animal
Welfare Policy*. Oxford: Oxford University Press.

Carlsson, F., P. Frykblom, and C. J. Lagerkvist. 2007*a*. "Consumer Willingness-to-Pay
for Farm Animal Welfare: Mobile Abattoirs versus Transportation to Slaughter."
European Review of Agricultural Economics 34: 321–44.

————2007*b*. "Farm Animal Welfare—Testing for Market Failure." *Journal of
Agricultural and Applied Economics* 39: 61–73.

Carter-Long, Lawrence. 2002. "A Dog's Worth." *SATYA* June/July.

Cassandra, B. Tucker, D. Fraser, and D. M. Weary. 2001. "Tail Docking Dairy Cattle:
Effects on Cow Cleanliness and Udder Health." *Journal of Dairy Science* 84: 84–7.

Chang, J. B., J. L. Lusk, and F. B. Norwood. 2009*a*. "How Closely Do Hypothetical
Surveys and Laboratory Experiments Predict Field Behavior?" *American Journal of
Agricultural Economics* 91: 518–34.

————2010. "The Price of Happy Hens: A Hedonic Analysis of Retail Egg
Prices." *Journal of Agricultural and Resource Economics* 35 (3): 406–23.

Cheng, K. J., T. A. McAllister, J. D. Popp, A. N. Hristov, Z. Mir, and H. T. Shin. 1998.
"A Review of Bloat in Feedlot Cattle." *Journal of Animal Science* 76: 299–308.

The Chicken of Tomorrow. 1948. Parts 1 and 2. <http://www.youtube.com/watch?
v=SG7TIQk1UiM> and <http://www.youtube.com/watch?v=Wc_AO2CVrmY&
feature=related>.

Chilton, S. M., D. Burgess, and W. G. Hutchinson. 2006. "The Relative Value of Farm
Animal Welfare." *Ecological Economics* 59: 353–63.

Clutton-Brock, Juliet. 1989. *A Natural History of Domesticated Mammals*. Austin, TX:
First University of Texas Press Paperback Printing.

Coase, R. H. 1960. "The Problem of Social Cost." *Journal of Law and Economics* 3: 1–44.

Copenhagen Consensus Center. 2008. "The World's Best Investment: Vitamins for
Undernourished Children, According to Top Economists, Including 5 Nobel Laure-
ates." May 30. <www.copenhagenconsensus.com/Admin/Public/DWSDownload.aspx?
File=%2fFiles%2fFiler%2fCC08%2fPresse++result%2fCopenhagen_Consensus_2008_
Results_Press_Release.pdf>. Accessed November 17, 2009.

Cowen, Tyler. 1996. "What Do We Learn From the Repugnant Conclusion?" *Society*
106: 754–75.

Cronon, W. 1991. *Nature's Metropolis: Chicago and the Great West*. New York: W. W. Norton
& Company.

Curtis, Stanley E. 2007. "Commentary: Performance Indicates Animal State of Being:
A Cinderella Axiom?" *The Professional Animal Scientist*. 23: 573–83.

————Rodney Baker, Mark Estienne, and Brendan Lynch, John McGlone, and Bjarne
Pedersen. 2009. "Scientific Assessment of the Welfare of Dry Sows Kept in Individual
Accommodations." CAST Issue Paper No. 42, March.

Dalai Lama. 2001. *An Open Heart*. New York: Little, Brown, and Company.

Danbury, T. C., C. A. Weeks, J. P. Chambers, A. E. Waterman-Pearson, and S. C. Kestin. 2000. "Self-Selection of the Analgesic Drug Carprofen by Lame Broiler Chickens." *The Veterinary Record* 146 (March 11): 307–11.

Dangour, Alan D., Sakhi K. Dodhia, Arabella Hayter, Elizabeth Allen, Karen Lock, and Ricardo Uauy. 2009. "Nutritional Quality of Organic Foods." *The American Journal of Clinical Nutrition* 90(3): 680–5. <http://www.food.gov.uk/multimedia/pdfs/organicreviewappendices.pdf>.

Davis, James C. 2004. *The Human Story*. New York: HarperCollins.

Davis, J. G., T. L. Stanton, and T. Haron. 2002. "Feedlot Manure Management." Livestock Series No. 1.220, Colorado State University Cooperative Extension.

Davis, Steven L. 2003. "The Least Harm Principle May Require that Humans Consume a Diet Containing Large Herbivores, Not a Vegan Diet." *Journal of Agricultural and Environmental Ethics* 16: 387–94.

Davis, H. P., W. H. Smith S. Dickinson, W. C. Coffey, and H. W. Nisonger. 1928. *Livestock Enterprises*. Chicago, IL: J.B. Lippincott Company, pp. 284, 391.

Dawkins, Marian Stamp. 1993. *Through Our Eyes Only?* New York: W.H. Freeman Spectrum.

——C.A. Connelly, and T.A. Jones. 2004. "Chicken Welfare is Influenced More by Housing Conditions than by Stocking Density." *Nature* 427: 342–4.

Dawkins, Richard. 1999. *The Selfish Gene*. Oxford University Press. New York.

——2006. *The God Delusion*. London, UK: Bantam Press.

Dawrst, Alan. 2009. "The Predominance of Wild-Animal Suffering Over Happiness: An Open Problem." Working Papers: Essays on Reducing Suffering. <http://www.utilitarian-essays.com/>. Accessed July 1, 2009.

Delaware Poultry Industry, Inc. No Date. "Collection of Statistics and Industry Facts." <http://www.dpichicken.org/download/2007allfacts.doc>. Accessed April 11, 2009.

De Mol, R. M., W. G. P. Schouten, E. Evers, H. Drost, H. W. J. Houwers, and A. C. Smits. 2006. "A Computer Model for Welfare Assessment of Poultry Production Systems for Laying Hens." *Netherlands Journal of Agricultural Science* 54: 157–68.

Descartes, Rene. 1637. *A Discourse on the Method of Rightly Conducting the Reason and Seeking Truth in the Sciences*. Rutland, VT: Orion Publishing Group.

Deyoe, G., W. A. Ross, and W. H. Peters. 1954. *Raising Livestock*, 2nd edn. New York: McGraw-Hill.

Dhuyvetter, Kevin C., Mike D. Tokach, and Steve S. Dritz. 2007. "Farrow-to-Finish Swine Cost–Return Budget." Kansas State University Agricultural Experiment Station Bulletin No. MF-292, October.

Diamond, Jared. 1997. *Guns, Germs, and Steel*. New York: W. W. Norton & Company, Inc.

Dickinson, D. L. and D. Bailey. 2002. "Meat Traceability: Are US Consumers Willing to Pay for It?" *Journal of Agricultural and Resource Economics* 27: 348–64.

Dixon, M. E. 2009. "Can They Suffer?" *Main Line Today*, July 28. <http://www.mainlinetoday.com/Main-Line-Today/August-2009/Can-They-Suffer/>.

Dostoevsky, Fyodor. [1879] 1990. *The Brothers Karamozov*. Tr. Richard Pevear and Larissa Volokhonsky. New York: Farrar, Straus, and Giroux, p. 319.

Driver, Doh. No Date. "Interviews: Paul Shapiro." *VegFamily*. <http://www.vegfamily.com/interviews/paul-shapiro.htm>.

Economic Research Service. 2003. "Farm Structure: Questions and Answers." United States Department of Agriculture, Washington, DC, September 3.

——2004. "Food Safety Innovation in the United States: Evidence from the Meat Industry." Agricultural Economic Report No. 56 (AER-831). United States Department of Agriculture, Washington, DC, April.

——2005. "Hogs: Background." Paper given to the Briefing Rooms, United States Department of Agriculture, Washington, DC.

——2006. *Red Meat Yearbook.* Washington, DC: United States Department of Agriculture.

——2007. "The Changing Economics of US Hog Production." Economic Research Report No. 45 (ERR-52). United States Department of Agriculture, Washington, DC, December.

——2008. "Meat Price Spreads: Documentation." Agricultural Center Report No. Ag. 101. United States Department of Agriculture, Washington, DC, November 19.

——2009. "Food Availability: Spreadsheets." United States Department of Agriculture, Washington, DC. <http://www.ers.usda.gov/Data/FoodConsumption/FoodAvail-Spreadsheets.htm>. Accessed December 9, 2009.

The Economist. 2006a. "Sick with Excess of Sweetness." December 23.

——2006b. "The Pursuit of Happiness." December 23.

——2007. "Noble or Savage?" December 22.

——2009a. "I Am Just a Poor Boy Though My Story's Seldom Told." April 4.

——2009b. "Egypt's Pigs: What a Waste." May 9.

——2009c. "Fair to Foul and Back Again" July 2.

EEC (European Economic Community). 1991. "Commission Regulation No. 1274/91: Introducing Detailed Rules for Implementing Regulation (EEC) No. 1907/90 on Certain Marketing Standards for Eggs." Brussels, Belgium, May.

Edmonds, David. 2009. "Philosophy's Great Experiment." *Prospect Magazine*, 156 (March).

EFSA (European Food Safety Authority). 2007. "Report of the Task Force on Zoonoses: Data Collection on the Analysis of the Baseline Study on the Prevalence of Salmonella in Holdings of Laying Hen Flocks of Gallus." *The EFSA Journal* 97.

Eicher, S. D., H. W. Cheng, A. D. Sorrells, and M. M. Schutz. 2006. "Short Communication: Behavioral and Physiological Indicators of Sensitivity or Chronic Pain Following Tail Docking." *Journal of Dairy Science* 89: 3047–51.

Elliott, Ian. 2010. "EU Ag. Groups Want to Link Welfare and Trade." *Feedstuffs*, February 8: 6.

EPA (Environmental Protection Agency). 2007. "Lifecycle Production Phases—Dairy." Ag. No. 101, September 11.

Espejo, L. A., M. I. Endres, and J. A. Salfer. 2006. "Prevalence of Lameness in High-Producing Holstein Cows Housed in Freestall Barns in Minnesota." *Journal of Dairy Science* 89: 3052–8.

Europa. "Animal Welfare on the Farm." (Collection of web pages regarding legislation by the European Commission.) <http://ec.europa.eu>. Accessed May 13, 2009.

Farm Animal Welfare Council. 1979. Press Statement. London. December 5.

——1995. "Report on the Welfare of Turkeys." PB No. 2033.

Farm Foundation. 2006. "Future of Animal Agriculture in North America." Oak Brook, IL: Farm Foundation. <http://www.farmfoundation.org/projects/documents/AnimalWelfare_000.pdf>. Accessed August 17, 2007.

Farm Sanctuary Website. 2009. "The Issues: Proposed Legislation and Public Policy." <http://www.trendtrack.com/texis/test/viewrpt?event=47aOaa2797>. Accessed July 23, 2009.

FAS (Foreign Agricultural Service). 2006. "Per Capita Consumption of Food in Select Countries." United States Department of Agriculture, Washington, DC. <http://www.fas.usda.gov/dlp/circular/2006/06–03LP/bpppcc.pdf>. Accessed December 9, 2009.

Federal Register. 2007. *Federal Register* 72 (199, October 16): 58,631–7

Feedstuffs Foodlink. 2008. "New Jersey Supreme Court Upholds State's Animal Welfare Standards." August 5. <http://www.feedstuffsfoodlink.com/ME2/dirmod.asp?sid=&nm=&type=news&mod=News&mid=9A02E3B96F2A415ABC72CB5F516B4C10&tier=3&nid=4A22477770A340B98DEEFB34E6CD69D3>.

Fehr, Ernst and Simon Gächter. 2000. "Fairness and Retaliation: The Economics of Reciprocity." *Journal of Economic Perspectives* (14):159–81.

——and K. Schmidt. 1999. "A Theory of Fairness, Competition and Cooperation." *Quarterly Journal of Economics* 114: 817–68.

Flandrin, Jean-Louis. 1996. *Food: A Culinary History from Antiquity to the Present*. Ed. Lawrence Kritzman. New York: Columbia University Press.

Food Safety and Inspection Service. 2006. "Fact Sheet: Meat Preparation. Veal From Farm to Table." United States Department of Agriculture, Washington, DC, October 17.

Francione, Gary. 2008. "Gary Francione Interview: Parts 1 and 2." *Vegan Freaks* (Podcast Interview) June 20 and 26. <http://www.youtube.com/watch?v=EO8NXatlQhc>.

——2009. "The Abolitionist Approach." A Series of Podcasts.

Frank, J. 2008. "Is There an 'Animal Welfare Kuznets Curve'?" *Ecological Economics* 66: 478–91.

Fraser, D. 2002. "Assessing Animal Welfare at the Farm and Group Level: The Interplay of Science and Values." *Animal Welfare* 12: 433–43.

Frasseto, Michael. 2008. *The Great Medieval Heretics*. New York: Blue Bridge.

Frédéric, Bastiat. 1848. "What is Seen and What is Not Seen." *Selected Essays on Political Economy*. New York: Irvington-on-Hudson, ch. 1. <www.econlib.org>.

Free, Ann Cottrell. 2000. "A Tribute to Ruth Harrison." *Animal Welfare Institute Quarterly* 49 (4, Fall). <http://www.awionline.org/ht/display/ContentDetails/i/2128/pid/2535>.

Fregonesi, J. A., D. M. Veira, M. A. G. von Keyserlingk, and D. M. Weary. 2007. "Effects of Bedding Quality on Lying Behavior of Dairy Cows." *Journal of Dairy Science*. 90: 5468–72.

Frewer, L. J., Kole, A., Van de Kroon. S. M. A, and C. De Lauwere. 2005. "Consumer Attitudes Towards the Development of Animal-Friendly Husbandry Systems." *Journal of Agricultural and Environmental Ethics* 18(4): 345–67.

Gambone, J. C. 1972. "Kansas: A Vegetarian Utopia: The Letters of John Milton Hadley, 1855–1856." *Kansas Historical Quarterlies* Spring.

Gandhi, Mohandas K. 1957. *An Autobiography: The Story of My Experiments With Truth*. Boston, MA: Bacon Press, pp. 426–7.

Gardner, Bruce. 2002. *American Agriculture in the Twentieth Century*. Cambridge, MA: Harvard University Press.

Gaskell, G., E. Einsiedel, W. Hallman, S. H. Priest, J. Jackson, and J. Olsthoorn. 2005. "Social Values and the Governance of Science." *Science* 310: 1908–9.

Gazzaniga, Michael S. 2008. "Spheres of Influence." *Scientific American Mind*, June/July: 32–9.

Gigerenzer, G. 2007. *Gut Feelings: The Intelligence of the Unconscious*. New York: Penguin Books.

Ginsberg, C. and A. Ostrowski. 2007. "The Market for Vegetarian Foods." *The Vegetarian Resource Group*. <http://www.vrg.org/nutshell/market.htm>. Accessed August 15, 2007.

Gladwell, Malcom. 2005. *Blink*. New York: Little, Brown and Company.

Glass, C. A., W. G. Hutchinson, and V. E. Beattie. 2005. "Measuring the Value to the Public of Pig Welfare Improvements: A Contingent Valuation Approach." *Animal Welfare* 14 (1, February): 61–9.

Gough, Julian. 2008. "The Sacred Mystery of Capital." *Prospect Magazine*. Afterword, July. <http://www.prospect-magazine.co.uk/article_details.php?id=10255>.

Grandin, Temple. 2007. "Bedding Type Affects Welfare." *Feedstuffs*, October 15: 9.

——and C. Johnson. 2005. *Animals in Translation: Using the Mysteries of Autism to Decode Animal Behavior*. New York: Scribner.

Gracia, A., M. L. Loureiro, and R. M. Nayga, Jr. 2009. "Valuing Animal Welfare Labels with Experimental Auctions: What Do We Learn from Consumers?" Paper presented at the International Association of Agricultural Economists Conference, Beijing, China, August 16–22.

Gregory, Gene (President of United Egg Producers). 2009. Personal Communication. February 6.

Griener, Jennifer. 2009. "Attempts to Ban Antibiotic Use in Pigs, Misguided." *Pork Magazine*, January 1. <http://www.porkmag.com/news_editorial.asp?pgID=768&ed_id=7004>.

Hagelin, J., H. E. Carlsson, and J. Hau. 2003. "An Overview of Surveys on How People View Animal Experimentation: Some Factors that May Influence the Outcome." *Public Understanding of Science* 12: 67–81.

Haidt, J. 2001. "The Emotional Dog and its Rational Tail: A Social Intuitionist Approach to Moral Judgment." *Psychological Review* 108: 814–34.

Hall, C. and V. Sandilands. 2007. "Public Attitudes to the Welfare of Broiler Chickens." *Animal Welfare* 16: 499–512.

Hancock, Dale and Tom Besser. 2006. "*E. Coli* 0157:H7 in Hay- or Grain-fed Cattle." Paper presented at the College of Veterinary Medicine, Washington State University, October 12. <http://www.puyallup.wsu.edu/dairy/nutrient-management/data/publications/E%20coli%20O157%20in%20hay-%20or%20grain-fed%20cattle%20Hancock%20and%20 Besser%2011%2006.pdf>. Accessed June 12, 2010.

Harrison, Ruth. 1966. *Animal Machines*. New York: Bailantine Books.

Hatkoff, Amy. 2009. *The Inner World of Farm Animals*. New York: Stewart, Tabori, & Chang.

Hayek, Friederick. 1988. *The Fatal Conceit: The Errors of Socialism*. Chicago, IL: University of Chicago Press, p. 76.

Hayes, Dermont J. and Helen H. Jensen. 2003. "Lessons from the Danish Ban on Feed-Grade Antibiotics." *Choices Magazine*, August.

Henderson, S. N., J. T. Barton, A. D. Wolfenden, S. E. Higgins, J. P. Higgins, W. J. Kuenzel, C. A. Lester, G. Tellez, and B. M. Hargis. 2009. "Comparison of Beak-Trimming Methods on Early Broiler Breeder Performance." *Poultry Science* 88: 57–60.

HFA (Humane Farming Association). No Date. "Animal Cruelty Conviction at Wiles Hog Factory." <http://www.hfa.org/campaigns/wiles_update11_07.pdf>. Accessed May 13, 2009.

——No Date. "Petition for Enforcement of Ohio Animal Cruelty Statutes at Wiles Farm."

Hills, Alison. 2005. *Do Animals Have Rights?* Cambridge, UK: Icon Books Ltd.

Hoffmann, S., Fischbeck, P., Krupnick, A., and McWilliams, M. 2007. "Using Expert Elicitation to Link Foodborne Illnesses in the United States to Food." *Journal of Food Protection* 70(5): 1220–9.

Holldobler, Bert and Edward O. Wilson. 1994. *Journey to the Ants*. Cambridge, MA: Belknap Press, Harvard University Press.

Hoppe, Robert A., Penni Korb, Erik J. O'Donoghue, and David E. Banker. 2007. "Structure and Finances of US Farms: Family Farm Report." Economic Information Bulletin No. EIB-24, June.

Horan, Richard, Jason Shogren, and Erwin Bulte. 2008. "Competitive Exclusion, Diversification, and the Origins of Agriculture." Paper presented at the American Agricultural Economics Association Annual Meeting, Orlando, Florida, July 27–9.

Hotzel, M. J., E. J. C. Lopes, P. A. V. de Oliveira, and A. L. Guidoni. 2009. "Behaviour and Performance of Pigs Finished on Deep Bedding With Wood Shavings or Rice Husks in Summer." *Animal Welfare* 18: 65–71.

Houpt, Katherine Albro. 2005. "Well-Being Assessment: Physiological Criteria." *Encyclopedia of Animal Science*. New York: Marcel Dekker, p. 205.

Hsee, Christopher and Yuval Rottenstreich. 2004. "Music, Pandas, and Muggers: On the Affective Psychology of Value." *Journal of Experimental Psychology: General* 133(1): 23–30.

HSUS (Humane Society of the United States). No Date. "About Cattle." Factory Farming Campaign. See generally <<http://www.humanesociety.org/issues/campaigns/factory-farming/>>.

——No Date. "About Chickens." Factory Farming Campaign.

——No Date. "About Pigs." Factory Farming Campaign.

——No Date. "A HSUS Report: The Welfare of Animals in the Dairy Industry."

——No Date. "A HSUS Report: The Welfare of Animals in the Turkey Industry."

——2006a. "A HSUS Report: The Welfare of Animals in the Egg Industry."

——2006b. "Help Farm Animals . . . Follow the Three R's." Factory Farming Campaign.<http://www.hsus.org/farm/resources/pubs/rrr.html>.

——2008a. "A HSUS Report: The Welfare of Animals in the Broiler Industry."

——2008b. "A HSUS Report: The Welfare of Animals in the Veal Industry."

——2009. "Factory Farming: Resources. About Farm Animals: Pigs." <http://www.hsus.org/farm/resources/animals/pigs/pigs.html>. Accessed January 19, 2009.

Huang, Kuo, S. 1985. "US Demand for Food: A Complete System of Price and Income Effects." Technical Bulletin No. 1714. United States Department of Agriculture Economic Research Service.

Huang, Kuo, S., and Biing-Hwan Lin. 2000. *Estimation of Food Demand and Nutrient Elasticities from Household Survey Data*. Technical Bulletin No. 1887. United States Department of Agriculture Economic Research Service, August.

Hueth, Brent, Maro Ibarburu, and James Kliebenstein. 2007. "Marketing Specialty Hogs: A Comparative Analysis of Two Firms from Iowa." *Review of Agricultural Economics* 29 (4, Winter): 720–33.

Ibrahim, Darian M. 2007. "A Return to Descartes: Property, Profit, and the Corporate Ownership of Animals." Arizona Legal Studies Discussion Paper No. 06–23.

Infoplease. 2010. "Per Capita Consumption of Principal Foods." <www.infoplease.com/ipa/A0104742.html>. Accessed June 20, 2010.

Jacob, Jacqueling P. and F. Ben Mather. 1998. "The Home Broiler Chicken Flock." IFAS Extension No. PS42. University of Florida.

Johansson, P. 1992. "Altruism in Cost–Benefit Analysis." *Environmental and Resource Economics*. 6: 605–13.

Kagel, J. H., R. C. Battalio, and L. Green. 1995. *Economic Choice Theory: An Experimental Analysis of Animal Behavior*. Cambridge, UK: Cambridge University Press.

Kanter, C, K. Messer, and H. Kaiser. 2009. "Does Production Labeling Stigmatize Conventional Milk?" *American Journal of Agricultural Economics*. 91(4): 1097–109.

Kastner, Jeffrey. 2001. "Animals on Trial." *Cabinet* 4, Fall. <http://www.cabinetmagazine.org/issues/4/animalsontrial.php>.

Katanbaf, M. N., E. A. Dunnington, and P. B. Siegel. 1988. "Restricted Feeding in Early and Late-Feathering Chickens." *Poultry Science* 68: 344–51.

Kestin, S. C., T. G. Knowles, A. E. Tinch, and N. G. Gregory. 1992. "Prevalence of Leg Weakness in Broiler Chickens and Its Relationship with Genotype." *Veterinary Record* 131: 190–94.

Khaitovich, P., H. Lockstone, M. Wayland, T. Tsang, S. Jayatilaka, A. Guo, J. Zhou, M. Somel, L. Harris, E. Holmes, S. Paabo, and S. Bahn. 2008. "Metabolic Changes in Schizophrenia and Human Brain Evolution." *Genome Biology* 9: R124.

Kilian, Emily. 2008. "Pork Industry Culture to Change." *Feedstuffs* 21 (80, May 26): 9.

Kliebenstein, James, Sean Hurley, Ben Larson, and Mark Honeman. 2004. "Cost of Organic Production: A Seasonal Analysis and Needed Price Premium for Continuous Production." Paper presented at the Agricultural and Applied Economics Association (AAEA) Annual Meeting, Denver, CO, August 1–4.

——, D. Stender, G. Huber, and J. Mabry. 2006. "Costs, Returns, Production and Financial Efficiency of Niche Pork Production in 2006," Iowa Pork Industry Center, October 2007.

Knowles, Toby G., Steve G. Kestin, Susan M. Haslan, Steven N. Brown, Laura E. Green, Andrew Butterworth, Stuart J. Pope, Dirk Pfeiffer, and Christine J. Nicol. 2008. "Leg Disorders in Broiler Chickens: Prevalence, Risk Factors, and Prevention." *PLoS ONE* 3(2): e1545.

Koch, Christof. 2008/2009. "What Is It Like to Be a Bee?" *Scientific American Mind* December/January: 18–19.

Krestin, S. C., T. G. Knowles, A. E. Tinch, and N. G. Gregory. 1992. "Prevalence of Leg Weakness in Broiler Chickens and its Relationship with Genotype." *Veterinary Record* 131: 190–4.

Kritzman, Lawrence. Ed. 1996. *Food: A Culinary History from Antiquity to the Present*. New York: Columbia University Press.

Lagerkvist, C. J., F. Carlsson, and D. Viske. 2006. "Swedish Consumer Preferences for Animal Welfare and Biotech: A Choice Experiment" *AgBioForum*. 9(1): 51–8.

Lawrence, John D. and Glenn Grimes. 2007. "Production and Marketing Characteristics of US Pork Producers, 2006." Staff General Research Paper 12828. Department of Economics, Iowa State University.

Lay, Donald C. Jr. 2002. "Management Tips to Reduce Pre-Weaning Mortality." Paper presented at the Forty-Sixth Annual North Carolina Pork Conference, February 19–21.

LayWel. 2004. "Welfare Implications of Changes In Production Systems for Laying Hens." Specific Targeted Research Project (STReP) Report No. SSPE-CT-2004–502315.

Leach, K. A., S. Dippel, J. Huber, S. March, C. Winckler, and H. R. Whay. 2009. "Assessing Lameness in Cows Kept in Tie-Stalls." *Journal of Dairy Science* 92: 1567–74.

Lehrer, Johan. 2010. How We Decide. New York: Mariner Books.

Levitt, S.D. and S.J. Dubner. 2009. *SuperFreakonomics*. New York: HarperCollins, p. 16.

Levy, David M. and Sandra J. Peart. 2001. "The Secret History of the Dismal Science: Part 1." *Economics, Religion, and Race in the 19th Century*, January 22. <www.econlib.org>.

Lichtenberg, E. 2004. "Some Hard Truths about Agriculture and the Environment." *Agricultural and Resource Economics Review* 33 (1, April): 24–33.

Liljenstolpe, C. 2008. "Evaluating Animal Welfare with Choice Experiments: An Application to Swedish Pig Production." *Agribusiness* 24: 67–84.

Lind, L. W. 2007. "Consumer Involvement and Perceived Differentiation of Different Kinds of Pork: A Means–End Chain Analysis." *Food Quality and Preference* 18 (4, June): 690–700.

List, J. A. 2002. "Preference Reversals of a Different Kind: The 'More is Less' Phenomenon." *American Economic Review* 92: 1636–43.

——2003. "Does Market Experience Eliminate Market Anomalies?" *Quarterly Journal of Economics* 118: 41–71.

——and C. A. Gallet. 2001. "What Experimental Protocol Influence Disparities Between Actual and Hypothetical Stated Values?" *Environmental and Resource Economics* 20(3): 241–54.

Little, J. M. and R. Berrens. 2004. "Explaining Disparities between Actual and Hypothetical Stated Values: Further Investigation using Meta-Analysis." *Economics Bulletin* 3 (6): 1–12.

LMIC (Livestock Marketing Information Center). 2008. "Retail Food Price Information." April. <http://www.lmic.info/>.

Loos, Trent. 2008. Speech given to the Student Section of the American Agricultural Economics Association, July 28 <http://asp.okstate.edu/baileynorwood/Misc2/zTrentLoos.htm>.

——2009. "From My Lips to Theirs." *Feedstuffs*. 8 (81, February 23): 9.

Loomis, John. 2006. "Importance of Including Use and Passive Use Values of River and Lake Restoration." *Journal of Contemporary Water Research and Education* 134: 4–8.

Lundeen, Tim. 2009. "Free-range Chickens More Prone to Disease." *Feedstuffs* January 26: 9.

Lusk, J. L. 2009. "The Effect of Proposition 2 on the Demand for Eggs in California." Working paper, Department of Agricultural Economics, Oklahoma State University.

——2010. "The Effect of Proposition 2 on the Demand for Eggs in California." *Journal of Agricultural and Food Industrial Organization* 8(1): article 3.

——and John D. Anderson. 2004. "Effects of Country of Orgin Labeling on Meat Producers and Consumers." *Journal of Agricultural and Resource Econmics*. Western Agricultural Economics Association 29 (02): 185–205.

Lusk, J. L. and F. B. Norwood. 2008. "Public Opinion about the Ethics and Governance of Farm Animal Welfare." *Journal of the American Veterinary Medical Association* 233: 1121–6.

————2009*a*. "Some Economic Benefits and Costs of Vegetarianism." *Agricultural and Resources Economics Review* 38 (2): 109–24.

————2009*b*. "Bridging the Gap between Laboratory Experiments and Naturally Occurring Markets: An Inferred Valuation Method." *Journal of Environmental Economics and Management* 58: 236–50.

————2009*c*. "Egg Prices and Sales for Select US Regions." Proprietary data.

————Forthcoming. "Speciesism, Altruism, and the Economics of Animal Welfare." *European Review of Agricultural Economics*.

————and J. F. Shogren. 2007. *Experimental Auctions: Methods and Applications for Economic and Marketing Research*. Cambridge, UK: Cambridge University Press.

————T. Nilsson, and K. Foster. 2007. "Public Preferences and Private Choices: Effect of Altruism and Free Riding on Demand for Certified Meat." *Environmental and Resource Economics* 36: 499–521.

Lymberry, Philip. 2002. "Laid Bare . . . The Case Against Enriched Cages in Europe." Report Written for Compassion in World Farming Trust, Hampshire, UK.

Lynch, P. Brendan, Laura Boyle, Finola Leonard, Annabel Tergny, and Patrick Brophy. 2000. "Studies of Housing on Pregnant Sows in Groups and Individually." Teagast Project No. 4563.

Lyubomirsky, Sonja. 2008. *The How of Happiness*. New York: The Penguin Press.

MacDonald, James and Penni Korb. 2008. "Agricultural Contracting Update, 2005." Economic Research Service. Economic Information Bulletin No. 35, April.

————Erik J. O'Donoghue, William D. McBride, Richard F. Nehring, Carmen L. Sandretto, and Roberto Mosheim. 2007. "Profits, Costs, and the Changing Structure of Dairy Farming." Economic Research Report No. ERR-47.

Manchester, William. 1993. *A World Lit Only By Fire*. New York: Little, Brown, and Company.

Marsh, John M. 1994. "Estimating Intertemporal Supply Response in the Fed Beef Market." *American Journal of Agricultural Economics* 76(3): 444–53.

Massachusetts Body of Liberties. 1641. Available from the Hanover Historical Texts Project, Hanover, MA.

Matheny, Gaverick. 2003. "Least Harm: A Defense of Vegetarianism from Steven Davis' Omnivorous Proposal." *Journal of Agricultural and Environmental Ethics* 16: 505–11.

————2006. "Utilitarianism and Animals." Peter Singer (ed.), *In Defense of Animals*. Malden, MA: Blackwell Publishing.

————and C. Leahy. 2007. "Farm-Animal Welfare, Legislation, and Trade." *Law & Contemporary Problems* 70: 325–58.

Matthews, L. R. and J. Ladewig. 1994. "Environmental Requirements of Pigs Measured by Behavioural Demand Functions." *Animal Behavior* 47: 713–19.

McCarthy, M., M. de Boer, S. O'Reilly and L. Cotter. 2003. "Factors Influencing Intention to Purchase Beef in the Irish Market." *Meat Science* 65 (3, November): 1071–83.

McFarlane, J.M., K. E. Boe, and S. E. Curtis. 1988. "Turning and Walking by Gilts in Modified Gestation Crates." *Journal of Animal Science* 66: 326–31.

McGeown, D., T. C. Danbury, A. E. Waterman-Pearson, and S. C. Kestin. 1999. "Effect of Carprofen on Lameness in Broiler Chickens." *The Veterinary Record* 144 (June 12): 668–71.

McGlone, J. J., E. H. von Borell, J. Deen, A. K. Johnson, D. G. Levis, M. Meunier-Salaun, J. Morrow Reeves, J. L. Salak-Johnson, and P. L. Sundberg. 2004. "Review: Compilation of the Scientific Literature Comparing Housing Systems for Gestating Sows and Gilts Using Measures of Physiology, Behavior, Performance, and Health." *The Professional Animal Scientist* 20: 105–17.

McKenna, Carol. 2001. *The Case Against Veal Crates*. New York: Compassion in World Farming.

McWilliams, James. 2010. "A Myth of Grass-Fed Beef." Freakonomics Blog, January 27.

Mead, P. S., L. Slutsker, V. Dietz, L. F. McCaig, J. S. Bresee, C. Shapiro, P. M. Griffin, and R. V. Tauze. 1999. "Food-Related Illness and Death in the United States." *Emerging Infectious Diseases* 5: 607–25.

Meager, Rebecca K. 2009. "Observer Ratings: Validity and Value as a Tool for Animal Welfare Research." *Applied Animal Behaviour Science* 119 (1–2, June): 1–14.

Meat and Poultry Wire Service News. 2009. "Norway: Higher Mortality in Organic Hen Production." March 10.

Menand, L. 2001. *The Metaphysical Club*. New York: Farrar, Straus, and Giroux.

Mench, Joy A. 1998. "Thirty Years after Brambell: Whither Animal Welfare Science?" *Journal of Applied Animal Welfare Science* 1(2): 91–102.

Messens, W., K. Grijspeerdt, K. De Reu, B. De Ketelaere, K Mertens, F. Bamelis, B Kemps, J. De Baerdemaeker, E. Decuypere and L. Herman. 2007. "Eggshell Penetration of Various Types of Hens' Eggs by *Salmonella Enterica* Serovar Enteritides." *Journal of Food Production* 70(3): 623–8.

Michell, H. 1935. "Notes on Prices of Agricultural Commodities in the United States and Canada, 1850–1934." *The Canadian Journal of Economics and Political Science* 1(2): 269–79.

Milberger, Sharon, Ronald M. Davis, and Amanda L. Holm. 2009. "Pet Owners' Attitudes and Behaviors Related to Smoking and Second hand Smoke. A Pilot Study." *Topacco Control* 18: 156–8.

Miller, Marlys. 2009. "Death on a Factory Farm." *Pork Magazine* March 3.

Montanari, Massimo. 1996. "Food: A Culinary History from Antiquity to the Present." Lawrence Kritzman (ed.), *Peasants, Warriors, Priests: Images of Society and Styles of Diet*. New York: Columbia University Press.

Montgomery, David. 2003. "Animal Pragmatism: Compassion Over Killing Wants to Make the Anti-Meat Message a Little More Palatable." *Washington Post* September 8: C01.

——2007. "Animal Pragmatism." *Washington Post*, September 8: C01.

Montgomery, Sy. 2006. *The Good Good Pig*. New York: Ballantine Books.

Morales, Alex. 2009. "Paul McCartney Calls for Meat-Free Day to Cut Cow Gas." Bloomberg.com June 15.

Moran, D and A McVittie. 2008. "Estimation of the Value the Public Places on Regulations to Improve Broiler Welfare." *Animal Welfare* 17: 43–52.

Morrison, Willie. 2010. "Buying Organic? Not So Fast." *Feedstuffs* February 15: 5.

Morwitz, Vicki G. and G. J. Fitzsimons. 2004. "The Mere-Measurement Effect: Why Does Measuring Intentions Change Actual Behavior?" *Journal of Consumer Psychology* 14(1–2): 64–74.

Moynagh, J. 2000. "EU Regulation and Consumer Demand for Animal Welfare." *AgBioForum* 3(2–3): 107–44.

Mozoyer, Marcel and Laurance Roudart. 2006. *A History of World Agriculture*. New York: Monthly Review Press.

Muirhead, Sarah. 2009. "Eco-Friendly Foods Not Always What They Seem." *Feedstuffs* 47 (81, November 16): 9.

Napolitano, F., G. Caporale, A. Carlucci, E. Monteleone. 2007. "Effect of Information about Animal Welfare and Product Nutritional Properties on Acceptability of Meat from Podolian Cattle." *Food Quality and Preference* 18 (2, March): 305–12.

NASS (National Agricultural Statistics Service). 2002. "2002 Census of Agriculture." US Department of Agriculture. <http://www.nass.usda.gov/census/census02/volume1/us/st99_1_058_058.pdf>. Accessed August 16, 2007.

——2008. "Poultry Slaughter: 2007 Annual Summary." United States Department of Agriculture Report No. Pou.2–1(08), February.

——2009a. "2008 Chicken Inventory and Egg Production." Montana and the US Press Release. United States Department of Agriculture and the Montana Department of Agriculture, February 27.

——2009b. "Livestock Slaughter: 2008 Annual Summary." United States Department of Agriculture Report No. Mt.An.1–2–1(09), March.

National Chicken Council. 2005. "National Chicken Council Animal Welfare Guidelines and Audit Checklist." April 5.

National Science Foundation. 2009. "Peer Pressure Plays Major Role in Environmental Behavior." National Science Foundation Press Release, Washington, DC, June 29.

Nepomnaschy, P. A., Kathleen B. Welch, D. S. McConnell, Bobbi S. Low, Beverly I. Strassman, and Barry G. England. 2006. "Cortisol Levels and Very Early Pregnancy Loss in Humans." *Proceedings of the National Academy of Sciences of the United States of America* 103(10): 3938–42. Published online February 22, doi: <10.1073/pnas.0511183103>.

Nestle, M. 2006. *What to Eat*. San Francisco, CA: North Point Press.

Norwood, F. Bailey. 2009. Personal Conversation With Paul Shapiro, May 11.

Norwood, F. Bailey and Jayson L. Lusk. 2008. *Agricultural Marketing and Price Analysis*. Upper Saddle River, NJ: Prentice Hall.

——2009. "The Farm Animal Welfare Debate." *Choices* 24(3).

——Forthcoming. "A Calibrated Auction-Conjoint Valuation Method: Valuing Pork and Eggs Produced under Differing Animal Welfare Conditions." *Journal of Environmental Economics and Management*.

——and Robert W. Prickett. 2007. "Consumers Share Views on Farm Animal Welfare." *Feedstuffs*. 79 (42, October 8): 14–16.

——Bharath Arunachalam, and Shida Rastegari Henneberry. 2009. "An Empirical Investigation into the Excessive-Choice Effect." *American Journal of Agricultural Economics* 91(3): 810–25.

Olynk, N. J., G. T. Tonsor, and C. A. Wolf. Forthcoming. "Consumer Willingness to Pay for Livestock Credence Attribute Claim Verification." *Journal of Agricultural and Resource Economics*.

Ophuis, P. A. M. O. 1994. "Sensory Evaluation of 'Free Range' and Regular Pork Meat Under Different Conditions of Experience and Awareness." *Food Quality and Preference* 5(3): 173–8.

Organic Foods Production Act of 1990. 2000. Amended (7 U.S.C 6501–6522), December 21.

Osbourne, James. 2009. "Animal Rights Terrorism on the Rise in the US." *FoxNews.com* June 3.

O'Shaughnessy, Patrice. 1997. "Warm & Damp: Fuzzies Talk Terror." *New York Daily News*, December 7.

Ovaskainen, Ville and Matleena Kniivila. 2005. "Consumer versus Citizen Preferences in Contingent Valuation: Evidence on the Role of Question Framing." *Australian Journal of Agricultural and Resource Economics* 49: 379–94.

Parkhurst, Carmen R. and George J. Mountney. 1988. *Poultry Meat and Egg Production*. New York: Van Nostrand Reinhold Company Limited.

Parry. Vicienne. 2009. "Pork Porkies." *Prospect Magazine* June: 20–1.

PETA. No Date. "Animal Rights Uncompromised: *PETA* on 'Pets.'". <http://www.peta.org/campaigns/ar-petaonpets.asp>. Accessed May 12, 2009.

——No Date. "PETA Offers $1 Million Reward to First In Vitro Meat." <http://www.peta.org/feat_in_vitro_contest.asp>. Accessed July 19, 2008.

PETA Kills Animals. No Date. <http://www.petakillsanimals.com/index.cfm>. Accessed May 28, 2009.

Peters, G., H. V. Rowley, S. Weidemann, R. Tucker, M. Short, and M. Schulz. 2010. "Red Meat Production in Australia: Life Cycle Assessment and Comparison with Overseas Studies." *Environmental Science & Technology* 44: 1327–32.

Phelps, Norm. 2007. *The Longest Struggle*. New York: Lantern Books.

Pinker, Steven. 1997. *How the Mind Works*. New York: W. W. Norton & Company.

Plain, Ron. 2006. "Retail Price of Pork Up." *US Weekly Hog Outlook: The PigSite* August 19.

Plott, C.R. and K. Zeiler. 2005. "The Willingness to Pay–Willingness to Accept Gap, the Endowment Effect, Subject Misconceptions, and Experimental Procedures for Eliciting Valuations." *American Economic Review* 95: 530–45.

Posner, R. A. 2004. "Animal Rights: Legal, Philosophical, and Pragmatic Perspectives." C. R. Sunstein and M. C. Nussbaum (eds.), *Animal Rights*. Oxford, UK: Oxford University Press.

Powell, Adam, Stephen Shennan, and Mark Thomas. 2009. "Late Pleistocene Demography and the Appearance of Modern Human Behavior." *Science* 324(5932): 1298.

Prickett, R., F. Bailey Norwood, and J. L. Lusk. 2010. "Consumer Preferences for Farm Animal Welfare: Results from a Telephone Survey of US Households." *Animal Welfare* 9: 335–47.

Progressive Newswire. 2008. "Unanimous Decision of New Jersey Supreme Court Results in Precedent-Setting Victory for Farm Animals." *CommonDreams.org* July 31.

Prosch, Allen, Ron Christensen, and Bruce Johnson. 2001. "Leasing and Valuing Swine Facilities." *Neb Guide*. Nebraska-Lincoln: Institute of Agriculture and Natural Resources. University of Nebraska-Lincoln.

Quinn, Char. 2009. "Animal Welfare versus Animal Rights in the Vail Valley." Special to the Daily. *Vail Daily* May 25.

Raper, S. C. B., T. M. L. Wigley, and R. A. Warrick. 1996. "Global Sea-Level Rise: Past and Future." In J. D. Milliman and B. U. Haq (eds.), *Sea-Level Rise and Coastal Subsidence: Causes, Consequences and Strategies*. Dordrecht, The Netherlands: Kluwer Academic Publishers, pp. 11–45.

Rauch, R. and J. S. Sharp. 2005. "Ohioans' Attitudes about Animal Welfare." Topical Report. Department of Human and Community Resource Development, Ohio State University, January.

Ray, John. 1985. "Bacon on the Side." *The Times* January 19.

Reach & Discover. 2004. *Refurbishing Roosts* 2: 2.

Read, Leonard. 1958. *I, Pencil: My Family Tree as Told to Leonard E. Read*. Irvington-on-Hudson, NY: The Foundation for Economic Education, Inc.

Reason Foundation. 2009. "Agricultural Subsidies: Corporate Welfare For Farmers." *Reason Hit & Run* January 27.

Regan, Tom. 2004. *The Case for Animal Rights*. Berkeley, CA: University of California Press.

Research Triangle Institute. 2007. "GIPSA Livestock and Meat Marketing Study." Study prepared for the Grain Inspection, Packers and Stockyard Administration, United States Department of Agriculture, Raleigh, NC, January.

Rising, Malin. 2009. "Zoo 'Attack' Shows Apes Can Plan." *Tulsa World* March 15: A14.

Rhodes, Tracy, Michael Appleby, Kathy Chinn, Lawrence Douglas, Lawrence Firkins, Katherine Houpt, Christa Irwin, John McGlone, Paul Sundburg, Lisa Tokach, and Robert Willis. 2005. "Task Force Report: A Comprehensive Review of Housing for Pregnant Sows." *Journal of the American Veterinary Medical Association* 227: 1580–90.

Robbins, Lionel. 1932. *An Essay on the Nature and Significance of Economic Science*. London, UK: Macmillan and Company.

Roberts, Russell. 2007. "Does the Trade Deficit Destroy American Jobs?" <http://www.invisibleheart.com/Iheart/TradeDeficitJobs.pdf>.

—— 2009*a*. "Charles Platt on Working at Wal-Mart." EconTalk Podcast, June 15. <www.econtalk.org>.

——2009*b*. "Roberts (and Hanson) on Truth and Economics." EconTalk Podcast, Library of Liberty and Economics, January 26. <www.econtalk.org>.

Robin, R. 2005. "In-Vitro Meat." *New York Times* December 11.

Rodenburg, T. B., F. A. M. Tuyttens, K. de Reu, L. Herman, J. Zoons, and B. Sonck. 2008*a*. "Welfare Assessment of Laying Hens in Furnished Cages and Non-Cage Systems: Assimilating Expert Opinion." *Animal Welfare* 17: 363–73.

————————————————2008*b*. "Welfare Assessment of Laying Hens in Furnished Cages and Non-Cage Systems: An On-Farm Comparison." *Animal Welfare* 17: 363–73.

Rogers, Lesley J. and Gisela Kaplan. 2004. "All Animals Are *Not* Equal." In Cass R. Sunstein and Marth C. Nussbaum (eds.), *Animal Rights: Current Debates and New Directions*. New York: Oxford University Press.

Rollin Bernand E. 1995. *Farm Animal Welfare*. Ames, IO: Iowa State University Press.

——2004. "Animal Agriculture and Emerging Social Ethics for Animals." *Journal of Animal Science* 82: 955–64.

Rossi, Johnny. 2006. "Implanting Beef Cattle." Bulletin No. 1302. Cooperative Extension Service, University of Georgia.

Rowley, Hazel V., Stephen Wiedemann, Robyn Tucker, Michael D. Short, and Matthias Schulz. 2010. *Environmental Science Technology* 44: 1327–32.

Rumsfeld, Donald. 2002. Press Conference Statement, NATO Headquarters, Belgium, February 12.

Rutten, Tim. 2009. "LA's Animal Terrorists: Opinion." *Los Angeles* Times March 22.

Sabate, Joan. 2003. "The Contribution of Vegetarian Diets to Health and Disease: A Paradigm Shift." *American Journal of Clinical Nutrition* 78: 502S–7S.

Sacremento Bee. 2008. "Ballot Watch: Proposition 2. Standards for Confining Farm Animals." September 27.

Sanotra, G. S., J. D. Lund, A. K. Ersboll, J. S. Petersen, and K. S. Vestergaard. 2001. "Monitoring Leg Problems in Broilers: A Survey of Commercial Broiler Production in Denmark." *World's Poultry Science Journal* 57: 55–69.

Savory, John C. and Katalin Maros. 1993. "Influence of Degree of Food Restriction, Age, and Time of Day on Behavior of Broiler Breeder Chickens." *Behavioural Processes* 29: 179–90.

Schardt, David. 2003. "Unforgettable Foods: What's More Likely to Make You Sick?" *Nutrition Action Healthletter* January–February. <http://findarticles.com/p/articles/mi_mo813/is_1_30/ai_97296397>. Accessed September 4, 2008.

Schnettler, B., R. Vidal, R. Silva, L. Vallejos, and N. Sepúlveda. 2009. "Consumer Willingness to Pay for Beef Meat in a Developing Country: The Effect of Information Regarding Country of Origin, Price and Animal Handling Prior to Slaughter." *Journal of Food Quality and Preference* 20: 156–65.

Schwartz, Barry. 2007. "When Words Decide." *Scientific American Mind* August/September: 37–43.

Seibert, Lacey and F. Bailey Norwood. 2011. "Production Costs and Animal Welfare For Four Stylized Hog Production Systems." *Journal of Applied Animal Welfare Science* 14(1): 1–17.

Seligman, Martin E. 1995. *The Optimistic Child*. New York: HarperPerennial.

Shields, Sara. 2009. "Understanding Mortality Rates of Laying Hens." Ham and Eggonomics Blog, September 22.

Silverman, A. P. 1978. "Rodents' Defence Against Cigarette Smoke." *Animal Behaviour* 26: 1279–81.

Simonson, Itamar and Aimee Drolet. 2004. "Anchoring Effects on Consumers' Willingness-to-Pay and Willingness-to-Accept." *Journal of Consumer Research* 31 (December): 681–90.

Sinclair, Upton. [1906] 1995. *The Jungle*. New York: Barnes and Noble.

Singer, Peter. 1975. *Animal Liberation*. New York: HarperCollins.

——2000. *Writings on an Ethical Life*. New York: HarperCollins.

——2004. "Ethics Beyond Species and Beyond Instincts: A Response to Richard Posner." In C. R. Sunstein and M. C. Nussbaum (eds.), *Animal Rights: Current Debates and New Directions*. New York: Oxford University Press.

——and Jim Mason. 2006. *The Way We Eat: Why Our Food Choices Matter*. Emmaus, PA: Rodale Publishers.

Smith, Adam. 1776. *An Inquiry into the Nature and Causes if the Wealth of Nations.* London: Methuen & Co. Ltd.

———[1817] 1991. *A Theory of Moral Sentiments.* Book 4. New York: Prometheus Books, p. 348.

Smith, Craig A., Kelly N. Roeszier, Thomas Ohnesorg, David M. Cummins, Peter G. Farlie, Timothy J. Doran, and Andrew H. Sinclair. 2009. "The Arian Z-linked Gene DMRTI is Required for Male Sex Determination in the Chicken." *Nature* 461 (7261): 267–71.

Smith, Lewis. 2008. "Pig Organs Available to Patients in a Decade." *Times Online* November 7.

Smith, Rod. 2007. "Free-Range Hens Experience Stress." *Feedstuffs* December 17, p. 2.

———2008a. "Study: Initiative to be Disastrous." *Feedstuffs* May 26, p. 1.

———2008b. "California Passes Prop 2." *Feedstuffs* November 10, p. 1.

———2009a. "Prop 2 Opens Door to Push for Veganism." *Feedstuffs* March 9, p. 3.

———2009b. "Prop 2 Fuels 'Freight Train.'" *Feedstuffs* March 30, p. 1.

———2009c. "Veal Producers Ahead of Schedule in Housing Transition." *Feedstuffs* June 25, p. 3.

———2009d. "Cage-Free Transition Has Consequences." *Feedstuffs* 42 (81, October 22) p. 8.

Soler, Jean. 1996. *Food: A Culinary History from Antiquity to the Present.* Ed. Lawrence Kritzman. New York: Columbia University Press.

Spears, Tom. 2007. "Study Finds Animals Have Personality." *CanWest News Service* November 26.

Specter, Michael. 2003. "The Extremist." *The New Yorker* April 14.

Spencer, C. 2000. *Vegetarianism: A History.* New York: Four Walls Eight Windows.

Stark, R. 1996. *The Rise of Christianity.* Princeton, NJ: Princeton University Press.

Stark, Rodney. 2004. "For the Glory of God: How Monotheism Led to Reformations." *Science, Witch-Hunts, and the End of Slavery.* Princeton, NJ: Princeton University Press.

Stassen, Steve. 2003. "Farrowing Crates vs. Pens vs. Nest Boxes." Greenbook. Energy and Sustainable Program, Minnesota Department of Agriculture, Minnesota.

Steinbeck, John. 1962. *Travels With Charley: In Search of America.* New York: Penguin Books.

Sumner, Daniel, Thomas Rosen-Molina, William Matthews, Joy Mench, and Kurt Richter. 2008. "Economic Effects of Proposed Restrictions on Egg-Laying Hen Housing in California." Paper presented at University of California Agricultural Issues Center, July.

Sumner, D. H., H. Gow, D. Hayes, W. Matthews, F. B. Norwood, J. T. Rusen-Moling, and W. Thurman. Forthcoming. "Economic and Market Issues on the Sustainability of Egg Production in the United States." *Poultry Science.*

Sunstein, C. R. 2004. "Introduction: What Are Animal Rights?" in C. R. Sunstein and M. C. Nussbaum (eds.), *Animal Rights: Current Debates and New Directions.* New York: Oxford University Press.

Surowieck, J. 2004. *The Wisdom of Crowds.* New York: Doubleday.

Tannahill, Reay. 1988. *Food in History.* New York: Three Rivers Press.

Task Force Report. 2005. "A Comprehensive Review of Housing for Pregnant Sows." *Journal of the American Veterinary Medical Association* 10 (227, November 15): 580–1590.

Taylor, Robert E. 1992. *Scientific Farm Animal Production*, 4th edn. New York: Macmillan Publishing Company.

Telezhenko, E., L. Lidfors, and C. Bergsten. 2007. "Dairy Cow Preference for Soft or Hard Flooring when Standing or Walking." *Journal of Dairy Science* 90: 3716–24.

Time Magazine. 1926. "Crops." September 20.

The Times. 1909. "Pig Keeping Methods and Swine Feeder." Letter to the Editor, February 22.

Tonsor, G.T. 2009. "Consumer Preferences for Mandatory Labeling of Animal Welfare Attributes." Working paper, Michigan State University.

——and C. A. Wolf. Forthcoming. "Drivers of Resident Support for Animal Care Oriented Ballot Initiatives." *Journal of Agricultural and Applied Economics*.

——N. Olynk, and C. Wolf. 2009. "Consumer Preferences for Animal Welfare Attributes: The Case of Gestation Crates." *Journal of Agricultural and Applied Economics* 41: 713–30.

——C. Wolf, and N. Olynk. 2009. "Consumer Voting and Demand Behavior Regarding Swine Gestation Crates." *Food Policy* 34: 492–8.

Tristram, Stuart. 2006. *Bloodless Revolution*. New York: WW Norton & Co.

Tucker, C. B. and D. M. Weary. Winter 2001–Spring 2002. "Tail Docking in Dairy Cattle." *Animal Welfare Information Center Bulletin* 11: 3–4.

Turner, Jacky, Leah Garces, and Wendy Smith. 2005. *The Welfare of Broiler Chickens in the European Union*. Guildford, UK: Compassion in World Farming Trust.

UC Davis. 2008. "To Cage or Not to Cage: Research Coalition Seeks Answers." News and Information. February 13.

Umberger, W. J., D. M. Feuz, C. R. Calkins, and K. Killinger-Mann. 2002. "US Consumer Preference and Willingness-to-Pay for Domestic Corn-Fed Beef versus International Grass-Fed Beef Measured Through an Experimental Auction." *Agribusiness: An International Journal* 18: 491–504.

United Egg Producers Certified. 2008. "United Egg Producers Animal Husbandry Guidelines for U.S. Egg Laying Flocks."

United States Census Bureau. 2000. "Average Number of Children Per Family and Per Family With Children, by State." <http://www.census.gov/population/socdemo/hh-fam/tabS-F1–2000.pdf>. Accessed November 27, 2009.

——2009. "Population Numbers." <http://www.google.com/publicdata/home>.

——2010. "Historical Estimates of World Population. International Programs." <http://www.census.gov/ipc/www/worldhis.html> Accessed on March 13, 2010.

United States Department of Agriculture. 2007. "Reference of Dairy Cattle Health Management Practices in the United States, 2007." Animal and Plant Health Inspection Service, Veterinary Services, National Animal Health Monitoring System, February.

University of Arkansas Libraries: Special Collections. 1951. "Chicken-of-Tomorrow Contest Papers." Correspondence and Publications. Manuscript Collection 580.

University of Missouri-Columbia. 2009. "Vitamin D Deficiency Related to Increased Inflammation in Healthy Women." *ScienceDaily* April 14. <http://www.sciencedaily.com/releases/2009/04/090408140208.htm>. Accessed April 27, 2009.

Unnevehr. 1996. "The Final Rule on Pathogen Reduction and HACCP." *Federal Register* 61 (144, July 25): 38805–55.

Unti, Bernard. 2004. *Protecting All Animals: A Fifty-Year History of the Humane Society of the United States*. Washington, DC: Humane Society Press.

Ursinus, W. W., F. Schepers, R. M. de Mol, M. B. M. Bracke, J. H. M. Metz, and P. W. G. Groot Koerkamp. 2009. "COWEL: A Decision Support System to Assess Welfare of Husbandry Systems for Dairy Cattle." *Animal Welfare* 18: 545–52.

Videra, J. 2006. "Religion and Animal Welfare: Evidence from Voting Data." *Journal of Socio-Economics* 35: 652–9.

Wallaces Farmer. 2009. "Iowa Feedlots 'Feel the Heat' As Cattle Death Loss Hit Last Week." July 10.

Ware, Katheryn. 2006. "Open Letter Against University of Minnesota's Invitation to Peter Singer." *Euthanasia.com.* <http://www.euthanasia.com/ware.html>. Accessed February 8, 2010.

Warrick, Joby and Stith, Pat. 1995. "Success Makes Murphy the Real 'Boss Hog.'" *The News & Observer* February 22. <http://www.pulitzer.org/archives/7538>.

Wasserman, Debra. 2006. *Simply Vegan*, 4th edn. Baltimore, MD: The Vegetarian Research Group.

Watanabe, Shigeru. 2009. "Pigeons Can Discriminate 'Good' and 'Bad' Paintings By Children." *Animal Cognition* 13(1): 75–85.

Watson, Lois. 2008. "Surprise Finding on Battery Hens." *Stuff.co.nz* September.

Weber, Christopher L. and H. Scott Matthews. 2008. "Food-Miles and the Relative Climate Impacts of Food Choices in the United States." *Environmental Science & Technology* 42: 3508–15.

Weekly Veal Market Summary. 2009. Agricultural Marketing Service, United States Department of Agriculture, March 5.

Webster, A. F. J. 2001. "Farm Animal Welfare: The Five Freedoms and the Free Market." *The Veterinary Journal* 161(3): 229–37.

Widowski, Tina and Carien Vandenburg. 2003. "Lighting Sources for Broiler Breeders." Poultry Industry Council Factsheet No. 139. Guelph, ON, January.

Wiepkema, P. R. 1997. "The Emotional Vertebrate." In M. Dol, S. Kasanmoentalib, S. Lijmbach, E. Rivas and R. van den Bos (eds.), *Animal Consciousness and Animal Ethics*. Assen: Van Gorcum, pp. 93–102.

Wierup, Martin. 2001. "The Swedish Experience of the 1986 Year Ban of Antimicrobial Growth Promoters, with Special Reference to Animal Health, Disease Prevention, Productivity, and Usage of Antimicrobials." *Microbial Drug Resistance* 7(2): 183–90.

Wilson, Lowell L., George Greaser, and Jayson Harper. 1994. "Agricultural Alternatives: Veal Production." Cooperative Extension Paper No. R3M995ps. Pennsylvania State University, College of Agricultural Sciences.

Wirth, H. 2007. "The Animal Welfare Movement and Consumer-Driven Change." *Farm Policy Journal* 4: 1–10.

Wise, S. M. 2004. "Animal Rights: One Step at a Time." In C. R. Sunstein and M. C. Nussbaum (eds.), *Animal Rights: Current Debates and New Directions*. Oxford: Oxford University Press, pp. 19–50.

Wolfson, David J. and Mariann Sullivan. 2004. "Foxes in the Hen House." In Cass R. Sunstein and Martha C. Nussbaum (eds.), *Animal Rights*. New York: Oxford University Press.

Wright, K. T. 1936. "Monthly Poultry Costs and Returns." *Journal of Farm Economics* 18(4): 706–10.

Wynne-Jones, Ros. 1997. "Refered and Reviled: Profile. Colin Blakemore." *The Independent* August 24.

Young, Beth. 2009. "Study Shows Moving Pigs Inside Has Huge Benefits." Press Release. University of Missouri Extension Service, November 23. <http://agebb.missouri.edu/news/DisplayStory.aspx?N=597/>.

Zimmerman Matthias R., Heiko Maischak, Axel Mithöfer, Wilhelm Boland, and Hubert H. Felle. 2009. "System Potentials: A Novel Electrical Long-Distance Apoplastic Signal in Plants, Induced by Wounding." *Plant Physiology* 149(3): 1593–600.

Index